Removal of MTBE With Advanced Oxidation Processes

The mission of the Awwa Research Foundation (AwwaRF) is to advance the science of water to improve the quality of life. Funded primarily through annual subscription payments from over 1,000 utilities, consulting firms, and manufacturers in North America and abroad, AwwaRF sponsors research on all aspects of drinking water, including supply and resources, treatment, monitoring and analysis, distribution, management, and health effects.

From its headquarters in Denver, Colorado, the AwwaRF staff directs and supports the efforts of over 500 volunteers, who are the heart of the research program. These volunteers, serving on various boards and committees, use their expertise to select and monitor research studies to benefit the entire drinking water community.

Research findings are disseminated through a number of technology transfer activities, including research reports, conferences, videotape summaries, and periodicals.

Removal of MTBE With Advanced Oxidation Processes

Prepared by:
Michael Kavanaugh, **Zaid Chowdhury**, and **Sunil Kommineni**
Malcolm Pirnie, Inc., Phoenix, AZ 85008

Sun Liang and **Joon Min**
Metropolitan Water District of Southern California,
La Verne, CA 91750

Jean-Philippe Crouè and **Nina Corin**
University of Poiters, Laboratoire de Chimie de l'Eau
et de l'Environment, France

Gary Amy and **Erich Simon**
University of Colorado, Boulder, CO 80309

William Cooper
University of North Carolina, Wilmington, NC

Paul Tornatore and **Michael Nickelsen**
Haley and Aldrich, Inc., Rochester, NY 14623

Sponsored by:
Awwa Research Foundation
6666 West Quincy Ave.
Denver, CO 80235-3098

Published by:

Awwa Research Foundation

Advancing the Science of Water®

IWA Publishing

DISCLAIMER

This study was funded by the Awwa Research Foundation (AwwaRF). AwwaRF assumes no responsibility for the content of the research study reported in this publication or for the opinions or statements of fact expressed in the report. The mention of trade names for commercial products does not represent or imply the approval or endorsement of AwwaRF. This report is presented solely for informational purposes.

Published by IWA Publishing, Alliance House, 12 Caxton Street, London SW1H 0QS, UK

Telephone: +44 (0) 20 7654 5500; Fax: +44 (0) 20 7654 5555; Email: publications@iwap.co.uk; Web: **www.iwapublishing.com**

AwwaRF report number 90933F. Originally published by AwwaRF for its subscribers in 2003. IWA Publishing version published 2004.

ISBN 1 84339 861 3

CONTENTS

TABLES

FIGURES

FOREWORD

The Awwa Research Foundation is a nonprofit corporation that is dedicated to the implementation of a research effort to help utilities respond to regulatory requirements and traditional high-priority concerns of the industry. The research agenda is developed through a process of consultation with subscribers and drinking water professionals. Under the umbrella of a Strategic Research Plan, the Research Advisory Council prioritizes the suggested projects based upon current and future needs, applicability, and past work; the recommendations are forwarded to the Board of Trustees for final selection. The foundation also sponsors research projects through the unsolicited proposal process; the Collaborative Research, Research Applications, and Tailored Collaborations programs; and various joint research efforts with organizations such as the U.S. Environmental Protection Agency, the U.S. Bureau of Reclamation, and the Association of California Water Agencies.

This publication is the result of one of these sponsored studies, and it is hoped that its findings will be applied in communities throughout the world. The following report serves not only as a means of communicating the results of the water industry's centralized research program but also as a tool to enlist the further support of the nonmember utilities and individuals.

Projects are managed closely from their inception to the final report by the foundation's staff and large cadre of volunteers who willingly contribute their time and expertise. The foundation serves a planning and management function and awards contracts to other institutions such as water utilities, universities, and engineering firms. The funding for this research effort comes primarily from the Subscription Program, through which water utilities subscribe to the research program and make an annual payment proportionate to the volume of water they deliver and consultants and manufacturers subscribe based on their annual billings. The program offers a cost-effective and fair method for funding research in the public interest.

A broad spectrum of water supply issues are addressed by the foundation's research agenda: resources, treatment and operations, distribution and storage, water quality and analysis, toxicology, economics, and management. The ultimate purpose of the coordinated effort is to assist water suppliers to provide the highest possible quality of water in an economical and reliable manner. The true benefits are realized when the results are implemented at the utility level. The foundation's trustees are pleased to offer this publication as a contribution toward that end.

Edmund G. Archuleta, P.E.
Chair, Board of Trustees
Awwa Research Foundation

James F. Manwaring, P.E.
Executive Director
Awwa Research Foundation

ACKNOWLEDGMENTS

The authors would like to acknowledge the generous financial support of AwwaRF without which this study would not have been possible. An efficient Project Officer and some experienced Project Advisory Committee (PAC) members have overseen the development and execution of this work. The project team would like to thank the Project Officer and the PAC members for their kind support and guidance:

Project Officer: Ms. Traci Case, AwwaRF
PAC Members: Dr. Harish Arora, O'Brien and Gere
Dr. Betrand Dussert, Vivendi-US Filter Incorporated
Dr. James Malley, University of New Hampshire
Dr. Lucik Melikian, Los Angeles Department of Water and Power

Several utility members across the U.S. participated in this project by supporting the bench and pilot-scale testing activities. The authors would like to acknowledge the support of the following water utility participants:

Mr. Robert Harvey, City of Santa Monica
Mr. Herman Miller, Suffolk County Water Authority
Dr. Robert Cheng, Long Beach Water Department

The authors would like to acknowledge the timely and constant support and services provided by the other team members including:

Dr. Urs von Gunten, EAWAG, Switzerland
Dr. Jonathan Pressman, Malcolm Pirnie, Inc.
Mr. Aaron Dotson, Malcolm Pirnie, Inc.
Mr. Aziz Ahmed, Malcolm Pirnie, Inc.
Ms. Patricia Hausser, Malcolm Pirnie, Inc.

Finally, the authors would like to thank the following people for providing bench/pilot-scale testing equipment and operational assistance:

Dr. Bill Cairns, Trojan Technologies
Mr. Alan Royce, Trojan Technologies
Mr. Daniel Gosselin, Trojan Technologies
Dr. Kuang-Ping Chiu, Trojan Technologies
Dr. Chi-Su Chou, Pulsar Environmental Remediation Technologies
Mr. James Bender, Pulsar Environmental Remediation Technologies

The authors would like to note that Dr. Joon Min, who was with Metropolitan Water District of Southern California at the time of this study, is now an employee of Carollo Engineers, Inc.

EXECUTIVE SUMMARY

MTBE USE AND OCCURRENCE

Methyl tertiary butyl ether (MTBE) has been added to gasoline as an octane enhancing replacement for lead and as an oxygenate to lower the ozone and carbon monoxide emission levels. The percentage of MTBE that was added to fuels increased from approximately 3% by volume in 1979 (when it was first introduced) to 11-15% in 1998 (USEPA, 1998c). MTBE was added approximately to 30% of the gasoline sold in United States in 1998 (USEPA, 1998c). As a result of this widespread usage and its unique physiochemical properties, a portion of this gasoline additive found its way into the nation's groundwater and surface water supplies and has been reported as a contaminant in some supplies.

The main sources of MTBE contamination of groundwater supplies are leaking underground storage tanks and pipelines, spills and contaminated sites and releases from manufacturing and storage sites. The primary sources of MTBE contamination in urban surface water supplies are releases from gasoline-powered recreation watercrafts and atmospheric deposition through precipitation of industrial or vehicular emissions. MTBE concentrations in the order of 100-200 µg/L were detected in drinking water wells in areas with leaky underground storage tanks (Gullick and LeChavellier, 2000). A national survey conducted by U.S. Geological Survey (USGS) found MTBE to be one of the most predominant organic contaminants, next only to chloroform in urban groundwaters (Squillacc et al., 1997). Detailed information on MTBE use and occurrence in natural waters is provided in Chapters 1 and 2.

REGULATORY UPDATE

The U.S. Environmental Protection Agency (USEPA) has not yet set any drinking water standards for MTBE. However, in 1997, the USEPA Office of Water issued a health advisory on MTBE. This advisory was issued to provide guidance to water suppliers whose source waters had been impacted by MTBE. This advisory recommended keeping the concentrations of MTBE in the range of 20 to 40 µg/L or less to likely avert unpleasant taste and odor effects (USEPA, 1997). The USEPA has included MTBE in its Contaminant Candidate List and may set a secondary maximum contaminant level (SMCL) in the future.

Several states, including Maine, New Jersey, California and New Hampshire have set primary maximum contaminant levels (MCLs) for MTBE. Maine and New Jersey have set MCLs for MTBE of 35 µg/L and 70 µg/L, respectively. In December 2000, California and New Hampshire adopted MCLs of 13 µg/L. The State MCLs are based on protection of human health and not on aesthetic concerns. California also set an SMCL of 5 µg/L based on organoleptic characteristics of MTBE.

Some recent studies indicated that the beneficial effects of fuel oxygenates are fewer for newer vehicles that use more advanced pollution control equipment. With increasing evidence of MTBE in drinking water supplies, the USEPA recommended complete elimination of MTBE from gasoline fuels by 2004. Several states including California, New York and Connecticut adopted legislations prohibiting the use of MTBE as a fuel oxygenate. Presently, more studies are being conducted to substitute MTBE with ethanol and other oxygenates to meet the 1990 Clean Air Act Amendments requirements.

TREATMENT TECHNOLOGIES

MTBE in drinking water sources can be removed using one of the following treatment alternatives:

- Advanced oxidation processes (AOPs)
- Air stripping
- Granular activated carbon (GAC) adsorption
- Resin adsorption

Air stripping, GAC and resin adsorption are inefficient for MTBE removal in certain cases due to MTBE's relatively low Henry's constant, low organic-carbon partition coefficient and high solubility in water.

ADVANCED OXIDATION PROCESSES

AOPs involve two stages of oxidation: 1) the formation of strong oxidants (e.g., hydroxyl radicals) and 2) the reaction of these oxidants with organic contaminants in water. However, the term AOP refers specifically to processes in which oxidation of organic contaminants occurs primarily through reactions with hydroxyl radicals (·OH) (Glaze et al., 1989). In water treatment applications, AOPs usually refer to a specific subset of processes that involve ozone (O_3) and/or hydrogen peroxide (H_2O_2), and/or ultraviolet light (UV). It should be noted that two primary mechanisms of UV dose are through conventional continuous wave application and pulsed application. In the context of this document, the application of high energy electron beam (E beam) is also considered an AOP process. All of these processes can produce hydroxyl radicals, which can react with and destroy a wide range of organic contaminants, including MTBE. Although a number of the processes noted above may have other mechanisms for destroying organic contaminants, in general, the effectiveness of an AOP is proportional to its ability to generate hydroxyl radicals. The various chemical and physical mechanisms through which AOP technologies produce hydroxyl radicals are discussed in Chapter 3 along with a literature review on prior studies on MTBE removal using AOPs.

When ozone and peroxide are simultaneously applied to water they react to form hydroxyl radicals. In ozone/peroxide treatment, MTBE is oxidized by radical reactions and by direct reaction with the oxidants (ozone and peroxide). In UV/peroxide treatment, hydroxyl radicals are generated when low-pressure (LP) or medium-pressure (MP), continuous wave UV light is applied to water that is dosed with hydrogen peroxide. In this continuous UV/peroxide treatment, oxidation of MTBE occurs predominantly through the radical route. Direct photolysis may also remove some MTBE. Unlike LP/MP-UV lamps, which provide continuous UV light, a pulsed-UV lamp produces high-intensity, pulsed polychromatic radiation in thousandths or millionths of a second. These UV pulses provide extremely intense flashes of radiation, which produce hydroxyl radicals to remove MTBE in water. In the E beam process, high-energy electrons are injected into the water to form hydroxyl radicals (·OH), and the reducing, aqueous electrons (e^-_{aq}) and hydrogen atoms (H·). These three highly reactive species can oxidize/reduce MTBE and its by-products.

OBJECTIVES AND APPROACH

The AOP technologies that were evaluated as part of this study include:

- Ozone/peroxide
- Continuous wave UV/peroxide
- Pulsed UV/peroxide and
- E beam

In addition to the AOP technologies that are listed above, there are other emerging technologies like cavitation, TiO_2 catalyzed UV oxidation and Fenton's reaction. These emerging technologies were not evaluated as part of this study.

The main objective of this study was to fill-in some of the research gaps with AOP treatment of MTBE. The specific objectives of this study were:

- To evaluate the impact of source water quality on MTBE and its by-products removal efficiencies. This evaluation was achieved through bench-scale and pilot-scale studies
- To develop design parameters for the most promising AOP(s) that are identified above
- To determine conceptual-level capital and operational costs for the selected AOPs
- Integrate the results from each AOP analysis into a comprehensive report that can be used by water treatment professionals

To evaluate the effect of source water quality, bench-scale tests were conducted on the aforementioned AOP processes using the following four waters:

- Suffolk County water, a low TOC (< 3 mg/L) and a low alkalinity (40-50 mg/L) water
- Santa Monica water, a low TOC (< 3 mg/L) and a high alkalinity (150-200 mg/L) water
- Miami Dade water, a high TOC (> 3 mg/L) and a low alkalinity (40-50 mg/L) water
- Long Beach water, a high TOC (> 3 mg/L) and a high alkalinity (150-200 mg/L) water

Based on the bench scale study findings, pilot scale tests were conducted for the aforementioned AOP technologies on Santa Monica water to further evaluate the impact of selected water quality parameters on the AOP performance and to develop design and operational parameters for economic evaluation of potential full-scale AOP facilities treating either surface or ground waters. For some treatment technologies pilot scale tests were also conducted on surface waters. Detailed discussions of the bench/pilot scale study experimental plans, methods and results for the ozone/peroxide, continuous wave UV/peroxide, pulsed UV/peroxide and E beam processes are in Chapters 4, 5, 6 and 7, respectively. A comparison of these promising AOPs with respect to treatment costs and implementation issues is presented in Chapter 8.

OZONE/PEROXIDE TREATMENT RESULTS

Bench-scale results of ozone/peroxide treatment demonstrated that MTBE removal efficiency is highly dependant on the source water quality. Bromate formation and the potential for minimizing the bromate formation by increasing the H_2O_2:O_3 ratio were found to be related to the hydroxyl radical scavenging capacity of the water. Ozone/peroxide treatment of waters with low scavenging capacity resulted in high MTBE removals and a high conversion of bromide to bromate. $O_3^-H_2O_2$ treatment for low alkalinity waters was found to be appropriate only when the initial bromide concentrations were low enough that the bromate formation stays below 10 μg/L (MCL for bromate). Waters with higher scavenging capacity needed higher oxidant dosages to achieve comparable MTBE removal.

Hydroxyl radical probe experiments with para-chlorobenzoic acid (pCBA) were used to assess hydroxyl radical exposure during reactions at differing oxidation levels and H_2O_2:O_3 ratios. These exposure estimations were shown to be useful in developing kinetic and mechanistic model predictions of MTBE degradation.

Pilot scale studies demonstrated approximately 90-100% MTBE removal when ozone alone or ozone with hydrogen peroxide was applied. For waters spiked with low concentrations of MTBE (< 200 μg/L), the type of source water did not significantly influence the removal efficiency of MTBE. In waters spiked with high MTBE concentrations (~2,000 μg/L), the application of ozone alone at dosages of 15-30 mg/L resulted in low MTBE removal (~86%). The addition of hydrogen peroxide (1:1 mass ratio as O_3:H_2O_2) with a contact time of 25 minutes improved the MTBE removal to greater than 95% for the waters containing high, initial MTBE (2,000 μg/L) concentrations.

In the absence of hydrogen peroxide, high levels of bromate were formed, well in excess of the regulatory limit. Addition of hydrogen peroxide significantly reduced the amount of bromate formed. Based on bench and pilot scale test results presented in Chapter 4, it can be inferred that ozone and peroxide dosages of 15 mg/L each (1:1 mass ratio), with a contact time of 25 minutes will be sufficient to lower the MTBE levels by 95% or higher in waters with influent MTBE of approximately 200 μg/L. The pilot tests with sequential ozone/peroxide and BAC treatment indicated higher overall MTBE and by-products removal efficiencies compared to ozone/peroxide treatment alone. Additional details about ozone/peroxide bench and pilot scale testing and relevant technology cost information is discussed in Chapter 4.

The costs for ozone/peroxide treatment (without BAC polishing treatment) of potable waters containing MTBE were found to vary between $1.08 per 1,000 gallons treated, for a 0.5 mgd system, to $0.48 per 1,000 gallons treated for a 10 mgd system. These costs were developed assuming ozone and hydrogen peroxide dosages of 15 mg/L each with a contact time of 25 minutes to lower the influent MTBE from 200 μg/L to 10 μg/L or less. It should be noted that in full-scale application, polishing treatment with BAC will be needed and this could increase the total cost of treatment by approximately $0.20 per thousand gallons of water treated.

CONTINUOUS WAVE UV/PEROXIDE TREATMENT RESULTS

Bench scale tests compared the removal of 200 μg/L MTBE using LP-UV/H_2O_2 and MP-UV/H_2O_2 from Santa Monica, Suffolk County, Miami Dade and Long Beach water samples. At an initial MTBE and H_2O_2 molar ratio of 1:85, both LP- and MP-UV/H_2O_2 removed MTBE most efficiently from the Suffolk County water that had low DOC and low alkalinity. For both LP- and MP-UV irradiation, the most retarded degradation of MTBE was observed in the Long Beach water that had high DOC, high specific UV absorbance, and high alkalinity. Intermediate MTBE removal rates were observed for Miami Dade and Santa Monica waters. MTBE was removed with MP-UV irradiation in all four waters even in the absence of H_2O_2. Since the degradation of MTBE with MP-UV irradiation generated similar major by-products in the presence or absence of

H_2O_2, it is still unclear if the degradation of MTBE in the absence of H_2O_2 is due to direct photolysis and/or radical pathways. Bench study results also showed rapid MTBE removal in the Suffolk County water with LP-UV irradiation in the absence of hydrogen peroxide. This was attributed to the presence of nitrates that may generate hydroxyl radicals during photolysis. The major findings regarding the formation of by-products are:

- Tertiary butyl formate (TBF) is formed in larger amounts than tertiary butyl alcohol (TBA)
- Acetone and formaldehyde are the major secondary by-products
- A higher concentration of H_2O_2 helps in removing the by-products more efficiently.
- More formaldehyde was formed in the high DOC waters
- More formaldehyde was observed in MP-UV/H_2O_2 treatment compared to LP-UV/ H_2O_2 treatment

Continuous UV/peroxide pilot tests were conducted using LP and MP lamps on Santa Monica groundwater and Lake Huron surface waters. The E_{EO} (kWh/m^3-order) values for MTBE removal in Santa Monica water spiked with 1.2 μM (or 100 μg/L) of MTBE by MP-UV/peroxide and LP-UV/peroxide treatment were respectively 2.2 kWh/m^3-order and 0.7 kWh/m^3-order. In tests conducted with Santa Monica water spiked with MTBE (~1.2 μM), TBA (~4.0 μM), nitrate (~36 mg/L as N) and bromide (~0.5 mg/L) to simulate poor quality groundwater, the E_{EO} values for MP-UV/peroxide and LP-UV/peroxide treatment were 2.58 kWh/m^3-order and 0.96 kWh/m^3-order, respectively. As indicated by the aforementioned results, the LP-UV/peroxide treatment had slightly lower E_{EO} values than MP-UV/peroxide treatment. In a test conducted with Santa Monica water spiked with 33 μM (or 2,900 μg/L) MTBE and 480 μM (or 16 ppm) H_2O_2, an E_{EO} of 0.64 kWh/m^3-order was measured for MP-UV/peroxide treatment of MTBE. This result illustrates the inverse relationship between E_{EO} and peroxide concentration.

Some algae growth was observed in the Santa Monica water that was stored in a large (3,000 gallon) storage tank at the Lake Huron Water Treatment Plant testing site. This algae growth may have resulted in an increase in TOC content and the associated hydroxyl radical scavenging capacity. The scavenging capacity for Santa Monica water was calculated based on the measured LP-UV/peroxide E_{EO} (of 1.44 kWh/m^3-order) with ~200 μM H_2O_2 and 1.2 μM MTBE. This scavenging capacity was found to be $1.4 \times 10^5\ s^{-1}$ or 2.5 times the value of $5.57 \times 10^4\ s^{-1}$ that was calculated for the Santa Monica water quality. The E_{EO} values measured for the Lake Huron surface water were lower than those for Santa Monica groundwater. This is attributed primarily to the lower scavenging capacity (no algae) as well as to the lower UV absorption coefficients for the Lake Huron water. The scavenging capacity of the Lake Huron surface water was calculated to be $4.7 \times 10^4\ s^{-1}$ based on a comprehensive water analysis.

It should be noted that the E_{EO} varies inversely with the optical pathlength of the photoreactor. Therefore, large, full-scale photoreactors with long pathlengths may result in lower E_{EO} values than those measured in this study. The E_{EO} values observed in this study were slightly lower than those reported in the literature for MP-UV/H_2O_2 treatment (Cater et al., 2000). The E_{EO} values reported in the literature were developed using batch reactors and they varied between 0.2 kWh/m^3-order to 7.5 kWh/m^3-order (Cater et al., 2000).

Consistent with the bench scale test results, the MP-UV/H_2O_2 treatment resulted in higher acetone and formaldehyde generation compared to LP-UV/H_2O_2 treatment. This indicates that the

MP-UV/peroxide treatment progressed farther in the oxidation pathway than the LP-UV/peroxide treatment.

The following design criteria were used in developing the UV/peroxide treatment costs for waters with quality comparable to Santa Monica groundwater (with scavenging capacity of approximately 5×10^4 s^{-1}):

- An E_{EO} of 0.35 kWh/m^3-order for MTBE removal using MP-UV/peroxide process, with an initial peroxide dose of 15 mg/L
- An E_{EO} of 0.13 kWh/m^3-order for MTBE removal using LP-UV/peroxide process, with an initial peroxide dose of 15 mg/L

Similar to other AOP technologies, it was assumed that MTBE and its by-products will be further removed by biological GAC treatment. Capital and operations and maintenance (O&M) costs were developed for MP-UV/peroxide and LP-UV/peroxide treatment of MTBE contaminated waters for three system sizes, 1 mgd, 3 mgd and 10 mgd. The cost per 1,000 gallons of treated water for MP-UV/peroxide treatment of MTBE varied from \$0.81 (for 1 mgd system) to \$0.41 (for 10 mgd system). For LP-UV/peroxide treatment, the costs per 1,000 gallons of treated water varied from \$0.55 (for 1 mgd) to \$0.32 (for 10 mgd). Polishing treatment with BAC is necessary and would cost an additional \$0.20 per 1,000 gallons. More details on these costs are provided in Chapter 5.

PULSED UV/PEROXIDE TREATMENT RESULTS

Bench-scale test results from pulsed UV/peroxide treatment showed that the background water quality plays an important role in the mechanism(s) involved in the reduction of MTBE. Effective removal of MTBE was observed for the four groundwaters that were tested. Regardless of the water characteristics, it was possible to remove MTBE to below 5–6 μg/L from 1,000–2,000 μg/L initial MTBE levels (99 percent reduction) when H_2O_2 was added with an irradiation time of 12 minutes. MTBE removal was also observed in experiments with no H_2O_2 addition. Generally, the water with high background NO_3^- inhibited the decay of MTBE much more than the water with high levels of TOC and alkalinity, which are known to scavenge hydroxyl radicals. Santa Monica water and Suffolk County water with high levels of NO_3^- exhibited slower MTBE removal rates when H_2O_2 was not added to the system. The addition of H_2O_2 increased the rates of MTBE decay. The influent water quality also affected the formation and fate of MTBE degradation by-products.

The pulsed UV/peroxide pilot tests were performed on Santa Monica groundwater and State Project (surface) Water (SPW). These pilot study results showed that the applied UV dose (frequency) was less sensitive than the hydrogen peroxide in terms of MTBE removal efficiency. Increasing the pulsed UV frequency from 4 Hz to 7 Hz did not significantly improve the MTBE removal efficiency under similar operating conditions. On the contrary, increasing the hydrogen peroxide significantly improved the overall treatment efficiency by generating more hydroxyl radicals to oxidize MTBE and the by-products. Santa Monica groundwater, which contained high levels of nitrate (~21 mg/L) exhibited overall lower level of removal efficiency than the SPW surface water, because nitrate strongly absorbs UV light thus decreasing the amount of hydroxyl radicals generated during the pulsed UV process. Depending on the operational conditions used, excess amount of nitrite may be formed in the treated water especially when the level of initial hydrogen peroxide is low. Thus, in pulsed UV/peroxide treatment, adequate amount of hydrogen peroxide (in the range of 10 mg/L) and optimized retention time (10-12 minutes) are necessary to remove MTBE and to keep the levels of nitrite low. The BAC post-treatment complemented the pulsed UV/peroxide treatment by enhancing the overall removal efficiency for both MTBE and its by-products.

Based on bench and pilot test results it was inferred that a pulsed UV dose of 2,000 kJ with a UV reactor residence time of 10 minutes is necessary to reduce the influent MTBE levels from 200 μg/L to 10 μg/L or less. Test data also indicated a peroxide dose of 10 mg/L is necessary to achieve the MTBE oxidation. Costs for pulsed UV/peroxide treatment with these design criteria were found to vary between $0.84 per 1,000 gallons treated (for a 0.5 mgd system) to $0.45 per 1,000 gallons treated (for a 10 mgd system). Similar to other AOP technologies, additional cost will be incurred from BAC polishing treatment. More details on pulsed UV/peroxide tests, results and costs are provided in Chapter 6.

ELECTRON BEAM TREATMENT RESULTS

The E beam process was shown to be effective in removing MTBE from spiked Santa Monica groundwater. The presence of TBA (in influent water spiked with TBA) resulted in lower MTBE removal efficiencies. For the E beam treatment doses that were tested [<10 kiloGray (kGy), 10 kGy = 1,000 kilorads or 1 megarad of radiation chemical dose], the effect of nitrate ion and bromide ion on MTBE removal were difficult to assess; however, the presence of either nitrate ion or bromide ion, appeared to increase the concentration of the reaction by-product formaldehyde. In the experiment that simulated the poor quality water (influent water spiked with MTBE, TBA, nitrate and bromide), the removal of MTBE appeared relatively unaffected, but the concentration of the reaction by-products formaldehyde and methylglyoxal were higher than for the other experiments. Presence of bromide in the influent water at concentrations of 0.5 mg/L had no effect on the removal of either the MTBE or TBA. However, bromate ion was observed at a concentration of 3.6 μg/L (well below the MCL) even after dosing the water with about 10 kGy of E beam dose. Assuming an E beam dose of 10 kGy, the costs for treating 1,000 gallons of water varied between $1.97 (for 0.5 mgd system) to $0.85 (for 10 mgd system). E beam treatment would also require polishing BAC treatment. E beam test plan, methods, results and costs are discussed under Chapter 7.

BIOLOGICAL ACTIVATED CARBON POLISHING TREATMENT RESULTS

Polishing treatment using BAC lowered the residual MTBE and by-products to non-detect levels in the ozone/peroxide and pulsed UV/peroxide treated waters. However, for waters with influent MTBE concentrations of 2,000 μg/L or greater, longer (more than 25 minutes) retention times in the BAC may be necessary to remove MTBE and its by-products to non-detect levels. The pilot tests with sequential ozone/peroxide or pulsed UV/peroxide and BAC treatment indicated higher overall MTBE and by-products removal efficiencies compared to the AOP treatments alone.

Although the pilot scale test results demonstrated that BAC alone may effectively remove MTBE over a wide range of influent concentrations, depending on the adsorption/desorption phase, the capacity for the BAC may fluctuate and thus may not provide consistent removal efficiencies. Also, fluctuations in MTBE concentration can adversely impact the BAC performance. Thus, in employing BAC in the overall treatment process, the primary AOP treatment should be optimized to provide sufficient removal of both MTBE and the degradation by-products, and BAC should be used as a polishing treatment for removing residual peroxide, MTBE and the oxidation by-products. A loading rate of 2 gpm/ft^2 and a retention time of 20 minutes in BAC was found to be necessary to remove MTBE and its by-products in the AOP treated waters. Costs for BAC polishing treatment varied between $0.20 per 1,000 gallons treated (for a 0.5 mgd system) to $0.16 per 1,000 gallons treated (for a 10 mgd system). More details on BAC treatment are provided in Chapters 4 and 6.

COMPARISON OF AOP TECHNOLOGIES

The above-discussed AOP technologies were compared using the following criteria:

- Treatment costs
- Qualitative factors (e.g., technology reliability, flexibility)
- Influent and treated water quality considerations

Based on the comparative analysis, it was concluded that all the AOP technologies that were evaluated in this study are capable of removing MTBE at 95% or higher efficiencies. But, ozone/peroxide and continuous UV/peroxide appear to be the most feasible technologies for AOP treatment of MTBE in drinking water sources. Water utilities are recommended to perform pilot or full-scale testing to understand their water quality impact on the AOP treatment processes, before selecting and implementing any AOP treatment. More details on comparison of AOP technologies are provided in Chapter 8.

CHAPTER 1
BACKGROUND AND INTRODUCTION

MTBE USE

MTBE is a fuel additive made by combining methanol and isobutylene. The methanol is typically derived from natural gas; isobutylene can be derived as a by-product of the petroleum refinery process. Since 1979, MTBE has been used in the United States as an octane-enhancing replacement for lead, primarily in mid-and high-grade gasoline at concentrations as high as 8 percent (by volume). Since the middle of the 1980s, it has been widely used throughout the country for this purpose. It is also used as a fuel oxygenate at higher concentrations (11 to 15 percent by volume) as part of the U.S. EPA's programs to reduce ozone and carbon monoxide levels in the most polluted areas of the country.

The Oxygenated Fuel (Oxyfuel) and Reformulated Gasoline (RFG) Programs were initiated by the U.S. EPA in 1992 and 1995, respectively, to meet the requirements of the 1990 Clean Air Act Amendments (CAAA). The Oxyfuel Program required 2.7 percent oxygen (by weight) in gasoline during fall and winter months to reduce the carbon monoxide emissions in non-attainment areas. In order to meet this requirement, gasoline producers had to use oxygen-containing compounds termed "fuel oxygenates" (e.g., MTBE, ethanol). Although the total number of areas participating in the Oxyfuel Program changed from year to year, in 1998, the CAAA required 28 areas to participate (USEPA, 1998a). Some of these areas also used RFG in the non-winter months.

The RFG Program required 2-percent oxygen (by weight) year round in the most polluted metropolitan areas to reduce ozone and smog. When MTBE was used to meet the RFG requirements its concentration in gasoline was 11 percent (by volume). The ten metropolitan areas that were required by the CAAA to use RFG were:

- Baltimore
- Chicago
- Hartford
- Houston
- Los Angeles
- Milwaukee
- New York
- Philadelphia
- Sacramento
- San Diego

Another 22 areas in 13 states and the District of Columbia voluntarily participated in the RFG Program: Arizona, Connecticut, Delaware, Kentucky, Maine, Maryland, Massachusetts, New Hampshire, New Jersey, New York, Rhode Island, Texas, and Virginia (USEPA 1998a). California had a separate statewide program with more stringent standards than the federal RFG program.

Although MTBE was not the only oxygenate alternative, it was preferred due to its availability and blending properties. In the year 1998, approximately 4.5 billion gallons of MTBE was used in gasoline, making it the second most produced organic chemical manufactured in the US (USEPA, 2000). Approximately 30 percent of all the gasoline in the US contains oxygenates for compliance with the RFG requirements. An additional 4 percent contains MTBE for compliance with the Oxyfuel requirements. MTBE, which is the most common fuel oxygenate, is used in more than 80 percent of oxygenated fuels. Ethanol, which is the second most common fuel oxygenate, is used in about 15 percent of the oxygenated fuels. Other oxygenates, which constitute the remaining 5 percent of the market, include tertiary amyl methyl ether (TAME), ethyl tertiary butyl ether (ETBE), di-isopropyl ether (DIPE), and tertiary butyl alcohol (TBA).

The cost of transportation and the convenience of use favors MTBE over ethanol. Because MTBE is more compatible with gasoline, it can be blended at the refinery and distributed with gasoline through pipelines. Ethanol, on the other hand, must be shipped separately from gasoline and added at the distribution terminal soon before use. If ethanol-blended gasoline is exposed to water or even water vapor (as in pipelines), ethanol will bring the water into solution and make the gasoline unusable. In addition, if ethanol-blended gasoline is stored for an extended period, the ethanol will begin to separate from the gasoline. As a result, ethanol is often manufactured close to the point of use or shipped by rail, increasing the cost of its use.

Concerns With MTBE Use

Although MTBE is known to be less toxic than many other gasoline constituents, concerns were raised about the potential for acute effects from inhaling it at service stations and long-term effects from drinking water contaminated with it. MTBE is also considered a potential human carcinogen.

In comparison to petroleum products, MTBE poses additional problems when it escapes into the environment through gasoline releases; typically from under-ground storage tank systems, above-the ground storage facilities, or pipelines. MTBE is capable of traveling through soil rapidly, is much more soluble in water than most other petroleum constituents, and is more resistant to biodegradation. As a result,. it often travels farther than other gasoline constituents, making it more likely to impact public and private drinking water wells. Because of its affinity for water and, consequently, its tendency to form large plumes, petroleum releases with MTBE can be more difficult to remediate than petroleum releases that do not contain MTBE.

REGULATORY UPDATE

In December 1997, the USEPA Office of Water issued a health advisory on MTBE. This advisory summarized the MTBE health effects and its organoleptic properties (USEPA, 1997). It did not impose any regulatory requirements on the providers of public drinking water; instead, it recommended concentrations that would be acceptable to most consumers of public drinking water supplies. These recommendations were based primarily on taste and odor thresholds. The health advisory stated that drinking water containing MTBE concentrations in the range of 20 to 40 μg/L [or part per billion (ppb)] would likely avoid unpleasant taste and odor effects for a large majority of people (USEPA, 1997). The document also noted that there is a large variation in the concentrations that people are able to detect MTBE. The various factors influencing detection limits include individual sensitivities, water quality, and water temperature. An independent study conducted by a group of researchers using the American Society for Testing and Materials (ASTM) protocol found 15 μg/L as the geometric mean of the MTBE detection thresholds for a large group of consumer panelists (Stocking et al., 2001).

In response to the growing concerns regarding MTBE in water, the USEPA appointed an independent Blue Ribbon Panel of experts to investigate the use of oxygen additives in gasoline. The Panel called for significant reduction in the use of MTBE in gasoline and recommended that Congress and EPA take action to lift the oxygen mandate and clean up MTBE contamination

(USEPA, 1999). Per the Panel recommendations, the USEPA is taking steps to strengthen its underground storage tank, safe drinking water, remediation, and research programs. Upon release of the Panel's report in September 1999, EPA immediately began working with and encouraging Congress to pass legislation that responded to the Panel's recommendations. The USEPA has provided Congress with a framework for legislation that will give it the authority to significantly reduce or eliminate MTBE. USEPA is also taking action to control MTBE under the Toxic Substances Control Act (TSCA) as a backup to the needed Congressional action. A TSCA rulemaking is procedurally burdensome and may take several years to complete (USEPA, 1997).

In response to occurrence of MTBE exceeding several hundred μg/L in drinking water wells and 15 μg/L in surface waters, the USEPA developed an action level for MTBE based on its adverse taste and odor properties in water at 20 – 40 μg/L. In parallel, states have established varying levels and standards for MTBE (ranging from 5 μg/L in California to >200 μg/L in Wyoming). In late 1998, California announced a SMCL of 5 μg/L based on the aesthetic properties of MTBE; SMCLs in California are enforceable. While the health effects of MTBE are still under debate, California proposed a health-based Public Health Goal (PHG) for MTBE of 13 μg/L. Subsequently, Office of Environmental Health Hazard Assessment (OEHHA) of California adopted a PHG of 13 μg/L for MTBE in drinking water in March 1999 based on carcinogenic effects observed in experimental animals. In December 2000, California adopted an MCL of 18 μg/L for MTBE in drinking water.

Some recent studies indicated that the beneficial effects of fuel oxygenates are fewer for newer vehicles that use more advanced pollution control equipment. With increasing evidence of MTBE in drinking water supplies, the USEPA recommended complete elimination of MTBE from gasoline fuels by 2004. Several states including California, New York and Connecticut adopted legislations prohibiting the use of MTBE as a fuel oxygenate. Presently, more studies are being conducted to substitute MTBE with ethanol and other oxygenates to meet the 1990 Clean Air Act Amendments requirements.

MTBE TREATMENT TECHNOLOGIES

Due to limitations of other drinking water treatment technologies (e.g. activated carbon and air stripping) to consistently and cost effectively reduce MTBE to low concentrations in drinking water, AOPs have received increased interest. Available oxidation technologies for drinking water treatment include: ozonation (O_3), ultraviolet (UV) photolysis, electron beam (E-beam) and several combinations such as O_3/H_2O_2 and UV/H_2O_2.

MTBE removal by ozonation can proceed through two mechanisms, molecular pathway or radical pathway. Of these two reactions, oxidation through molecular pathway occurs slowly (second-order rate constant is < 1 L mole^{-1} sec^{-1}) (Buxton et al., 1988). By contrast, oxidation through the radical pathway is fast (1.6×10^9 L mole^{-1}sec^{-1}) (Buxton et al., 1988). UV photolysis of MTBE can occur upon direct or indirect photolysis (Zepp, 1988). Direct photolysis involves light absorption by MTBE followed by decomposition of MTBE. Indirect photoreactions of MTBE occur when ozone or H_2O_2 is added prior to irradiation. E-beam irradiation produces approximately equal concentrations of oxidizing and reducing species. Irradiation of aqueous solutions results in the formation of the aqueous electron, hydrogen atoms, and hydroxyl radical. The aqueous electron reacts with MTBE at a rate of 1.75×10^7 L mole^{-1} sec^{-1}, while the rate constant for the reaction of MTBE with hydrogen atoms, was found to be $\leq 8.0 \times 10^4$ L mole^{-1} sec^{-1} (Cooper and Tornatore, 1999).

PROJECT GOALS

With the recent emergence of MTBE contamination as an important drinking water quality issue, the water industry has been engaged in searching for appropriate technologies such as AOPs to remove MTBE from drinking water. The purpose of this project was to perform a careful evaluation of the technical and economic feasibility of AOPs for MTBE removal. Specifically, the objectives of this project were:

- Identify and fill data gaps related to the implementation and operation of AOPs with respect to MTBE removal. These data gaps primarily include the by-product formation and control and the effect of water quality parameters on AOP operation to achieve a desired MTBE removal efficiency.
- Select and optimize the design of the most promising AOP(s) as a function of water quality parameters. This will be achieved through bench-scale studies and pilot-scale studies.
- Determine conceptual-level engineering costs for these selected AOPs.

The deliverable from this study was to integrate the results from each AOP analysis into a comprehensive report. The results from each of the AOP analyses will be compared using a variety of charts and tables, prepared and formatted using similar assumptions, (e.g., costs per quantity of water treated; MTBE removal efficiency).

PROJECT APPROACH

Bench-scale and pilot-scale testing was performed to assess the ability of some promising AOPs including O_3/H_2O_2, low-pressure (LP)-UV/H_2O_2, medium-pressure (MP)-UV/H_2O_2, pulsed-UV/H_2O_2 and E-beam technologies to remove MTBE and its by-products from drinking water. This study was initiated by a comprehensive literature survey, with the objective of finalizing bench-scale testing conditions. During the rigorous bench-scale testing of the aforementioned AOP processes, the effect of water quality parameters (e.g., organic carbon, alkalinity, nitrate, bromide), process operational factors (e.g., H_2O_2 dose, oxidant dose, contact time), UV configurations (e.g., pulsed UV, LP-UV, MP-UV), and their influence on the rate of MTBE reduction and the formation of by-products were evaluated.

Bench-Scale Testing

The bench-scale testing that was conducted, as part of this study evaluated the effectiveness of O_3/H_2O_2, UV/H_2O_2 and E-beam under changing water qualities. The research tested each of the AOPs listed above, except E-beam, with the following four water types:

1. low TOC and low alkalinity;
2. low TOC and high alkalinity;
3. high TOC and low alkalinity; and
4. high TOC and high alkalinity

The goal of this testing was to generate design parameters for further evaluation in pilot-scale testing. Water quality impact on oxidation rates and by-product generation were evaluated.

Table 1.1 summarizes the naturally occurring waters that are representative of the four types of waters listed above. By using the same waters for all of the bench-scale testing, the study was able to compare the effectiveness of each of the AOPs under the changing water quality parameters. All of the waters were natural field waters with spiked concentrations of MTBE and TBA. Water quality data from AWWA water database indicated that the four waters selected were representative of most surveyed groundwaters and surface waters. Impacts of other ionic constituents in water such as nitrate and bromide on MTBE destruction and by-product formation was also investigated.

Table 1.1
Summary of the test water types

	Low TOC (< 3 mg/L)	High TOC (> 3 mg/L)
Low Alkalinity (40 – 50 mg/L)	Suffolk County Water Authority	Miami-Dade
High Alkalinity (150 – 200 mg/L)	City of Santa Monica	Long Beach Water District

Bench-scale tests for different technologies were done using different batches of water from the above-mentioned locations. Water quality of the individual batches of water that were used for testing is summarized in the respective sections.

Pilot-Scale Testing

Pilot-scale test plans and experimental matrices were developed giving due consideration to the bench-scale study findings. The goals of the pilot-scale study were to: (1) confirm MTBE removal efficiencies observed in bench-scale studies, (2) identify by-products of MTBE oxidation, (3) evaluate the effect of selected water quality parameters on the AOP performance, (4) determine the effectiveness of biological filters for by-product removal and (5) develop design and operating parameters for economic evaluation of potential full-scale facilities treating either surface or ground waters. Pilot-scale tests of all the AOP technologies were conducted on Santa Monica, CA groundwater to enable an even comparison of the processes.

Pilot-scale tests for the different AOP technologies were performed at different time-periods using different qualities of Santa Monica ground water and State Project surface water. Water quality of the individual batches is summarized in the respective chapters (Chapter 4-7).

CHAPTER 2
MTBE OCCURRENCE AND TREATMENT TECHNOLOGIES

This chapter discusses the occurrence of MTBE in the US and provides an overview of the various treatment technologies that are available to remove MTBE from drinking water supplies. A summary of the MTBE properties is also provided in order to understand the principles behind the various treatment technologies.

MTBE OCCURRENCE

Although scattered incidents of localized water contamination by MTBE have been reported since the early 1980s, the first report to suggest that oxygenate contamination of water might be occurring on a widespread basis came as a result of the USGS National Water Quality Assessment (NAWQA) program. This was the most comprehensive national occurrence survey of MTBE in water resources conducted to date (Squillace et al., 1997; EPA, 1998; Moran et al., 1999; Gullick and LeChavellier, 2000). For samples taken in 1993-94, MTBE was detected (> 0.2 µg/L) much more frequently in shallow groundwater in urban areas (27 percent of sites) than in shallow groundwater in rural areas (1.3 percent) or in deeper groundwater from major aquifers (1.0 percent) (Squillace et al., 1997; EPA, 1998). Of the 60 VOCs measured in urban groundwater, MTBE was the second most frequently detected, ranking just behind chloroform. Only 3 percent of the urban wells and springs sampled contained MTBE concentrations >20 µg/L. In contrast, the maximum MTBE concentration in both the shallow rural groundwater and the deeper groundwater samples was 1.3 µg/L. MTBE was detected in all eight of the urban land areas studied but in only three of the 21 agricultural areas (Squillace et al., 1997; EPA, 1998).

Other regional studies have since supplemented the USGS NAWQA reports. MTBE occurrence surveys for California source water supplies have been reported by the Metropolitan Water District of Southern California (MSDSC) in Los Angeles, the University of California at Davis, and the Lawrence Livermore National Laboratory in Livermore (Dale et al., 1997; Gullick and LeChavellier, 2000). Drinking water supplies in Maine were also surveyed. The USGS has also studied New England groundwater, Long Island (N.Y.) and New Jersey groundwater, fractured-bedrock aquifers in south-central Pennsylvania, and MTBE occurrence in storm water in metropolitan areas. EPA has summarized the results of some of these studies in a report (EPA, 1998).

American Water System of the American Water Works Company performed a survey of surface and subsurface drinking water supplies. The MTBE concentrations found in the water supplies examined in this study were nearly all below 20 µg/L. This study also found that conventional water treatment processes (coagulation/sedimentation/filtration) may not be able to remove MTBE at low levels (Gullick and LeChavellier, 2000). AwwaRF is sponsoring a project to develop a national database of MTBE concentrations in surface water and groundwater drinking water supplies. Performed by the MWDSC and the USGS, this project is scheduled to be completed in 2002 (Delzer et al., 2001).

Studies have also examined the MTBE occurrence at underground tank sites with fuel releases and demonstrated the potential for substantial MTBE contamination. Buscheck (1998) reported detection of MTBE in more than 83 percent of the groundwater plumes monitored at more than 700 service station sites in four states. About 45 percent of the sites had concentrations in excess of 1,000 μg/L. Similarly, MTBE was detected at 78 percent of the 236 leaking underground fuel tank sites in California studied by the Lawrence Livermore National Laboratory; 70 percent of the 236 sites had MTBE present at concentrations >20 μg/L and 10 percent at concentrations >10,000 μg/L (Happel et al., 1998).

PHYSICAL AND CHEMICAL CHARACTERISTICS OF MTBE

The effectiveness of MTBE treatment methods is directly related to its (MTBEs) physical and chemical characteristics. MTBE is a polar organic compound that has a chemical formula of $CH_3OC(CH_3)_3$. At room temperature, it is a volatile, flammable, colorless liquid with a terpene-like odor (Squillace et al., 1997). The taste and odor thresholds for MTBE in water range from 2.5 μg/L to 680 μg/L, and 2 μg/L to 190 μg/L, respectively, depending on the individual being tested, level of chlorination, temperature of water, and other factors (API, 1994; ARCO, 1993; Young et al., 1996; Shen et al., 1998; Malcolm Pirnie, 1998; Dale et al., 1997).

It has been suggested that the treatment of MTBE in drinking water using conventional treatment processes (i.e., air stripping and GAC) poses challenges relative to other organic contaminants, particularly benzene, toluene, ethylbenzene, and xylenes (BTEX) compounds, because of its unique physical and chemical properties. A summary of the physical and chemical characteristics of MTBE relative to BTEX compounds is presented in Table 2.1 (OSTP Report, 1997; Crittenden et al., 1997; California MTBE Research Partnership, 2000). Comparing the characteristics of MTBE with other BTEX compounds helps in showing how the selection of the treatment and remediation technologies may differ if MTBE is present in the water:

- MTBE is about 30 times more soluble than benzene in water. Pure MTBE can reach an equilibrium concentration in water of approximately 5 percent, i.e., 48,000 mg/L. The relatively low organic carbon partitioning coefficient (K_{OC}) and the high water solubility of MTBE make GAC adsorption less effective for MTBE relative to BTEX compounds.
- As the data in Table 2.1 shows, MTBE's Henry's constant is approximately an order of magnitude lower than those of BTEX compounds. Since the air stripping potential of a compound is primarily determined by its Henry's constant (H), air stripping of MTBE will tend to be more difficult and more costly relative to those other compounds.
- Furthermore, MTBE has been found to be more resistant to biodegradation than BTEX compounds. However, several researchers have been investigating various methods to enhance the biodegradation rates for MTBE (e.g., Hanson et al., 1999; Salanitro et al., 1999).
- MTBE is three times more volatile than benzene as indicated by the vapor pressure data. When MTBE is present in the soil as a result of petroleum release, it may separate from the rest of the petroleum, reaching the groundwater faster than the BTEX compounds and dissolving rapidly. Once in the groundwater, MTBE travels at about the same rate as the groundwater since it much less likely to adsorb to soil or organic carbon (EPA, 1998b).

Table 2.1
Physical and chemical characteristics of MTBE and BTEX compounds

Physical and chemical properties	MTBE	Benzene	Toluene	Ethyl-benzene	o-Xylene
Molecular weight (g/mole)	88	78	92	106	106
Vapor density at 1 atm and 10°C	3.80	3.36	3.97	4.57	4.57
Specific gravity at 25°C	0.744	0.88	0.8669	0.867	0.8802
Boiling point (°C)	53.6-55.2	80.1	110.6	136.25	144.4
Water solubility (mg/L) at 25°	43,000-54,300	1730	534.8	161	175
Vapor pressure at 25°C (mm Hg)	245-276	76 – 95	28.4	9.53	6.6
Henry's constant at 25°C (dimensionless)	0.018	0.23	0.272	0.336	0.212
Log K_{oc}	1.091-1.049	1.18-2.16	1.56-2.25	2.94 1.98-3.04	1.68-1.83
Log K_{ow}	1.20	2.36	2.73	3.24	3.10

Date Sources: OSTP Report, 1997; Crittenden et al., 1997; California MTBE Research Partnership, 2000

OVERVIEW OF MTBE TREATMENT TECHNOLOGIES

As mentioned earlier, the physical and chemical properties of MTBE play an important role in the treatment of MTBE above ground. Compared with other components of gasoline, MTBE is more difficult to remove from contaminated water. The more commonly considered strategies for removal of MTBE from drinking water sources include air-stripping, adsorption to activated carbon or other sorbents, and advanced oxidation processes. Brief discussions of these technologies are provided in the following sections.

Advanced Oxidation Processes

AOPs destroy MTBE and other organic contaminants directly in the water through chemical oxidation, as opposed to simply transferring them from the liquid phase into a gas phase (as in the case of air stripping) or solid phase (as in the case of GAC and resins). Removal of organic compounds from water by AOPs is primarily accomplished through the reaction of organic contaminants with highly reactive hydroxyl radicals (·OH) that can be produced through a variety of mechanisms. Compared to more established drinking water treatment alternatives such as air stripping and GAC, AOPs are generally considered emerging technologies (with fewer full-scale installations and less well understood). Currently, there are some full-scale applications where organic contaminants (e.g., PCE and NDMA) are being removed from drinking water using AOPs (California MTBE Research Partnership, 2000). There are, however, no full-scale installations for AOPs for removal of MTBE from drinking water. Thus, thorough pilot- and full-scale testing of an AOP system is required before it can be implemented for MTBE removal in drinking water applications.

Some of the challenges with respect to the implementation of AOPs in drinking water treatment are associated with the formation and fate of oxidation by-products (e.g., TBA and TBF), non-selective radical oxidation, radical scavenging, and bromate formation (for ozone-based AOPs). Although it is possible to overcome these challenges, cost for AOP treatment may be high as a result of the greater energy usage, greater chemical dosage, and/or secondary treatment polishing steps.

Some of the established and emerging AOPs potentially applicable for the removal of MTBE from drinking water include:

- Ozone/hydrogen peroxide
- Continuous wave UV/hydrogen peroxide
- Pulsed UV/hydrogen peroxide
- Electron beam
- TiO_2-catalyzed UV
- Sonication/hydrodynamic cavitation
- Fenton's reaction

Among the various AOP technologies listed above, ozone/peroxide and continuous wave UV/peroxide are well understood and have been demonstrated at several field-scale sites to successfully remove MTBE from water to meet drinking water standards. In addition to these two relatively new AOPs, E-beam and pulsed UV/peroxide were considered for further consideration due to their technical feasibility for removing MTBE from drinking water. E beam and pulsed UV/peroxide technologies are still in developmental stages for removal of organic contaminants in drinking water applications; however, they have been widely demonstrated for disinfection and remediation applications.

Air Stripping

The tendency for a compound to be removed from water by air stripping is characterized by its Henry's constant. Because MTBE's Henry's constant is several times lower than those of other organic compounds commonly treated through air stripping (e.g., TCE and Benzene), air stripping of MTBE is more difficult and more costly than these other compounds. However, air stripping is a proven technology that has been used successfully to remove MTBE from drinking water. To optimize the performance of air strippers, the contact between air and water is maximized while energy costs associated with equipment design are minimized. This optimization provides the highest rate of mass transfer between water and air at the lowest operating cost. The most common mass transfer design for air stripping system is randomly packed towers. Packed tower air strippers are being used successfully for drinking water treatment in La Crosse, Kansas, and Rockaway Township, New Jersey (California MTBE Research Partnership, 2000). In addition to packed towers, established and emerging air stripping technologies potentially applicable for MTBE treatment in drinking water include low profile air strippers, bubble diffusion strippers, spray towers, and aspiration air strippers.

Packed tower aeration was found to be superior to the other air stripping technologies from a cost perspective, regardless of hydraulic capacity, removal efficiency requirements, or initial MTBE concentrations (California MTBE Research Partnership, 2000). At higher flow rates (>600 gpm) and higher removal efficiency (>95 percent), packed towers are less expensive

(California MTBE Research Partnership, 2000). However, for lower flow rates (<100 gpm), low profile air strippers become cost competitive with packed towers (California MTBE Research Partnership, 2000). In addition, lower profile air strippers are generally easier to install, maintain, and modify for changing flow and water quality conditions than packed towers. Thus, for hydraulic capacities less than 100 gpm, which may be a small drinking water application, a low profile air stripper can be used for MTBE removal. For drinking water applications requiring hydraulic capacities greater than 100 gpm the packed tower aeration technology may be used.

Granular Activated Carbon Adsorption

Carbon adsorption technology is implemented by passing contaminated water through a vessel containing GAC. Intermolecular attraction between the dissolved organic contaminant (adsorbate) and the GAC surface (adsorbent) results in adsorptive forces that physically attracts the adsorbate to the GAC as the water passes through the vessel. As such, the adsorbate remains attached to the GAC matrix while the water leaves the system with a decreased concentration. The adsorption potential of a type of GAC for a given contaminant depends on numerous factors, including the GAC structures (i.e., pore size distribution) and the physical and chemical characteristics of the adsorbate. Based on the MTBE isotherms currently available, coconut shell GAC appears to have better adsorption characteristics than coal-based GAC (California MTBE Research Partnership, 2000).

Simplicity and stable operation are the primary advantages of GAC relative to other water treatment technologies. In addition, GAC is well established for removing organic compounds from water. However, field application specifically for MTBE removal in large-scale drinking water systems is currently limited (California MTBE Research Partnership, 2000). GAC systems are easy to implement due to wide commercial availability for other applications. Numerous vendors can supply GAC and the necessary equipment (e.g., vessels, piping, and pumps). Because of the simplicity of the equipment and materials, capital and installation costs are relatively low compared to more innovative technologies. Finally, there is no off-gas treatment required for GAC systems and the formation of by-products is limited to spent carbon that can either be thermally regenerated or discarded.

The effectiveness of GAC for the treatment of MTBE has been limited by its poor physical and chemical adsorption characteristics. In particular, MTBE's high solubility causes the compound to preferentially remain in solution rather than be adsorbed onto a solid surface. In addition, NOM and other SOCs compete with MTBE for the adsorption sites of GAC. Since, MTBE is only weakly adsorbed by GAC, other more preferentially adsorbed SOCs in the contaminated water can result in the desorption or the displacement of previously-sorbed MTBE from the GAC matrix.

The carbon usage rates and unit treatment costs are highly dependent on influent MTBE concentrations, background water quality, and the concentration of other SOCs. Cost analysis suggested that GAC is most cost-effective for the removal of lower MTBE concentrations, which results in lower carbon usage rates and, consequently, lower O & M costs. GAC is also more likely to be cost-effective for waters that have relatively low levels of NOM (e.g., some groundwaters). Computer modeling studies predict that carbon fouling from NOM can cause approximately 50% increase in carbon usage rates for the removal of 20 μg/L MTBE (California MTBE Research Partnership, 2000). Finally, GAC is more cost-effective for waters contaminated solely with MTBE since other SOCs will preferentially occupy adsorption sites, thereby increasing carbon usage rates. Adsorption modeling and cost estimates show that moderate loads of total BTEX (800 μg/L) can cause greater than 50 percent increase in carbon usage rates for MTBE removal for systems treating influent with 20 μg/L MTBE (California MTBE Research Partnership, 2000).

Synthetic Resin Sorbents

Synthetic resin sorbents, like GAC, rely on the process of sorption to remove organic compounds from water (Davis and Powers, 1999). The primary advantage of resin sorbents over GAC is their on-site regenerability and their resistance to competitive NOM sorption. Resins used for drinking water applications could be regenerated on-site through steam stripping or microwave irradiation.

Resin isotherm data for MTBE suggests that Ambersorb 563, a carbonaceous resin manufactured by Rohm and Haas (Philadelphia, PA), is currently the resin industry's most promising candidate for MTBE removal from water. Two independent studies found Ambersorb 563 to have superior sorption capacity for MTBE compared to Filtrasorb 400, a coal-based GAC widely used by the water industry (California MTBE Research Partnership, 2000).

Limited data are available on the effects of background water quality toward the MTBE removal efficiency of resins. Available data suggest that the performance of Ambersorb resins is unaffected by pH (6.5 to 8.5), temperature (10°C vs 25°C), oxidants (HOCL, H_2O_2, and O_3), the presence of NOM, and the presence of TBA. The Ambersorb resins have also been found to be unsusceptible to biofouling. However, *m*-xylene, which can be considered representative of BTEX compounds, has been found to decrease resin MTBE sorption capacity when it is present at relatively high concentrations (43.2 mg/L) (Davis and Powers, 1999).

CHAPTER 3
STATE OF KNOWLEDGE OF AOPs FOR MTBE REMOVAL

INTRODUCTION

Discussed in this chapter is the current state of knowledge of MTBE removal using AOPs. The mechanisms of MTBE oxidation, reactor details, bench and pilot-scale study summaries are discussed for each AOP technology. MTBE oxidation pathways and rates, water quality impacts on oxidation effectiveness and literature gaps are also addressed in this chapter.

AOP TECHNOLOGIES - CURRENT KNOWLEDGE

This section presents a summary of literature review on AOP treatment technologies (e.g., O_3/H_2O_2, UV/H_2O_2 and E-Beam) for removing MTBE and similar contaminants (e.g., ethers) and their byproducts (e.g., t-butyl formate (TBF), t-butyl alcohol (TBA), aldehydes, etc.). Specific emphasis is placed on the importance of the water quality matrix (pH, temperature, total organic carbon (TOC), alkalinity, and other co-contaminants) on AOP operational conditions and performance effectiveness.

MTBE Oxidation by Ozone/Peroxide

O_3/H_2O_2 Process Description

In an O_3/H_2O_2 system, H_2O_2 is used in conjunction with O_3 to enhance the formation of hydroxyl radicals. Since O_3 decomposes rapidly, it is typically produced on-site using a generator fed with dried compressed air or oxygen (Hoigné, 1998). The gas mixtures produced from air and oxygen by an ozone generator usually consist of 0.5 to 1.5% and 1 to 2% by volume O_3, respectively (Montgomery, 1985). The use of air to generate ozone requires dehumidification, which may be cost prohibitive relative to the use of pure liquid oxygen. In addition, larger quantities of ozone can be produced from a unit of liquid oxygen (14% O_2 by weight) compared to a unit of compressed air (2% O_2 by weight), which facilitates greater mass transfer of the ozone into the source water. Finally, ozone can be generated from liquid oxygen using less energy than it can be from compressed air.

When O_3 is added to water, it participates in a complex chain of reactions that result in the formation of radicals such as the hydroxyl radical ($\cdot OH$) and the superoxide radical ($O_2\cdot$) (Hoigné, 1998). Like O_3, these radical products ($\cdot OH$ and $O_2\cdot$) are oxidants that are capable of MTBE destruction. Of the radical intermediates formed in ozonated water, $\cdot OH$ is the most powerful MTBE oxidant, even more powerful than O_3 itself. Oxidation of ethers by O_3 is known to occur very slowly with second-order kinetic rate constants less than 1 $M^{-1}s^{-1}$ (Buxton et al., 1988). By contrast, oxidation of ethers by radical oxidants is extremely rapid. Hydroxyl radicals react with MTBE according to a rate constant of 1.6×10^9 $M^{-1}s^{-1}$ (Buxton et al., 1988).

H_2O_2 can be combined with ozone to enhance the formation of $\cdot OH$ in solution. H_2O_2 is a weak acid, which partially dissociates into the hydroperoxide ion (HO_2^-) in water. H_2O_2 reacts slowly with O_3, whereas the HO_2^- ion can rapidly react with O_3 to form $\cdot OH$ (Hoigné, 1998):

$$H_2O_2 + H_2O \longleftrightarrow HO_2^- + H_3O^+ \tag{3.1}$$

$$O_3 + HO_2^- \longrightarrow \cdot OH + O_2^- + O_2 \tag{3.2}$$

The reaction of MTBE with hydroxyl radical has been shown to result in by-products such as *tertiary*-butyl alcohol (TBA), *tertiary*-butyl formate (TBF), acetone, formaldehyde and acetaldehyde (Liang et al., 1999a).

$$\text{MTBE} + \cdot OH \longrightarrow \text{TBA, TBF, acetone, aldehydes} \longrightarrow CO_2 + H_2O \tag{3.3}$$

O_3/H_2O_2 Reactor Description

In O_3/H_2O_2 AOPs, the O_3 gas is fed through spargers, porous piping or plates, or Venturi-type injectors at dosages equivalent to 1-2 mg/L ozone per mg/L dissolved organic carbon (DOC); however, higher dosages are recommended for source waters with high alkalinity (>100 mg/L as $CaCO_3$) or natural organic matter (NOM) (Hoigné, 1998). O_3 transfer efficiencies from the gas to the water of up to 90-95% can be achieved (Montgomery, 1985). H_2O_2 is fed from an aqueous solution, at peroxide to ozone mass ratios ranging from 0.3:1 to 3:1 (Liang, 1999a, b; Applebury, 1999). The specific ratio will be a function of disinfection requirements, bromide concentration, contaminant concentration, and other water quality parameters.

Since an ozone residual can provide disinfection credit, a lower peroxide to ozone ratio is typically applied to source waters requiring disinfection (e.g., surface waters) in order to leave some ozone residual. However, researchers have shown that bromate formation is a strong function of the H_2O_2 to O_3 ratio, and that higher peroxide to ozone ratios can effectively reduce the concentration of bromate generated (Siddiqui et al., 1994; Liang et al., 1999a and 1999b; White, 1999). These counter-acting effects should be considered when trying to determine the optimal peroxide to ozone dosage ratio to apply for a specific water source with significant (>100 μg/L) influent bromide concentrations. For source waters requiring minimal disinfection (e.g., some groundwaters), a higher peroxide to ozone ratio can be applied to minimize bromate formation. In the case of waters requiring disinfection, alternative bromate formation mitigation measures may be required if a high peroxide to ozone ratio is used for disinfection credit or an alternative disinfectant (e.g., Cl_2) may be used to fully meet disinfection standards.

Two types of ozone contact configurations exist for application: 1) conventional 3 to 5 meter deep continuously stirred reactor basins and 2) long (>100 feet) plug flow reactors. In a conventional ozone reactor, ozone is bubbled through the base of the reactor and allowed to diffuse through the reactor until it either escapes through the top or is completely reacted. This results in high ozone concentrations at the base of the reactor, independent of the contaminant concentrations, which promote the reaction of ozone with other chemical constituents to form regulated by-products (e.g., bromate). These reactors are typically covered so that excess O_3 can be collected and directed to an off-gas decomposer. Automatic monitoring and control systems are used to regulate chemical feed rates, pH, and other parameters. In addition, a variety of safety, monitoring, and control systems are included to facilitate operation.

The second type of H_2O_2/O_3 contact system, referred to as HiPOx, has been commercialized by Applied Process Technology, Inc. (San Francisco, CA). HiPOx is a continuous, in-line plug flow reactor where O_3 and H_2O_2 are injected into the water stream in precisely controlled ratios at multiple ports along the flow reactor. The primary advantage of this system is that high dosages can be applied at the beginning of the flow reactor, where contaminant concentrations are high. As contaminant concentrations are reduced along the line, decreasing dosages can be applied; thereby, controlling formation of regulated by-products (e.g., bromate). Using multiple injection ports, the concentration of molecular ozone in solution can be maintained at a lower concentration, typically below 0.5 mg/L, than in a large continuously stirred reactor. This keeps the H_2O_2 / O_3 ratio high, which in turn increases the rate of molecular ozone being converted to the hydroxyl radical and increases the rate of hypobromite reduction to bromide (Staehelin and Hoigné, 1982; von Gunten and Oliveras, 1997). As a result, the HiPOx system has been shown to effectively reduce bromate formation, even under high concentrations of influent bromide (1,000 μg/L), by minimizing the molecular ozone available to oxidize the bromide to hypobromite and having excess hydrogen peroxide available to reduce any hypobromite produced (Applebury, 1999). In addition, this system can be operated without the loss of pressure experienced by bringing the source water in contact with the atmosphere, thereby reducing pumping costs.

O_3/H_2O_2 Bench and Pilot-Scale Studies

Karpel Vel Leitner et al. (1994) studied the reaction of ozone combined with hydrogen peroxide on gasoline additives such as MTBE in a dilute aqueous solution. Experiments conducted in a semi-continuous reactor with MTBE showed that the use of H_2O_2/O_3 is a more effective treatment process than ozone alone. Applied dosages of 3.0 and 1.7 mg ozone per mg of MTBE were found to result in 80% reduction of MTBE under ozone alone (at pH 8) and under H_2O_2/O_3, respectively. TBA, TBF, and formaldehyde were identified as the by-products of the H_2O_2/O_3 -MTBE reactions.

Dyksen et al., (1992) performed pilot tests using in-line application of ozone and hydrogen peroxide to evaluate process issues for the removal of organic chemicals such as trichloroethylene (TCE), tetrachloroethylene (PCE), *cis*-1,2 dichloroethylene (*cis*-1,2 DCE), and MTBE. The results indicated that H_2O_2/O_3 is more effective than ozone alone for removal of TCE, PCE, *cis*-1, 2 DCE, and MTBE. Non-detectable levels of MTBE (the method detection limit for MTBE was 0.2 μg/L) were recorded using an ozone dosage of 8 mg/L, a contact time of 3 to 6 minutes, and a hydrogen peroxide-to-ozone ratio of 0.5.

Liang et al., (1999a) conducted a pilot-scale study to investigate the effectiveness of ozone and H_2O_2/O_3 processes for MTBE removal in surface water. Using two treatment trains with a total flow capacity of 12 gpm, H_2O_2/O_3 was found to be more effective than ozone alone for MTBE removal under their tested conditions. The results indicated that O_3/H_2O_2, with 4 mg/L of ozone and 1.3 mg/L of hydrogen peroxide, can achieve average MTBE removals of approximately 78% for both source water supplies tested.

Liang et al., (1999b) also investigated the removal of MTBE from contaminated groundwater through the use of ozone and O_3/H_2O_2. Experiments conducted in a large-scale semi-batch reactor again demonstrated that H_2O_2/O_3 (at a H_2O_2 to O_3 ratio of 1.0) was consistently more effective in oxidizing MTBE than ozone alone, even at ozone doses as high as 10 mg/L. Applied ozone doses greater than 10 mg/L were necessary to reduce MTBE concentrations from approximately 200 and 2,000 μg/L to concentrations below the California secondary drinking water standard of 5 μg/L. However, at these dosages and contact times, both ozone and H_2O_2/O_3 were also found to completely oxidize MTBE oxidation by-products, such as TBF and TBA.

MTBE Oxidation by UV/Peroxide

UV/H_2O_2 Process Description

UV light is on the high-energy end of the light spectrum with wavelengths less than that of visible light (400 nm) but greater than that of X-rays (100 nm). UV radiation (hv) can destroy organic contaminants, including MTBE, through direct and indirect photolysis (Zepp, 1988). In direct photolysis, the absorption of UV light by MTBE places it in an electronically excited state, causing it to react with other compounds, and eventually degrade. In contrast, indirect photolysis of MTBE is mediated by hydroxyl radicals that are produced when ozone or peroxide are added to the source water either prior to or during UV irradiation.

The most common sources of UV light are continuous wave low-pressure mercury vapor lamps (LP-UV), continuous wave medium-pressure mercury vapor lamps (MP-UV), and pulsed-UV (P-UV) xenon arc lamps. Both LP-UV and MP-UV mercury vapor lamps produce a series of line outputs, whereas the xenon arc lamp produces a continuous output spectrum. For most traditional applications of UV irradiation with H_2O_2 or O_3, LP- and MP-UV have been used; however, MP-UV is receiving increasing attention because of its greater potential for direct photolysis. In addition, MP-UV lamps radiate over a wider range of wavelengths (200 to 400 nm) than LP-UV lamps, which better facilitates the formation of hydroxyl radicals when hydrogen peroxide is present; hydrogen peroxide absorbs more in the higher wavelengths (250 to 300 nm). Furthermore, while an LP-UV lamp is more electrically efficient than an MP-UV lamp, the latter produces a greater UV output per lamp. Thus, medium pressure UV systems can be expected to use fewer lamps, take up less space, and require less maintenance. Finally, after extensive Environmental Technology Verification (ETV) testing, the USEPA has recently credited Calgon Carbon's (Markham, Ontario, Canada) MP-UV lamp system (Sentinel™) with 3.9 $\log_{10}$ inactivation for *Cryptosporidium parvum* (EPA/600/R-98/160VS, 1999; Bukhari et al., 1999). Calgon Carbon (Markham, Ontario) has decided not to use pulsed UV lamps due to their short lifetimes and minor observed benefits relative to MP-UV (Crawford, 1999).

To describe the removal efficiency for organic contaminants using UV lamps, Calgon Carbon (Markham, Ontario, Canada) has defined the term Electrical Energy per Order of Removal (EE/O) as the kilowatt-hours (kWh) of electricity required to reduce the concentration of a compound (e.g., MTBE) in 1,000 gallons by one order of magnitude (or 90%) (Calgon AOT Handbook, 1996). The unit for EE/O is kWh/1,000 gal/order of removal and is defined at the optimum H_2O_2 or O_3 concentration. According to Calgon Carbon, the EE/O provides a convenient way to compare the effectiveness of removal of various organic compounds, using UV irradiation for a given source water (i.e., the EE/O will change depending on the water quality). The higher the EE/O value of a contaminant, the more difficult, and thus more costly, it is to remove that contaminant relative to those with lower EE/O values. For example, MTBE is more difficult to treat than benzene, with EE/O values of around 10 for MTBE and 2 to 5 for benzene. An EE/O of 10 for MTBE means that it would take ~10 kWh to reduce MTBE from 600 μg/L to 60 μg/L in 1,000 gal of water. It will take another 10 kWh to reduce the MTBE from 60 μg/L to 6 μg/L, and so on (Calgon AOT Handbook, 1996).

The effectiveness of the H_2O_2/UV process for the destruction of MTBE relies on several synergistic oxidation mechanisms. The oxidation of organics can occur by either direct photolysis or reactions with hydroxyl radicals. Hydroxyl radicals are produced from the photolytic dissociation of H_2O_2 in water by UV irradiation (Wagler and Malley, 1994; Calgon AOT Handbook, 1996). As in H_2O_2/O_3 systems, the degradation of MTBE is primarily due to the oxidation reactions initiated by the highly reactive hydroxyl radicals (Stefan et al., 2000; Cater et al., 2000):

$$H_2O_2/\text{UV process: } H_2O_2 \xrightarrow{h\upsilon} 2\,{\cdot}OH \qquad (\lambda < 300\text{ nm}) \qquad (3.4)$$

$$\cdot OH + \text{MTBE} \longrightarrow \text{Oxidation by-products} \qquad (3.5)$$

UV/H_2O_2 Reactor Description

Two basic UV reactor design configurations are used for the removal of organic contaminants from water. Calgon Carbon, Inc. (Markham, Ontario, Canada) currently uses both reactor designs for MTBE removal, depending on the flow rate (Crawford, 1999). For large-scale drinking water applications (>500 gpm), a tower design is typically utilized. In the tower configuration, multiple UV lamps are arranged horizontally within a single large reactor vessel with the contaminated water flowing perpendicularly past the UV lamps. For example, a tower system may consist of twelve 20 kW UV lamps arranged horizontally throughout the tower. Heat transfer for MP-UV lamps is typically less than one degree Celsius for every 4 kWh/1000 gallons. Therefore, no cooling systems are needed for the large-scale tower configuration.

In the small-scale systems (<500 gpm), Calgon Carbon (Markham, Ontario, Canada) employs reactors where a single UV lamp per reactor vessel is arranged vertically. For example, a smaller-scale system may consist of three individual reactor vessels in series, each containing one 30 kW UV lamp in a vertical position. For very small systems (<50 gpm), these higher watt lamps operate at a higher temperature and thus, require a cooling fan to effect heat transfer (Crawford, 1999). Safety interlocks are provided on Calgon UV reactors to protect personnel from both the UV radiation and high voltage supply (Calgon AOT Handbook, 1996).

US Filter (Santa Clara, California, USA) markets a LP-UV oxidation system, referred to as Ultrox, which can use either ozone or peroxide, or a combination of both as supplemental oxidants (Gruber, 1994). A typical Ultrox system can consist of a combination of the following four components: 1) a stainless steel reaction chamber with LP-UV lamps; 2) an air compressor/ ozone generator; 3) a hydrogen peroxide feed system; and 4) a catalytic ozone decomposition unit. As a first step in the treatment process, the contaminated source water is mixed with peroxide and then fed into the reaction chamber. The reaction chamber ranges in size from 325 to 3,900 gallons and is divided into a series of parallel sub-chambers, each housing a bank of LP-UV mercury vapor lamps (Gruber, 1994). As the water flows through each sub-chamber, it passes in front of each bank of UV lamps; the number of sub-chambers and the number of lamps depend on the size of the system and type of contaminant being destroyed. The Ultrox system employs low-intensity UV lamps, hence the surface temperatures of the quartz sheath surrounding each lamp rarely exceeds 90°F (Gruber, 1994). Ultrox systems can operate from flow ranges of 5 gpm to 1,200 gpm and higher flow rates are attainable with multiple treatment trains (Himebaugh, 1999).

To minimize problems associated with potential fouling of the UV lamp sleeves in cases where the influent water has high concentrations of fouling agents (e.g., iron, calcium, and magnesium), UV systems are equipped with automated cleaning devices. Pneumatically driven wipers, periodically clean quartz sleeves that separate the water from the UV lamps. Quartz sleeve cleaning devices are common in UV oxidation technologies, and the costs are generally included in the total costs of the system (Crawford, 1999). The frequency of UV lamp cleaning is a function of the presence of iron and other scaling agents in the water. However, the wiping mechanisms used today are well designed and allow for trouble-free operation for source waters containing moderate concentrations of iron (< 0.5 mg/L) and other fouling agents.

The two primary design variables that must be optimized in sizing a UV/peroxide AOP system are the UV power radiation per unit volume of water treated, more commonly referred to as UV dose, and the dose of hydrogen peroxide. UV dose, when applied to AOP, is a measure of the total lamp electrical energy applied to a fixed volume of water. The units are measured in kWh/1,000 gallons treated. This parameter combines flow rate, residence time and light intensity into a single term. The dose of UV light and peroxide/ozone required per unit volume of water treated will vary from one type of water to another. Design tests are typically performed to measure the UV dosage required to achieve the desired effluent concentration. The dosage to be applied is determined in an iterative manner by examining the effect on treatment of selected process variables such as pH, oxidant concentration and retention time.

The key design and operating parameters for a UV/peroxide system include the H_2O_2 dose, the UV lamp type and intensity, the reactor contact time, and the control systems (pH and temperature). The low molar extinction coefficient for H_2O_2 (Wagler and Malley, 1994) results in the use of MP-UV lamps for higher hydroxyl radical yields—UV doses typically range from 2.5 kWh/1000 gallons to 15 kWh/1000 gallons depending on water quality and contaminant concentrations (Crawford, 1999).

Hydrogen peroxide can be added either as a single slug dose or at multiple points in the system. The optimum dose of H_2O_2 is normally determined for each water source based on bench/pilot-scale testing, but, is commonly estimated at twice the COD and not less than 1 to 2 mg/L. Where COD data is not available TOC can be used as a surrogate to estimate the peroxide dosages. As previously noted, currently, there are no federal or state regulations for H_2O_2 residual in treated drinking water. Since peroxide can cause corrosion in the distribution system pipes and also react with residual chlorine, drinking water purveyors may not allow any residual peroxide in treated waters. Additionally, low concentrations of peroxide can increase the dissolved oxygen concentrations in the water and thereby encourage bacterial re-growth in the distribution system. Thus, if H_2O_2 is added at very high concentrations (>10 mg/L), polishing treatment with GAC or BAC is required. Consequently, Calgon Carbon (Markham, Ontario, Canada) commonly keeps H_2O_2 doses at less than 3-5 mg/L to minimize H_2O_2 residuals. Once the optimum H_2O_2 dose is determined, the EE/O for the target compound is used to determine the energy costs.

Pulsar Environmental Remediation Technologies, Incorporated (Auburn, California) markets modular, pulsed UV reactors, known as Riptide™ systems, for remediation applications. These Riptide™ reactors are available in 3 different sizes: 1-10 gpm (Riptide-8), 10-60 gpm (Riptide-20) and 60-400 gpm (Riptide-350) (Bender, 1998). The large Riptide™ reactor (Riptide-350) is a vertical chamber of 20-inch diameter and 6-ft tall (Bender, 1999). The Riptide™ system is comprised of a multi-pass reaction chamber containing a high-energy UV flashlamp. These Pulsar (Auburn, California) UV lamps radiate UV light in a broad spectrum ranging from 185 nm to 400 nm, in a radiation profile known as blackbody or continuum radiation (Bender, 1998). The

Pulsar (Auburn, California) lamps also radiate in the visible and infrared from 400 nm to 3000 nm, in accordance with the blackbody profile (Bender, 1998).

UV/H_2O_2 Bench and Pilot-Scale Studies

Wagler and Malley (1994) conducted bench-scale studies to determine the effectiveness of ultraviolet (UV) light, H_2O_2, and UV combined with H_2O_2 in removing MTBE from contaminated groundwater in New Hampshire. In general, treatment of a simulated groundwater with pH between 6.5 and 8.0 by UV alone or by H_2O_2 alone produced less than 10% removal of MTBE after 2 hours of exposure. In contrast, the combination of UV and H_2O_2 within the pH range of 5.5 to 10 produced more than 95% removal of MTBE after only 40 minutes of exposure time. This study confirmed that the hydroxyl radical formed in the H_2O_2/UV process is the primary oxidant responsible for the oxidation of MTBE. During these oxidation experiments, methanol, formaldehyde, TBA, and 1,1-dimethylethyl-formate were identified as by-products of the H_2O_2/UV process. Furthermore, H_2O_2/UV oxidation of an actual groundwater containing MTBE and other VOCs resulted in 83% removal of MTBE after 2 hours of contact time.

Chang and Young (1999) determined the kinetics of H_2O_2/UV degradation of MTBE by using a recirculating batch reactor with a low-pressure mercury UV lamp. With a spiked MTBE concentration of 10 mg/L, H_2O_2/UV treatment resulted in 99.9% removal. The major by-product identified was TBF. The second order rate constant for the MTBE/·OH reaction under the H_2O_2/UV treatment process was found to be $4.82 \times 10^9\ M^{-1}s^{-1}$. The mean second order rate constant for the reaction of TBF with ·OH was found to be $1.19 \times 10^9\ M^{-1}s^{-1}$. The yield for TBF formation from the MTBE/·OH reaction was calculated to be 27% under the conditions of this experiment.

Stefan et al., (2000) proposed a detailed reaction mechanism for the degradation of MTBE by UV/peroxide oxidation. According to this mechanism, the primary by-products of MTBE oxidation by ·OH are TBF, 2-methoxy-2-methyl propionaldehyde, formaldehyde, acetone, TBA and methyl acetate (Stefan et al., 2000). Other intermediates, such as carbonyl compounds (hydroxy-*iso*-butyraldehyde, hydroxyacetone, pyruvaldehyde) and organic acids (hydroxy-*iso*-butyric, formic, pyruvic, acetic, oxalic) were also detected (Stefan et al., 2000).

Cater et al., (2000) developed a kinetic model for predicting the EE/O values for MTBE oxidation by UV/peroxide. The EE/O values for pseudo-first-order degradation of MTBE were found to be between 0.2 and 7.5 kWh/m^3/order (Cater et al., 2000). These EE/O values were found to be a function of the concentrations of MTBE, peroxide and other organics such as benzene, toluene and xylene (BTX). The removal efficiency at initial MTBE concentration of 10 mg/L was not affected by the presence of BTX at concentrations of less than 2 mg/L. However, in the UV/peroxide treatment of MTBE, the removal efficiencies decreased significantly at BTX concentrations above 2 mg/L (Cater et al., 2000).

In preliminary bench-scale studies, peroxide-assisted Pulsar (Auburn, California) P-UV systems successfully reduced the MTBE from influent concentrations ranging from 40 to 2,000 ug/L to less than 5 ug/L (Bender, 1998). Presence of high turbidity, large particles and excess TDS can effect the performance of pulsed UV systems (Bender, 1999), similar to other UV-dependent AOPs. Pulsar Environmental (Auburn, California) is currently working with NSF International for certification of the Riptide™ system for use in drinking water applications (Bender, 1999).

In July 1998, a pilot treatment plant was constructed at the City of Santa Monica, Charnock well field to evaluate treatment technologies for removal of MTBE and TBA from drinking water. The treatment plant included several treatment processes including an H_2O_2/UV oxidation system (for MTBE, TBA destruction), several carbon adsorption systems (for by-product destruction and polishing), and a packed tower air stripper. Additional systems, including a granular media filter and bag filter with an oxidant injection system, were installed to evaluate iron and manganese removal for pre-treatment to the H_2O_2/UV system. Calgon Carbon (Markham, Ontario, Canada) provided the H_2O_2/UV system, which consisted of a tower reactor with three MP-UV lamps. The reactor was approximately 42-inches in diameter and 6-feet in height. The pilot facility began operation on July 31, 1998 at a design flow of 140 gpm but testing was conducted at flows as high as 350 gpm. The pilot testing was conducted over a period of approximately 12 months and included an optimization phase, a reliability phase, and a sensitivity phase for MTBE, TBA, and other by-product testing (Rodriguez, 1999).

Raw groundwater from Charnock well field was spiked at concentrations of approximately 1,000 μg/L MTBE and/or 200 μg/L TBA. MTBE and TBA were removed to less than 10 μg/L; however, the results of the pilot testing indicated that the costs were significantly higher than expected and the removal efficiencies were lower than predicted. The detection of by-products in the treated water (TBA, TBF, and acetone) mandated the use of an additional treatment unit, which would further increase the cost of treatment. Finally, residual H_2O_2 in the treated water would require installation of a carbon system for its removal. Testing conducted indicated that the H_2O_2/UV technology worked for MTBE removal, but the high energy requirements, complications caused by several sleeve/lamp failures, formation of by-products, and requirement for additional treatment processes significantly reduced the advantages of this technology (Rodriguez, 1999).

The recent emergence of UV irradiation as a disinfection technology has prompted the investigation of possible negative side effects of UV irradiation on drinking water containing NOM. In particular, there have been concerns regarding the potential effects of UV irradiation on by-product formation when UV is used in conjunction with chlorine addition. Chlorine is sometimes added at plant headworks as a pre-treatment step or, more likely, used as a primary or secondary disinfectant farther down the treatment train. The use of chlorine is associated with the formation of by-products such as trihalomethanes (THMs) and haloacetic acids (HAA_9: HAA_5 plus tribromo-, bromochloro-, bromodichloro- and dibromochloroacetic acid) from NOM normally present in water. These by-products are suspected human carcinogens and are regulated by the EPA. A number of researchers have investigated the potential of UV irradiation to change the composition or distribution of these by-products by promoting the photolysis of NOM (e.g., humic acids) into smaller molecules, which have higher potential for THM and HAA_9 formation (e.g., Stewart et al., 1993; Hengesbach et al., 1993).

Zheng et al., (1999a) recently showed that UV irradiation at a dosage of 100 mJ/cm^2 (the expected maximum UV dosage used for water disinfection applications) of water pre-chlorinated at 6 to 48 mg/L can result in a 1 to 7% decrease in the THM concentration and a change in the HAA_9 concentration ranging from –3% to 4%. These results suggest that under typical treatment conditions, where pre-chlorination is usually less than 5 mg/L, there will be negligible change in THM or HAA formation. It was suggested that UV irradiation might have caused a fraction of the Cl_2 residual to decompose into chlorine radicals (Cl·), which then reacted with THM precursors and converted them into HAA_9 precursors. In a parallel study, Zheng et al., (1999b) also studied the effect of UV irradiation prior to chlorine addition on the subsequent formation of THMs and HAA after chlorine is added. The study found that UV irradiation (dosage range from 0 to 3,000 mJ/cm^2) had insignificant effects on the formation of THMs after the addition of chlorine at a dosage of 5 mg/L to 10.5 mg/L. Similar results were observed for HAA formation. Thus, it could be inferred that the use of UV, especially at doses less than 100 mJ/cm^2, either prior to or after a chlorination process, has a minor impact on THM and HAA_9 formation.

Zheng et al., (1999b) investigated the effects of pre-chlorination with H_2O_2/UV on THM and HAA_9 formation. Subjecting drinking water to a UV dosage of 100 mJ/cm^2 and H_2O_2 dosages ranging from 3.6 to 51 mg/L, THM concentrations increased between 2% and 7% and HAA concentrations increased between 2% and 6%, compared to values observed without H_2O_2/ UV treatment. THM and HAA formation generally increased linearly for UV doses up to 2,000 mJ/cm^2. At a UV dosage of 2,000 mJ/cm^2, the THM concentrations increased between 37% and 146% and HAA_9 concentration increased between 39% and 128%. UV doses for MTBE removal applications are expected to be greater than 2000 mJ/cm^2 and thus increases in THM and HAA formation may be significant. Consequently, these preliminary results suggest that the combined effects of H_2O_2 and UV irradiation on THM and HAA_9 formation potentials should be considered in drinking water situations.

MTBE Oxidation by E-Beam

E-Beam Process Description

High-energy electron beam (E-beam) treatment refers to the use of ionizing radiation from an electron beam source to initiate chemical changes in aqueous contaminants. In contrast to other forms of radiation, such as infrared and ultraviolet, ionizing radiation from an E-beam is absorbed almost completely by the target compounds' electron orbitals, thus increasing the energy level of its orbital electrons. The energy level of radiation is sufficiently high to produce changes in the molecular structure of compounds, but is too low to induce radioactivity (Siddiqui et al., 1996a; HVEA, 1999). Electron beam processes use the portion of the electromagnetic spectrum between 0.01 eV and 10 eV (Siddiqui et al., 1996a).

Within 10^{-16} to 10^{-12} seconds, E-beam irradiation (∧∧∧) of water results in the formation of electronically excited species, including ions and free radicals, along the path of the electrons. The products of direct reactions of water molecules with the electron beam are formed in isolated volumes referred to as "spurs". As these spurs expand through diffusion, a fraction of the initial products escape into the bulk solution and transfer their energy to other aqueous chemical species, causing more reactions to occur (Nickelsen et al., 1992). After approximately 10^{-7} s, oxidizing species, such as hydroxyl radicals, and reducing species, such as aqueous electrons and hydrogen atoms, are formed from the E-beam irradiation of water (Nickelsen et al., 1992; Allen, 1961). The net reaction is shown below:

$$H_2O + \wedge\wedge\wedge \longrightarrow 2.7\ \cdot OH + 0.6\ \cdot H + 2.6\ e_{aq}^- + 0.45\ H_2 + 0.7\ H_2O_2 + 2.6\ H_3O^+ \qquad (3.6)$$

The combination of products that result from this reaction creates a unique environment where oxidizing and reducing reactions occur simultaneously (Allen, 1961). In particular, note that the oxidizing species, $\cdot OH$, and the reducing species, e_{aq}^-, are expected to be present in similar steady-state concentrations. These two species, along with another reducing species, the hydrogen atom ($\cdot H$), are the most reactive products of this reaction and control the rate of the electron beam process for MTBE destruction. Some of the reactions of these species with MTBE are as follows (Cooper and Tornatore, 1999):

$$MTBE + e_{aq}^{-} \longrightarrow (CH_3)_3C\cdot + {}^{-}OCH_3 \qquad (3.7)$$

$${}^{-}OCH_3 + H_2O \longrightarrow HOCH_3 + OH^{-} \qquad (3.8)$$

$$MTBE + \cdot H \xrightarrow[-H_2]{} \text{Reduction by-products} \qquad (3.9)$$

$$\cdot OH + MTBE \longrightarrow \text{Oxidation by-products} \qquad (3.10)$$

The aqueous electron reacts with MTBE according to a rate constant of $1.75 \times 10^7\ M^{-1}s^{-1}$, while the rate constant for the reaction of MTBE with hydrogen atoms was found to be less than $8.0 \times 10^4\ M^{-1}s^{-1}$ (Cooper and Tornatore, 1999). The reaction rate of MTBE with hydroxyl radicals is approximately 90 and 20,000 times faster than with aqueous electrons and hydrogen atoms, respectively (Buxton et al., 1988; Cooper and Tornatore, 1999). Therefore MTBE oxidation with E beam is likely to occur through hydroxyl radical route than the other routes.

Radiolysis is the chemical effect produced when materials (in this case water) are exposed to high-energy electrons in the case of the electron beam process. In aqueous solution, high-energy electrons injected into the solution result in the following reaction describing the pure water reactants that are formed 10^{-7} seconds after the electrons are injected into the water (Buxton et al., 1988):

$$H_2O \text{ -/\!\!\setminus\!\!/\!\!\setminus-> } [2.6]\ e^{-}_{aq} + [0.6]\ H\cdot + [2.7]\ \cdot OH + [0.7]\ H_2O_2 + [2.6]\ H_3 0^{+} + [0.45]\ H_2 \qquad (3.11)$$

or,

$$H_2O \text{ -/\!\!\setminus\!\!/\!\!\setminus-> } [0.26]\ e^{-}_{aq} + [0.06]\ H\cdot + [0.27]\ \cdot OH + [0.07]\ H_2O_2 + [0.26]\ H_3O^{+} + [0.045]\ H_2 \qquad (3.12)$$

Unlike photochemical reactions where one photon of light initiates one (molecular) reaction, a high-energy electron is capable of the formation of several thousand reactive species as it dissipates its energy. The efficiency of conversion of high-energy electrons to a chemical process is defined as G (shown in brackets). G is the number of radicals, excited states, ions or other products, formed or lost in a system absorbing 100 eV of energy (equation 3.10), or in SI units, $\mu mol\ J^{-1}$ (equation 3.11). (G values in equation 1, molecules $100eV^{-1}$, are converted into SI radiation chemical yields of $\mu mol\ J^{-1}$ by multiplying with 0.1036). The concentration of the three reactive species and hydrogen peroxide, in pure water is summarized in Table 3.1.

Of the radiolysis products formed in aqueous solution, the most reactive are the oxidizing, hydroxyl radical ($\cdot OH$), and the reducing, aqueous electron (e^{-}_{aq}) and hydrogen atom ($H\cdot$). Thus, the chemistry of primary interest in the high-energy electron beam process is of the three reactive species Table 3.2). The reactions of these three reactive species result in the destruction of the pollutants and other constituents in the water to be treated. (For the most current compendium of rate constants the reader is referred to the University of Notre Dame Radiation Laboratory web

Table 3.1
Estimated concentrations of transient reactive species in pure water at several doses using high energy electron irradiation

Dose (kGy)	Concentration (mM)			
	e^-_{aq}	H·	·OH	H_2O_2
1	0.27	0.06	0.28	0.07
5	1.4	0.3	1.4	0.4
10	2.7	0.6	2.8	0.7

Table 3.2
Properties of transient reactive species formed by the electron beam process

Reactive Species	·OH	e^-_{aq}	H·
Redox Properties	Strongly Oxidizing	Strongly Reducing	Reducing
Mode of Action	Electron Transfer H Abstraction OH Addition	Electron Addition	H Abstraction H Addition
G Values	2.7	2.6	0.6

site http://www.rcdc.nd.edu). The fact that ionizing radiation produces both oxidizing and reducing species simultaneously differentiates it from most other advanced oxidation technologies.

The data in Table 3.1 suggests that at relatively moderate doses the concentration of the reactive species is relatively high, in the milli-molar concentration range. In a recent study it was shown that the electron beam process is the most efficient process for generating the ·OH (Bolton et al., 1998). The two compounds that were studied were both destroyed primarily by the ·OH; however, because of the high concentration of e^-_{aq}, compounds that do not react with the ·OH, for example carbon tetrachloride, can also be effectively removed (destroyed) in aqueous systems with the electron beam process.

A major advantage of the electron beam process is that electrons can penetrate solid material. We have shown that the presence of clay, up to 3-5%, does not adversely affect the removal of the pollutant chemicals (Cooper et al., 1993a; Lin et al., 1995; Mak et al., 1996; 1997; Nickelsen et al., 1992; 1994). Another advantage of this penetrability is that closed systems can be engineered where the solution to be irradiated at no time comes in contact with the electron accelerator.

The electron beam process, because of the short reaction times of less than 0.1 seconds, is configured as a flow through treatment. In a typical treatment system the absorbed dose can be determined by measuring the temperature difference of the water stream before and after irradiation. The relationship is derived for pure water by the following relationships:

$$0.01 \text{ Gy} = 100 \text{ erg g}^{-1} \qquad (3.13)$$

or

$$10\ \text{kGy} = 1.0 \times 10^8\ \text{erg g}^{-1} \quad (3.14)$$

substituting in,

$$1\ \text{erg} = 2.39 \times 10^{-8}\ \text{cal} \quad (3.15)$$

then,

$$10\ \text{kGy} = 2.39\ \text{cal g}^{-1} \quad (3.16)$$

converting to °C,

$$10\ \text{kGy} = 2.39°\text{C} \quad (3.17)$$

or,

$$\Delta T = 4.18\ \text{kGy}\ °\text{C}^{-1} \quad (3.18)$$

The relationship between dose and temperature is one way to estimate relative energy consumed for the treatment of a compound(s). It provides an estimate of the temperature increase in the treated solution and this can then be related to energy (and cost) for the treatment.

E-Beam Reactor Description

In the E-beam process, a continuous stream of high-energy electrons irradiates contaminated water. The generation of high-energy electrons is accomplished through the use of an electron accelerator in which electrons emitted by a hot cathode (e.g., tungsten filament) are accelerated by means of a voltage differential (Nickelsen et al., 1998). The accelerated electrons are then deflected magnetically by a scanner to produce an E-beam which scans the water surface with a particular radiation pattern and frequency. Once in the water, the electrons react with water molecules to form reactive intermediates, such as hydroxyl radicals, hydrated electrons, and hydrogen atoms, as discussed above (Cooper, 1998). The shape and frequency of the E beam pattern is controlled to apply a uniform amount of electrons (dose) to the source water stream. A common unit of electron dose is the rad, defined as the energy absorption of 100 ergs per gram of material. The maximum depth of penetration of an E-beam is directly proportional to the energy of the incident electrons and inversely proportional to the density of the falling stream, the beam power, and the length of time the water is exposed to the electron beam. Most E-beam systems for drinking water treatment are designed such that the electron beam infiltrates less than a centimeter into the source water. For example, 1.5 MeV electrons have a depth of penetration of approximately 7 mm in water (Nickelsen et al., 1994). E beam irradiation is not significantly affected by the turbidity or color of the feed water.

Currently, there is only one E-beam configuration used to treat drinking water: a high energy electron beam scans in a raster pattern over a thin (approximately 4 mm) sheet of water. This configuration is designed to apply radiation doses up to several thousand krad and scanned at 200 Hz by 60 Hz to cover an area 1.2 m wide by 5 cm deep (Nickelsen et al., 1994). A key piece of equipment for the application of E-beam is the design of a water distribution system. For E-beam to be effective, the source water must be spread over a plate at a sufficiently shallow thickness to allow electrons to penetrate most of the water. If the water thickness is too deep because of higher flow rates or limitations of the distribution system, multiple water passes through the electron beam may be required to meet effluent goals. Because E-beam systems can potentially emit x-rays, the electron accelerator, the beam scanner, and the contact chamber are usually completely surrounded by lead of varying thickness to attenuate any emitted x-rays to less than 0.2 mRem/h (Nickelsen et al., 1998).

E-Beam Bench and Pilot-Scale Studies

Over 700 E-beam systems have been implemented in materials applications and food and drug industry disinfection applications worldwide (Tornatore, 1999). There are presently no electron beam processes in continuous drinking water service, but over 200 pilot and demonstration studies have been performed on over 60 different organic contaminants in remedial applications. Some of the targeted contaminants include benzene (Nickelsen et al., 1992; Nickelsen et al., 1994), phenol, TCE, and PCE (Lin et al., 1995; USEPA, 1998b), THMs and THM precursors (Cooper et al., 1999), and MTBE (Cooper, 1998). Of the 200 pilot and demonstration studies, more than 10 have been designed to evaluate performance on MTBE, TBA, TBF and formaldehyde (Tornatore and Cooper, 1999).

Many of these pilot-scale studies have been completed using the mobile E-beam treatment system designed and built by High Voltage Environmental Applications (HVEA). This system can have process flow rates of up to 40 gpm at an applied power of 20 kW (Cooper et al., 1999; HVEA, 1999). The HVEA also has a research station, formerly known as EBRF (Electron Beam Research Facility), at the Miami Dade Central District Wastewater Treatment Plant in Miami, Florida (HVEA, 1999). At this (EBRF/HVEA) facility, there is an E-beam reactor designed for large-scale research. The E-beam Reactor in Miami Dade is comprised of a horizontal 1.5 MeV insulated-core transformer (ICT) electron accelerator capable of delivering up to 60 mA of beam current (Nickelsen et al., 1994).

Test results from MTBE pilot and demonstration studies have shown the ability of E-beam systems to reduce MTBE concentrations from 1,000 μg/L to less than 5 μg/L (Tornatore, 1999). Tornatore and Cooper (1999) also conducted large-scale E-beam experiments for MTBE destruction with a flow rate of 100 gallons per minute in a recycle mode. MTBE was reduced to below the detection limit (87 μg/L) after cumulative doses of 665 and 2,000 krads applied to initial MTBE concentrations of 2,300 and 31,000 μg/L, respectively. At the equivalent energy dose, primary reaction intermediates (TBA, TBF) were also treated to low residual concentrations. A dose response curve for TBA and TBF has been developed which suggests that the energy required for removal of these compounds to below detection levels is approximately equal to the energy requirement for MTBE removal. TBA and TBF reactions use reducing chemistry, and thus, these reactions proceed concurrently with the process of MTBE destruction (Tornatore, 1999).

A series of experiments were performed with the electron beam technology at Orange County Water District (OCWD) to treat MTBE and a variety of other contaminants in a number of different water sources (Cooper et al., 1999). The results indicate that MTBE was readily treated in all experiments, and that treatment efficiency was dependent upon basic water chemistry, delivery system limitations, and the presence of competing organic and inorganic compounds.

MTBE removal efficiencies of greater than 99.5%, with final concentrations less than 5 μg/L, were reported in several experiments (Cooper et al., 1999). Primary reaction intermediates (TBA, TBF) were reduced concurrently with MTBE, suggesting that the oxidizing and reducing chemistry involved is efficient in treating MTBE and its by-products. Feed water chemistry was shown to have an impact on the treatment efficiency of MTBE and other compounds; waters lower in TOC and pH demonstrated higher removal efficiencies. Researchers concluded that the application of the electron beam process is best suited for high flow rate, single or multiple constituent treatment scenarios where complete oxidation or mineralization of contaminants is the desired endpoint (Cooper et al., 1999).

One aspect that differentiates the E-beam process from other AOP technologies is that with the constant production of both oxidizing and reducing species, when by-products are formed that do not react efficiently with the oxidizing ·OH and do with the e^-_{aq}, increased removal efficiency can result. An example of this would be be acetone. The ·OH reaction rate is relatively slow (1.3×10^8 $M^{-1}s^{-1}$) and in a predominantly oxidizing system acetone would accumulate. However, the reaction rate constant of acetone with the e^-_{aq} is rapid (6.5×10^9 M^{-1} s^{-1}) and destruction is initiated via a reducing pathway. Therefore, the presence of the e^-_{aq} would help to minimize the formation of acetone when waters containing MTBE are treated.

In summary, some key features of the E-beam process are:

1. It is non-selective in the destruction of organic chemicals because strongly reducing reactive species (e^-_{aq}/H·), and strongly oxidizing reactive species (·OH), are formed at the same time and in approximately the same concentrations in solution.
2. The formation of the reactive species is pH independent in the range 3-11.
3. The process is temperature independent.

MTBE OXIDATION PATHWAYS AND RATES

The reaction between ·OH and many organic contaminants occurs rapidly; however, this reaction by itself does not directly result in the mineralization of these contaminants but produces organic oxidation byproducts that can further react with ·OH. There are at least three independently proposed mechanisms in the literature for the complete oxidation of MTBE to carbon dioxide and water (Barreto et al., 1995; Kang and Hoffmann, 1998; Kang et al., 1999; Stefan et al., 2000). If the reaction rate for a particular by-product is slower, it may be the rate-limiting step in the complete mineralization of the target compound and that by-product will accumulate. Ideally, AOP systems are designed to completely mineralize the organic contaminants of concern to CO_2 and H_2O but this may require more energy and greater chemical dosages, and may ultimately prove to be cost prohibitive in certain applications.

While highly dependent on the specific water quality, Table 3.3 shows experimentally determined hydroxyl radical rate constants for MTBE and its various oxidation by-products (water quality varies, as discussed in the literature). A possible mechanism for the oxidation of MTBE involves either direct conversion to TBA or oxidation of the methyl group to form TBF. TBF can then be hydrolyzed to TBA and formaldehyde. TBA can lose a methyl group to form isopropyl alcohol, which can be further oxidized to acetone. In some AOPs, further oxidation of these by-products was found to be the rate-limiting step in the ultimate mineralization of MTBE. Finally, acetone may be converted to formaldehyde or formate, which can be mineralized to CO_2 and H_2O. As stated above, economic considerations can limit the complete mineralization of

Table 3.3
Hydroxyl radical oxidation rate constants for MTBE and its by-products

Compound	Reaction	Second order rate constants (L $mole^{-1}$ sec^{-1})	Reference
MTBE	$(CH_3)_3COCH_3 + \cdot OH \rightarrow (CH_3)_3COCH_2\cdot + H_2O$	1.6×10^9 4.82×10^9	Buxton et al. (1988) Chang and Young (1999)
TBF	$(CH_3)_3COCHO + \cdot OH \rightarrow ?$	1.19×10^9	Chang and Young (1999)
TBA	$(CH_3)_3COH + \cdot OH \rightarrow \cdot CH_2C(CH_3)_2OH + H_2O$	4.8×10^8	Buxton et al. (1988)
Isopropyl alcohol	$(CH_3)_2CHOH + \cdot OH \rightarrow (CH_3)_2COH + H_2O$	1.6×10^9	Buxton et al. (1988)
Acetone	$CH_3COCH_3 + \cdot OH \rightarrow ?$	1.1×10^8	Buxton et al. (1988)
Formaldehyde	$HCHO + \cdot OH \rightarrow ?$	1×10^9	Buxton et al. (1988)
Acetaldehyde	$CH_3CHO + \cdot OH \rightarrow CH_3CO + H_2O$	7.3×10^8	Buxton et al. (1988)
Glyoxal	$HCOCHO + \cdot OH \rightarrow \cdot COCHO + H_2O$	6.6×10^7	Buxton et al. (1988)

MTBE, leaving short-chained carboxylic acids, alcohols, aldehydes, and/or acetone in the effluent water (Karpel Vel Leitner et al., 1994; Liang et al., 1999a). These residual oxidation compounds are also produced by the partial oxidation of NOM (Hoigné, 1998). These compounds represent a source of concern in drinking water applications due to their high solubility and uncertain toxicity. In addition, the presence of these more easily degradable compounds may promote biological growth and fouling in the distribution system. Thus, these compounds are often further treated with a biological activated filter or some other polishing treatment step.

IMPACT OF WATER QUALITY PARAMETERS

There are many water quality parameters that may impact the effectiveness of MTBE destruction by AOPs. For example, for a given AOP system, nearly all dissolved organic compounds present in the source water will serve to reduce the removal efficiency of the target compound by consuming ·OH (Hoigné, 1998). Below is a brief discussion of each of these water quality parameters and the mitigation measures that can be taken to limit the impact of these parameters on increased costs of AOP to meet the required effectiveness.

Alkalinity

The detrimental impact of alkalinity on the effectiveness of AOPs has been extensively studied (AwwaRF, 1998). As mentioned previously, the hydroxyl radical is non-selective and thus, will react with organic or inorganic compounds other than the contaminants of concern. Both carbonate and bicarbonate will scavenge hydroxyl radicals to create carbonate radicals which in turn react with other organic or inorganic compounds present, albeit at a much slower rate (Hoigné and Bader, 1976; AwwaRF, 1998). The reaction for the scavenging of hydroxyl radicals by bicarbonate ions is shown below (Morel and Hering, 1993):

$$\cdot OH + HCO_3^- \rightarrow CO_3\cdot + H_2O \qquad (3.19)$$

The rate constants, k, for the reactions of the hydroxyl radical with carbonate and bicarbonate are 3.8×10^8 and 8.5×10^6 $M^{-1}s^{-1}$, respectively (Buxton et al., 1988). These rate constants are much slower than the reaction rate constant of hydroxyl radicals with MTBE (~10^9 $M^{-1}s^{-1}$). However, for these second order reactions, the actual reaction rate, r, is a function of both the rate constant and the concentration of the reactant, C: $r = k\,[C]$. Waters with medium to high alkalinities (>100 mg/L as $CaCO_3$) likely contain carbonate and bicarbonate ions at concentrations several orders of magnitude greater than MTBE and thus, the reaction of hydroxyl radicals with carbonate and bicarbonate can proceed as fast or faster than their reaction with MTBE. Consequently, MTBE-impacted waters high in bicarbonate ions may require a lowering of alkalinity (e.g., pH adjustment or carbon dioxide stripping) prior to treatment by AOP (AwwaRF, 1998) or higher doses of oxidants coupled with increased reaction time.

Total Organic Carbon and Natural Organic Matter

TOC measurement incorporates all organic compounds present in the water, both dissolved organic carbon (DOC) and particulate organic carbon (POC). Source waters for drinking contain TOC concentrations typically in the range from less than 1 mg/L to greater than 7 mg/L and include naturally occurring compounds and anthropogenic compounds (e.g., pesticides, gasoline components, and chlorinated compounds). NOM, a subset of TOC, is commonly used to describe large macromolecular organic compounds present in water. These macromolecules can include humic substances, proteins, lipids, carbohydrates, fecal pellets, or biological debris (Stumm and Morgan, 1996) and while not highly reactive, often contain reactive functional groups (Hoigné, 1998). Organic matter present in the water (whether anthropogenic or natural) will serve to scavenge hydroxyl radicals and thus limit the effectiveness of AOPs. The rate constants reported in the literature for hydroxyl radical reactions with NOM range from 1.9×10^4 to 1.3×10^5 $(mg/L)^{-1}s^{-1}$ (AwwaRF, 1998). Given the high molecular weight of NOM (5,000 to 10,000 g/mole) these rate constants are comparable to the reaction rate constant for MTBE, and thus, high concentrations of NOM can result in significant reduction of MTBE destruction potential. Also, high concentrations of NOM may be mitigated by the addition of higher dosages of oxidant and longer reaction times.

Nitrates and Nitrites

Hydroxyl radicals can be formed by several mechanisms, including UV photooxidation of hydrogen peroxide. Any constituent present in the water that adsorbs UV light will serve to decrease the formation of hydroxyl radicals and the subsequent destruction of MTBE. Nitrates and nitrites adsorb UV light in the range of 230-240 nm and 300-310 nm, and consequently, high nitrate (> 1 mg/L) or high nitrite (>1 mg/L) concentrations have been shown to limit the effectiveness of UV technologies (Calgon, 1996).

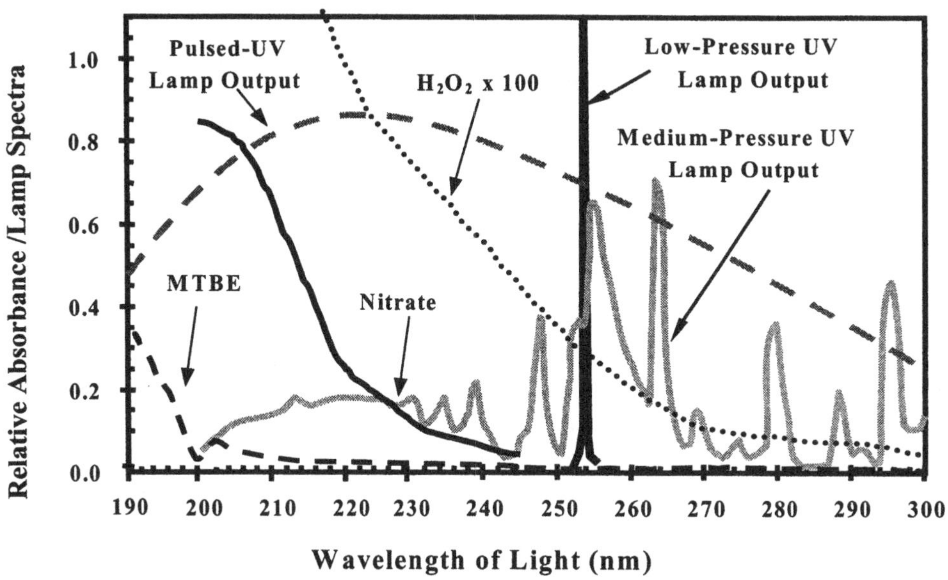

Figure 3.1 Absorption scans for MTBE, nitrate and hydrogen peroxide. Shown in the background are the lamp emission spectra.

Shown in Figure 3.1 are the absorbance scans from wavelengths between 190 nm to 300 nm for MTBE, nitrate and hydrogen peroxide, at concentrations that are typically seen in potable water AOP systems. The lamp spectral outputs for LP-UV, MP-UV and pulsed UV are shown in the background. As shown in Figure 3.1, nitrate absorbs light in the region that peroxide absorbs UV light. Absorption of UV light by peroxide results in the formation of hydroxyls radicals. Presence of nitrate in water interferes with UV-based AOPs.

Phosphates and Sulfates

While phosphates and sulfates are commonly present in low concentrations in source waters, these compounds have the potential to scavenge hydroxyl radicals. However, they are extremely slow in reacting with ·OH, and their scavenging effect can usually be neglected (Hoigné, 1998) for ozone/peroxide/UV systems.

Iron (II), Copper (I), and Manganese (II)

While not well understood, the presence of these reduced metals in combination with NOM and hydroxyl radicals may lead to the formation of iron or copper organic complexes or the oxidation of Mn (II) to form permanganate (Hoigné, 1998; Calgon, 1996). The formation of permanganate has been observed to occur with a rate constant of 3×10^3 to 2×10^4 $M^{-1}s^{-1}$ (Reckhow et al., 1991). In addition, the presence of iron (absorptivity 200-400 nm) and other scaling agents may result in fouling of the UV systems.

Turbidity

Systems relying on UV irradiation for the dissociation of H_2O_2 or O_3 exhibit a decrease in efficiency as turbidity increases. Turbidity lowers the transmittance of the source water and thus, lowers the penetration of the UV radiation into the source water.

Bromate Issues

During the use of AOPs that employ ozone, bromate can be formed when bromide is present in the source water (Krasner et al., 1993; von Gunten and Hoigné, 1994; Siddiqui et al., 1999). Bromide concentrations in raw drinking water can range from non-detect to several hundred µg/L to several mg/L in brackish water (Hoigné, 1998). The conversion of bromide to bromate can proceed as follows (Krasner et al., 1993; Siddiqui et al., 1994):

$$Br^- \longrightarrow BrO^- \longrightarrow BrO_2^- \longrightarrow BrO_3^- \qquad (3.20)$$

$$Br^- \longrightarrow Br\cdot \longrightarrow BrO\cdot \longrightarrow BrO_2^- \longrightarrow BrO_2\cdot \longrightarrow BrO_3^- \qquad (3.21)$$

When bromide concentrations in the source water exceed 0.1 mg/L, bromate was produced at concentrations above the Stage 1 D/DBP MCL of 10 µg/L (Liang et al., 1999a). However, prior studies of ozone-based AOPs, specifically H_2O_2/O_3 systems, have shown that varying the chemical ratio of O_3 to H_2O_2 is effective at minimizing the bromate formation (Song et al., 1997). In addition, research has demonstrated that bromate formation is reduced by approximately 20 percent at a slightly acidic pH (~6.5), when compared to the ambient pH (Liang et al., 1999a).

RESEARCH GAPS

AOPs destroy MTBE and other organic contaminants directly in the water through chemical transformation, as opposed to simply transferring them from the liquid phase into a gas phase (in the case of air stripping) or solid phase (in the case of GAC and resins). However, despite this advantage, there are significant limitations and challenges in the full-scale application of AOPs. In general, AOPs are much less well understood than air stripping and adsorption due to the complex chemical and physical processes involved in oxidation reactions. The implementation of AOPs and the determination of their effectiveness are difficult for several reasons. As with all treatment technologies, the effectiveness of AOPs will be largely determined by the specific water quality matrix of the contaminated water. However, in the case of AOPs, the effect of background water quality on contaminant removal is much less well understood than for other technologies. For example, the presence of high bromide concentrations or natural organic matter (NOM) can result in the formation of regulated oxidation by-products that may cause water quality to deteriorate beyond its initial state of contamination. Similarly, the presence of nitrates, NOM, and carbonates can interfere with the destruction of the target contaminant(s) and ultimately reduce the effectiveness of the selected AOP. In general, most of the technical difficulties associated with AOPs stem from the fact that oxidation processes are non-selective with the potential for significant interference. To compensate for these limitations, more energy or higher chemical dosages may be required, potentially resulting in higher costs.

As described above, there is still a significant amount of uncertainty regarding the technical and economic effectiveness of AOPs for removing MTBE from drinking water under a variety of water quality scenarios. More bench- and pilot-scale studies will help in understanding the MTBE and its by-products removal efficiencies that can be achieved under different water quality conditions and operational parameters. In addition, the following specific topics warrant further research.

Water Quality Impacts on Effectiveness of AOPs

The effectiveness of AOPs is directly related to water quality parameters such as pH, alkalinity, NOM, TOC, turbidity, and concentrations of other interfering compounds (e.g., nitrates and bromide). More studies on AOP treatment of MTBE are required to evaluate the impact of the above-listed water quality parameters. The evaluation criteria should include MTBE removal efficiency, oxidation by-product formation, DBP formation potential, and costs. For ozone-peroxide AOP, the effect of influent bromide concentration on bromate formation needs to be further evaluated. Similarly, a detailed analysis of the effect of influent water turbidity, TOC and nitrate concentrations on the effectiveness of AOPs relying on UV-light (LP, MP, pulsed) is warranted.

By-Product Formation and Control

One of the areas that require more research is the issue of by-product formation and control. Oxidation of MTBE to carbon dioxide and water involves many steps and during this complex sequence of reactions many oxidation by-products (e.g., TBA, TBF, acetone) are formed. If these by-products are not completely mineralized, they will be present in treated water at concentrations that may present health hazard or aesthetic impact in the treated water. A better understanding of by-product formation mechanisms and subsequent mitigation strategies will be necessary prior to the acceptance of AOPs by the regulatory community for drinking water applications. This includes research to determine the most cost-effective treatment option, such as biologically activated carbon, for by-product removal in drinking water applications.

Cost Evaluation As a Function of Water Quality and Contamination Scenario

Finally, future research is also necessary to evaluate the engineering costs for MTBE oxidation by AOPs. Capital and O&M costs for promising AOP processes have to be developed as a function of water quality, flow rate, influent MTBE concentration, and required removal efficiency. These cost evaluations must be performed under uniform design criteria (e.g., required removal efficiency) and operational assumptions (e.g., power rate). A unified costing approach will enable a direct comparison of the various AOPs for specific water qualities, and further allow comparison with alternative technologies for control of MTBE, TBA and related compounds.

CHAPTER 4
OZONE/PEROXIDE TREATMENT

BENCH SCALE TESTING

Introduction

The recent realization of the potential large-scale impacts of MTBE contamination on community drinking water supplies has spurred investigations into enhanced removal of this resistant compound. Since conventional treatment processes that are useful for many other volatile organic compounds, including air stripping, granular activated carbon, and ozone oxidation, have proved ineffective in efficient removal of MTBE (California MTBE Research Partnership, 2000; Davidson and Creek, 1998), research has focused on the use of various advanced oxidation processes (AOPs) to achieve this goal (Acero, von Gunten, Schmidt, and Haderlein, 2000b; Cater, Stefan, Bolton, and Safarzadeh-Amiri, 2000; Chang and Young, 2000; Cooper and Mezyk, 2001; Cooper, Tobien, Mezyk, Adams, Nickelsen, O'Shea, Inclan, Tornatore, Hajali, and Weidman, 2001; Kang and Hoffmann, 1998; Karpel Vel Leitner, Papailhou, Croué, Peyrot, and Doré, M., 1994; Liang, Palencia, Yates, Davis, Bruno, and Wolfe, 1999a; Stefan, Mack, and Bolton, 2000). AOPs rely on the generation of hydroxyl radicals, which in turn are known to rapidly react with MTBE ($k_{OH} = 1.6 \times 10^9\ M^{-1}s^{-1}$) (Buxton, Greenstock, Helman, and Ross, 1988) to form oxidation by products and potentially innocuous mineralized end products. Considering capital limitations and the presencc of preexisting infrastructure, the removal technology of choice will depend on the potential for enhanced removal of MTBE and will also depend heavily on the potential for minimal potentially hazardous by-product and oxidation product formation.

Previous investigations have shown the potential utility of the peroxone (O_3-H_2O_2) treatment process in MTBE removal (Acero, von Gunten, Schmidt, and Haderlein, 2000b; Liang, Palencia, Yates, Davis, Bruno, and Wolfe, 1999a; Stefan, Mack, and Bolton, 2000), however, easy application of this optimized process in waters of varying water quality is not straightforward. This is due in part because of the wide variations existent in water quality parameters in different waters. Variations in some water quality parameters affect a waters hydroxyl radical scavenging capacity during oxidation. Alkalinity affects oxidation reactions by inhibiting the O_3 decay cycle through removal of hydroxyl radicals while the effect of dissolved organic carbon (DOC) is more complicated, due to both inhibition and promotion of O_3 decay. Variations in these constituents could affect the hydroxyl radical exposure during a reaction and therefore impact MTBE removal efficiencies.

Objectives

What has been needed is an easy tool that can be used to predict MTBE removal efficiency and degradation product formation of this process for different water types, since contamination may occur in water of all types. Development of this tool for determining a water's capability to efficiently remove MTBE, or other pollutants that may be efficiently removed with hydroxyl radicals, paves the way to implementing this powerful treatment technique in full scale operations. The goal herein was to develop a method for characterizing the hydroxyl radical

Table 4.1
General water quality classifications for chosen waters

	Low Alkalinity ($<$100mg/L as $CaCO_3$)	High Alkalinity ($>$100mg/L as $CaCO_3$)
Low DOC ($<$1mg/L)	Suffolk County	Santa Monica
High DOC ($>$3mg/L)	Miami Dade	Long Beach

formation potential at typical oxidant doses, which could, in turn, be used in predictive modeling of compound oxidation. Furthermore, an assessment of the effects that alkalinity and DOC may have on MTBE removal and bromate formation during O_3-H_2O_2 treatment would help in predicting utility of this method for select water types. The hydroxyl radical probe compound chosen, para-chlorobenzoic acid (pCBA), can be used to indirectly predict hydroxyl radical exposure during a reaction via a simple HPLC analysis (Elovitz, and von Gunten, 1999; Elovitz, von Gunten, and Kaiser, 2000). This information can then be used to predict radical mediated micropollutant decay through comparison of previously determined rate constants, as demonstrated on anther micropollutant, atrazine (Acero, Stemmler, and von Gunten, 2000a).

Four waters, containing high/low DOC/Alkalinity, as shown in Table 4.1, were chosen to encompass normal variations in these driving parameters for natural waters. Other factors that could influence hydroxyl radical exposure are water, temperature, and pH (Elovitz, von Gunten, and Kaiser, 2000). Variation of these factors was not incorporated into this study since assessment of the potential for MTBE degradation on a natural water using the method in this investigation can be performed at its natural temperature and pH. Of note, however, is the previous demonstration that increasing reaction temperature or pH can increase the ratio of hydroxyl radical exposure to ozone exposure (R_{CT}) (Elovitz, von Gunten, and Kaiser, 2000), suggesting these parameters could be manipulated to enhance performance of this method.

Methods

Both O_3 dose and H_2O_2:O_3 ratio were varied according to the orthogonal matrix shown in Figure 4.1, with the 4 mg/L O_3 and 0.35:1 H_2O_2:O_3 serving as the baseline oxidation dose. Initial MTBE concentration was set at 200 µg/L for all MTBE degradation experiments. Hydroxyl radical probe experiments, performed to estimate hydroxyl radical concentrations/yields, followed the same matrix as in Figure 4.1, except pCBA was added as a probe compound instead of MTBE. Initial bromide concentrations were raised up to 100 µg/L prior to experimentation, except when natural water bromide levels were above this value, for assessment of bromate formation potential at the different oxidation levels tested. Previous research has shown that bromate formation is related to O_3 dose (Liang, Palencia, Yates, Davis, Bruno, and Wolfe, 1999a) and excess H_2O_2 during peroxone treatment may exhibit only limited bromate minimization (von Gunten and Oliveras, 1998). Temperature was held at natural ground water temperatures for selected waters during experiments using a water bath.

Oxidation experiments for waters exhibiting slow ozone decay were performed with 500 mL reaction volumes in true-batch reactors, which were amberized 500 mL Erlenmeyer flasks with glass sampling ports attached. Flask tops were cut and capped with silica stoppers such that no headspace remained when 500 mL water added. Time series samples were collected

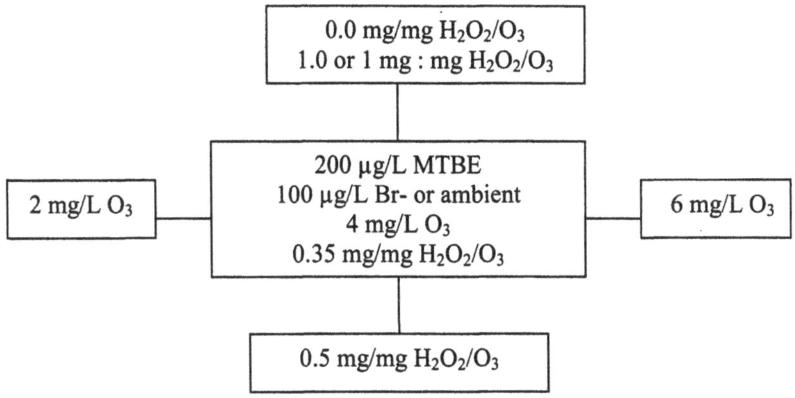

Figure 4.1 Orthogonal matrix schematic

for dynamic ozone, MTBE, and pCBA decay characterization. Oxidation experiments for waters exhibiting ozone decay too fast to measure accurately (complete decay within 20 seconds) were performed with 20 mL reaction volumes in 50 mL beakers. Only final concentrations were analyzed for these samples. Ozone, MTBE and pCBA samples were taken using 5 mL latex free SlipTip™ syringes. Ozone concentrations were assessed by adapting the published indigo trisulfonate quenching methods to reaction volumes used (Bader and Hoigné, 1982; Chiou, Mariñas, and Adams, 1995; Simon, 2001). MTBE and degradation product samples were quenched in sodium thiosulfate and analyzed via heated purge and trap GC-MS. PCBA sampling used excess sample collected for ozone decay measurements and were analyzed via HPLC (Elovitz and von Gunten, 1999). Bromide/bromate were analyzed via ion chromatography (Simon, 2001). Water quality parameters, including alkalinity, DOC, pH, turbidity, conductivity, cations, and hardness were determined following Standard Methods (American Public Health Association, American Water Works Association, and Water Environment Federation, 1998).

Results and Discussion

The four waters used in this investigation had initial water quality parameters measured prior to experimentation. Table 4.2 summarizes measured water quality parameters. The two waters with high initial DOC, Long Beach and Miami Dade waters, exhibited "instantaneous" O_3 decay, being too fast to dynamically measure, at all oxidant levels tested. This observation was suspected to be caused by promotion of O_3 decay by the DOC and the high natural temperature for the Long Beach water. The other waters tested, Santa Monica and Suffolk County waters, had in common lower DOC values and slower ozone decay kinetics, which allowed for time series decay measurements for O_3, MTBE, and pCBA.

Alkalinity can act as an inhibitor of the O_3 decomposition cycle by scavenging hydroxyl radicals and not forming superoxide, or other species contributing to O_3 decomposition. DOC action is more complex, since different functional groups of natural organic matter (NOM) can act in both promotion and inhibition of O_3 decay. Furthermore, NOM character from a specific source can change seasonally or as a result of isolated incidents, thus affecting it's impact on O_3 decay kinetics and potential hydroxyl radical exposure during a reaction (Elovitz, von Gunten, Kaiser, 2000). An estimate of the hydroxyl radical scavenging capacity of each water, shown in

Table 4.2
Initial water quality of sample waters

Water Quality Values	Water ID			
	Santa Monica	Long Beach	Miami-Dade	Suffolk County
Groundwater Temp. (°C)	20°	29.3°	20°	12°
Alkalinity (mg/l as $CaCO_3$)	325	145	45	23.8
DOC (mg/L)	0.45	4.38	3.81	0.24
SUVA (l/mg m)	2.6	5.55	2.07	1.44
Ambient pH	8.07	8.55	8.63	6.52
Bromide (μg/L)	226	133.9	35.6	96.51
Ca (ppm)	103.6	9.2	20.9	13.4
Mg (ppm)	54.7	0.6	3.8	5.7
Mn (ppm)	0.01	0.002	ND	ND
Fe (ppm)	ND	0.038	ND	ND
K (ppm)	3.5	0.82	3.2	0.99
Hardness (mg/L as Ca)	483.9	25.6	67.9	57
Turbidity (NTU)	0.24	0.81	0.18	0.05
Conductivity (uS)	1218	358	270	225
Initial MTBE (μg/L)	ND	ND	ND	3.5

ND = Not detected

Table 4.3, was calculated using published second order rate constants (first order with respect to hydroxyl radicals and first order with respective to reactive species) and initial DOC, bicarbonate and carbonate, and bromide concentrations. Due to the complex reaction chemistry involved, MTBE degradation was not directly related to this scavenging capacity estimation. Further evaluation of the individual impacts of each contributor to hydroxyl scavenging is needed to draw this correlation.

Compiled MTBE degradation experiments, shown for Santa Monica water in Figure 4.2, show expected trends of increased MTBE removal with increasing O_3 dose and increased H_2O_2:O_3 ratio. Also evident in this figure is the reduced reaction time required when H_2O_2 is added, showing that for some waters, increased MTBE removal amount and rate can be achieved with H_2O_2 addition. The other water exhibiting slow enough ozone decay to measure, Suffolk County water, had complete removal of MTBE at all but the lowest O_3 level, due in part to it's very low (hydroxyl radical) scavenging capacity (see Table 4.3). For the other waters tested, both containing high DOC, only final measurement of MTBE and degradant concentrations were possible. Analysis of MTBE degradation product formation, as shown in Figure 4.3 for Suffolk County water, shows that even once MTBE is completely removed during a reaction, secondary oxidation may only partially remove degradation products, due to there slower reaction rates with hydroxyl radicals and lower concentrations. Calculated molar recovery decreased over time suggesting that partial mineralization of MTBE during the course of oxidation experiments occurred.

Table 4.3
Estimated hydroxyl radical scavenging capacity of sample waters

	Santa Monica	Long Beach	Miami Dade	Suffolk County
DOC (mg/L)	0.446	4.38	3.81	0.243
HCO_3^- (M)	3.21E-03	1.40E-03	4.30E-04	2.37E-04
CO_3^{2-} (M)	3.55E-05	4.67E-05	1.73E-05	7.40E-08
Br (M)	2.83E-06	1.68E-06	1.25E-06	1.21E-06
Scavenging capacity (s^{-1})	5.57E+04	1.42E+05	1.07E+05	9.45E+03

	k_{OH} (M^{-1} s^{-1})	k_{OH} Reference
DOC (mg/L)	2.50E+04	—
HCO_3^- (M)	8.50E+06	Buxton et al., 1988
CO_3^{2-} (M)	4.00E+08	Buxton et al., 1988
Br (M)	1.10E+09	Pinkernell & von Gunten, 2001

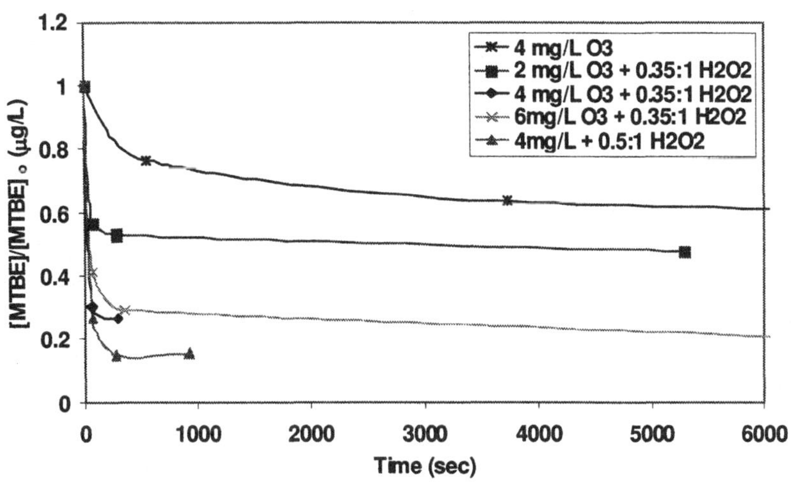

Figure 4.2 Compiled MTBE decay profiles for Santa Monica experiments

Not enough experiments were performed to accurately be able to predict degradant profiles from MTBE removal. This should theoretically be possible, however, potential for secondary oxidation reactions to occur once MTBE sufficiently removed complicates calculation. Also, prediction was further complicated by degradation products undergoing potential other reactions once formed, such as hydrolysis (Church, Pankow, and Tratnyek, 1999) or potential NOM mediated conversions.

Analysis of bromate formation in bromide-spiked waters at differing O_3 doses and H_2O_2:O_3 ratios can show the relative impacts of peroxone treatment on various water qualities. In Figure 4.4, the percent removal of MTBE was related to bromate formation for three of the tested waters; here, the intent is to portray tradeoffs between MTBE oxidation versus bromate formation, and to indicate potential mitigation of bromate formation by the presence of peroxide.

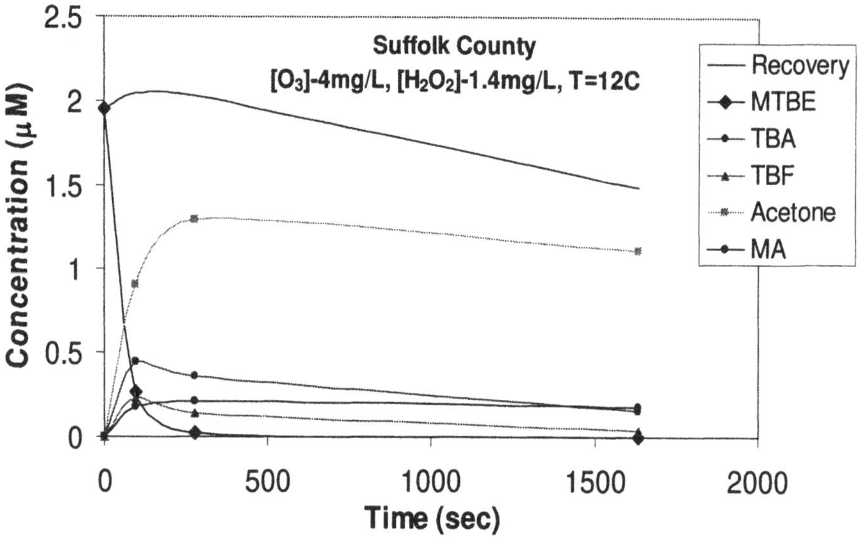

Figure 4.3 MTBE and degradant profile for a Suffolk County water oxidation experiment

The fourth water, Suffolk County, is not included in these comparisons since MTBE was completely removed at all but the lowest oxidation levels for this water. The open diamonds represent an increase in O_3 dose, keeping the H_2O_2:O_3 ratio constant at 0.35:1, with the line drawn through these points showing the relationship between ozone dose, MTBE removal, and bromate formation. The closed diamonds represent the changing H_2O_2:O_3 ratio at the constant O_3 dose of 4 mg/L.

Waters tested with low DOC (Santa Monica and Suffolk County) exhibited much higher bromide-bromate conversion than waters with high DOC. Also, greatest reductions in bromate formation with increased H_2O_2:O_3 ratio were seen in the waters tested with low DOC. The impact of H_2O_2 addition seems to be influenced by scavenging capacity of a water as well as promotion of O_3 decay by DOC. Waters with high DOC (Long Beach and Miami Dade) showed little impact of increased H_2O_2:O_3 ratio on bromate minimization, suggesting hydroxyl radical formation potential was already close to optimum due to promotion of O_3 decay by DOC. Of these two waters, the one with higher alkalinity and resultant hydroxyl radical scavenging, Long Beach water, exhibited the lowest bromate concentrations and least reduction of bromate formation with increased H_2O_2:O_3 ratio. These observations may be due to both very short O_3 exposures and hydroxyl radical exposures during reaction. The Miami Dade water, having less alkalinity contributing to hydroxyl radical scavenging, ended up with higher hydroxyl radical exposure. This corresponded to increased potential bromate formation via the hydroxyl radical pathway (Pinkernell, and von Gunten, 2001) as well as higher over all degradation efficiency of MTBE.

Model development required the separate experiments with the hydroxyl radical probe compound, pCBA, to indirectly determine the hydroxyl radical exposure during a reaction. Knowing the hydroxyl radical exposure (a function of the hydroxyl radical concentration versus time profile), the MTBE degradation predictions could be made and compared with real data for model comparison. The first modeling approach, with results from three Santa Monica water experiments presented in Figure 4.5, was to determine a waters R_{CT} value, the ratio of hydroxyl radical exposure to O_3 exposure, from pCBA decay measurements, and ozone exposure calculations. Knowing the R_{CT} for a water at a specific oxidation level, predictions of

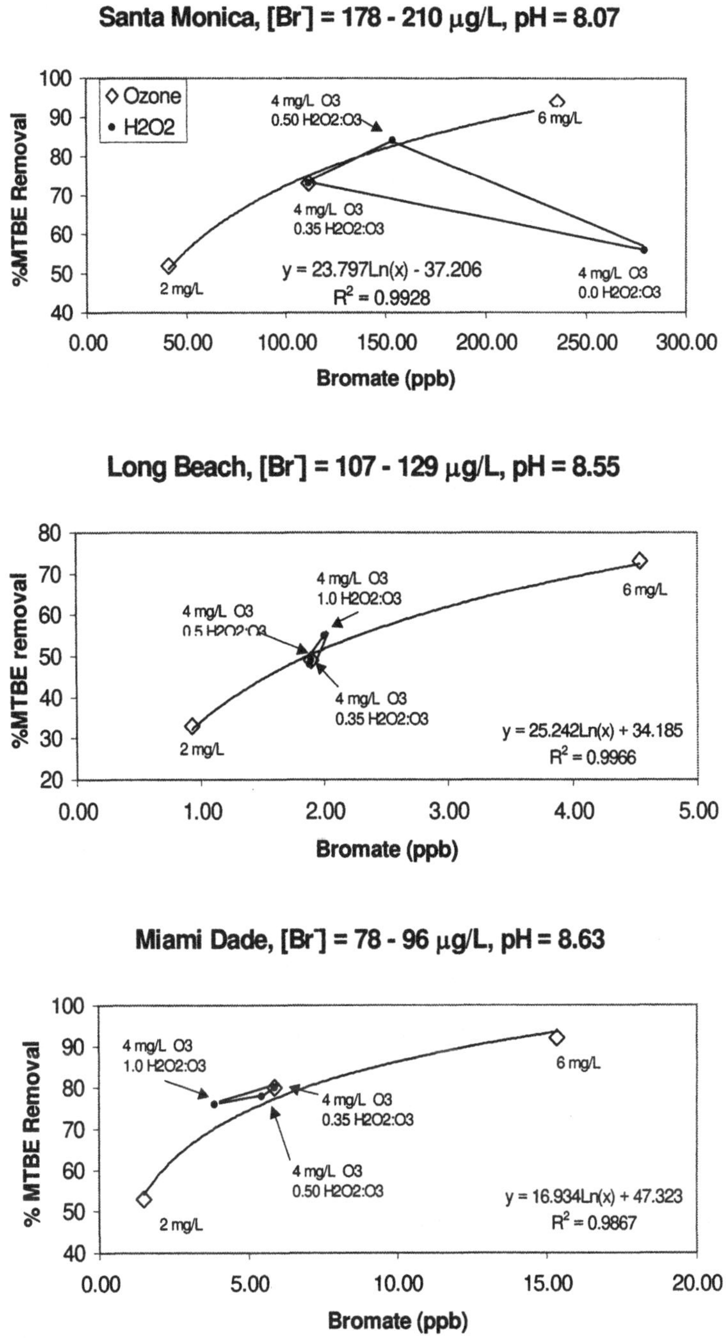

Figure 4.4 MTBE removal/bromate formation comparison at all oxidation levels

contaminant removal were shown to closely resemble real data obtained from separate experiments. R_{CT} determinations at higher oxidation levels for the Santa Monica water and at all oxidation levels for the Suffolk County water were not feasible using pCBA as the hydroxyl radical probe compound. This was due to high pCBA consumption, resulting from low scavenging capacities for these waters. Excessive amounts of pCBA would have been required for post oxidation detection, enough to significantly impact reaction kinetics from consumption of

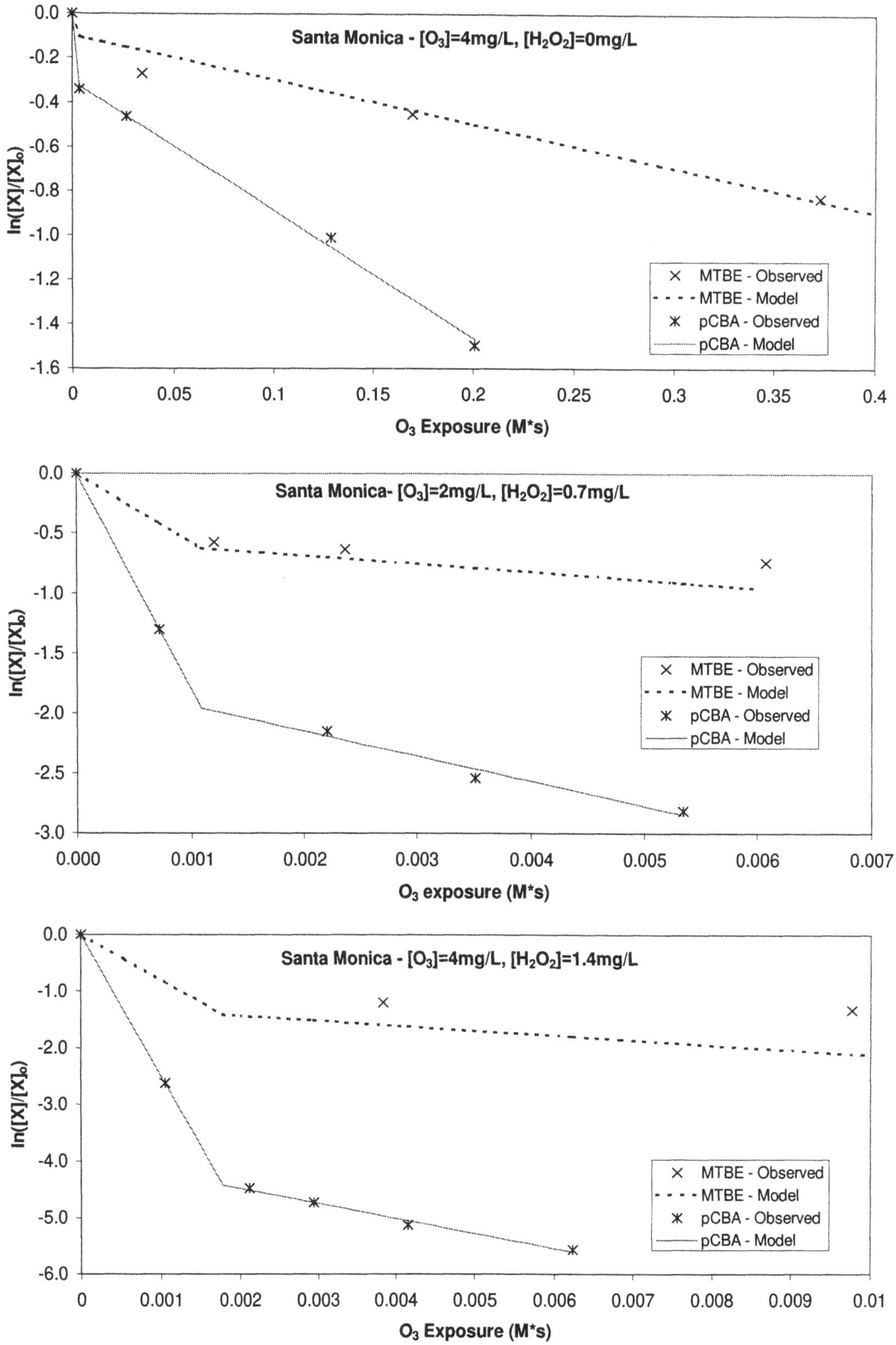

Figure 4.5 R_{CT} model predictions for Santa Monica water, with pCBA data included

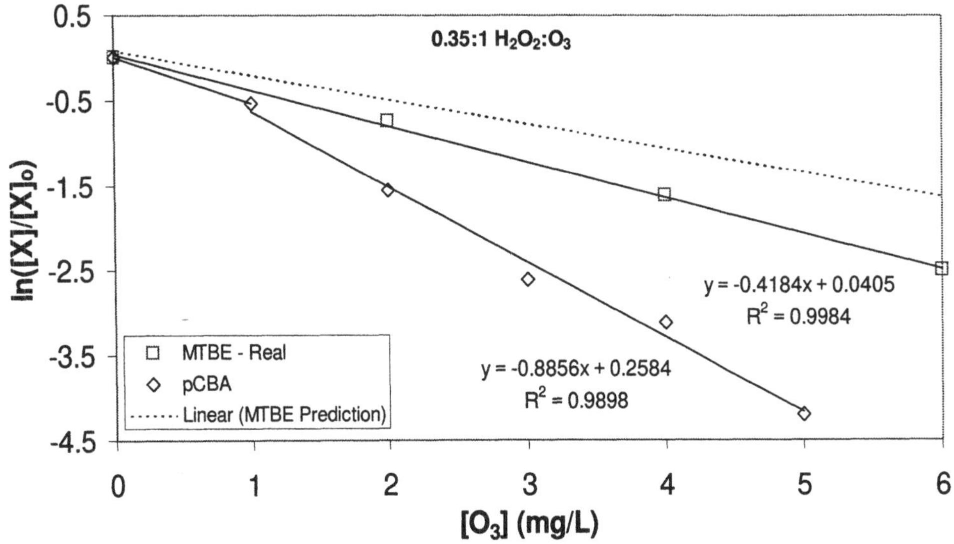

Figure 4.6 Miami Dade MTBE predictions based on final hydroxyl radical exposure estimates

hydroxyl radicals by pCBA. Therefore, an alternative hydroxyl radical probe compound needs to be discovered, one retaining the traits of easy detectability and low ozone reaction rate, but having a lower hydroxyl radical rate constant than pCBA.

For waters exhibiting instantaneous O_3 decay, a dynamic modeling approach was not feasible, thus an R_{CT} could not be calculated. A simpler approach was attempted by estimated the overall hydroxyl radical exposure during a reaction from the removal efficiency of pCBA. Since pCBA removal is kinetically related to reaction with hydroxyl radical, estimates of overall hydroxyl radical exposure could be made. Using the published hydroxyl radical rate constant for MTBE of $k_{OH} = 1.6 \times 10^9\ M^{-1}s^{-1}$ (Buxton, Greenstock, Helman, and Ross, A. B. (1988), along with estimated hydroxyl radical exposures from pCBA experiments, MTBE removal prediction could be made. Figure 4.6 shows the MTBE real and prediction lines for the Miami Dade water with 0.35:1 H_2O_2:O_3 ratio. This figure was typical, showing an under prediction of MTBE removal efficiency. Comparing real MTBE data from all experiments show a hydroxyl radical rate constant at $k_{OH} = 2.1 \times 10^9\ M^{-1}s^{-1}$ (%RSD = 13.8%), a value within reported literature range of $1.6 - 3.9 \times 10^9\ M^{-1}s^{-1}$. This observation suggests that the higher rate constant be used for MTBE when using this approach in predicting removal efficiency.

The last approach used to model MTBE degradation was mechanistic, utilizing the ACUCHEM™ software (Lassiter, 1986), which solves multiple differential equations input in the form of chemical equations. Reactions considered were between hydroxyl radicals and carbonate, bicarbonate, bromide, MTBE, pCBA, DOC, as well as O_3 decay and O_3 conversion to hydroxyl radical. Overall conversion of O_3 to hydroxyl radical is typically about 50% during an oxidation reaction, however, this value may change when H_2O_2 is added to a system. In order to accurately predict MTBE degradation, the conversion of O_3 to hydroxyl radical was needed to be calibrated at each oxidation level. This calibration was performed using pCBA decay results obtained during R_{CT} experiments. For the same reasons described previously, only the Santa Monica water had useable dynamic pCBA results for MTBE model calculations, as shown in Figure 4.7. Identifying

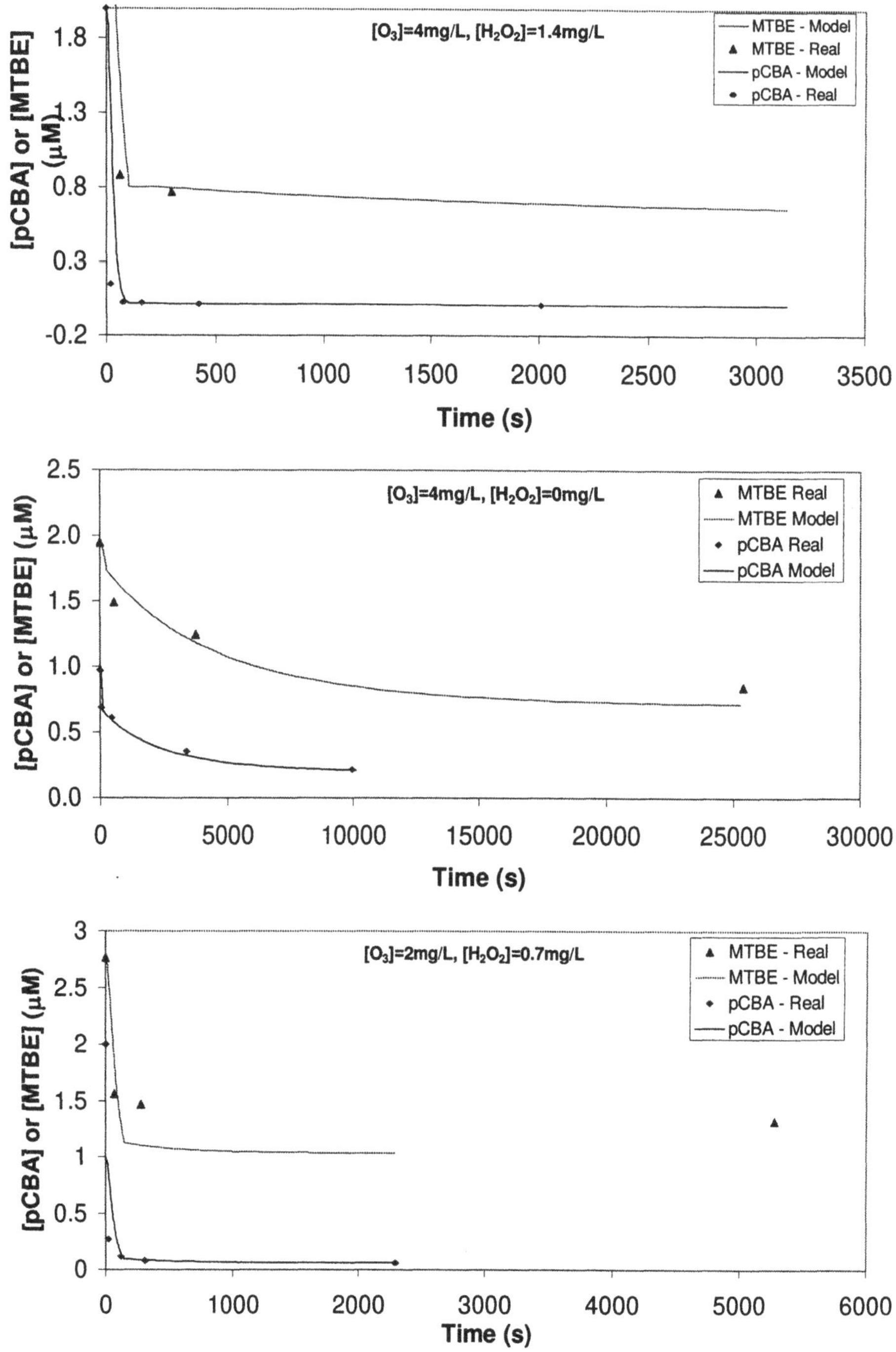

Figure 4.7 ACUCHEM™ model predictions for Santa Monica water

another probe compound for use with waters of low scavenging capacity, such as Suffolk County water, would make this approach more versatile. With pCBA and MTBE data coming from separate experiments, the small error within model predictions may be indicative of experimental error between replicate oxidations.

Summary

Results from experiments performed on the four diverse waters in this investigation demonstrated that MTBE removal efficiency is dependant on initial water quality parameters. Bromate formation and potential for minimization by increasing the H_2O_2:O_3 ratio are also affected by initial water quality, and appear to be related to hydroxyl radical scavenging capacity of a water. In fact, results suggest that the balance between bromate formation and MTBE removal relies on scavenging capacity. Waters with low scavenging capacity have higher likelihood of high MTBE removal with the O_3-H_2O_2 approach, with the drawback of high bromide conversion to bromate. Waters of this type need lower initial bromide concentrations to ensure bromate formation stays below 10 μg/L. Waters with higher scavenging capacity need higher oxidation levels to achieve comparable MTBE removal. MTBE degradation products are of concern, since even when 100% MTBE has been oxidized, degradation products were shown to be removed only slowly by secondary oxidation. These results suggest that investigation of other post-oxidation removal alternatives for these compounds would be wise if toxicity is a concern.

Utilization of a hydroxyl radical probe compound, such as pCBA, can be used to assess hydroxyl radical exposure at various oxidation levels easily. This information can then be used in several modeling approaches to predict MTBE removal without the need to perform costly MTBE analyses or work with the volatile compound. More experiments are needed to be able to make accurate predictions of MTBE oxidation product formation. Secondary oxidation of degradation products and post-experimental transformations complicate this estimation. An alternative hydroxyl radical probe compound to pCBA, having a lower hydroxyl radical rate constant while retaining the other qualities of an ideal radical probe, was determined to be needed to better model MTBE degradation in waters with low scavenging capacity.

PILOT SCALE TESTING

Introduction

Although the advanced oxidation process is effective in treating MTBE from the water, it may not be cost effective to treat the degradation byproducts such as TBA and aldehydes, since much longer retention time or higher oxidant dose would be required to do that. Consequently, one of the approaches in dealing with the complete removal of MTBE and the byproducts may be to use advanced oxidation process as the primary treatment method for MTBE and to use complementary approach to treat the MTBE degradation byproducts. The idea presented in the proposal was to use AOP, such as ozone and UV in combination with H_2O_2, followed by the biological GAC, which can be used to remove MTBE degradation byproducts such as TBA, TBF, and aldehydes.

The flow through pilot scale test matrix were developed based on the bench scale results to validate the performance of the ozone and ozone in combination with hydrogen peroxide for treating MTBE and MTBE degradation by-products. The experimental conditions were chosen to treat MTBE to below the secondary standard of 5 μg/L.

The La Verne Pilot Plant located at the Metropolitan's Weymouth Filtration Plant (La Verne, CA) was retrofitted to perform the MTBE pilot scale tests. The retrofit included 1) installation and acclimation of two biological GAC (BAC) columns, which were used as the post treatment for the MTBE degradation by-products from the advanced oxidation process, 2) modification of the ozone systems to deliver high concentration of applied ozone dose (up to 30 mg/L), 3) installation of a secondary GAC tank to remove any excess MTBE from the system, and 4) modification of a backwash system to provide dechlorinated water to wash the biological columns.

Experimental Plan and Objectives

The pilot scale ozone tests were designed to investigate the effect of water matrix (surface water vs. groundwater), applied ozone dose, and initial MTBE concentration. For the groundwater used, effect of the presence of TBA was also studied in removing MTBE during the advanced oxidation process since TBA is often found as a primary contaminant in the gasoline contaminated groundwater.

The main objective of the tests done during the pilot scale phase is to validate the performance of the advanced oxidation processes, specifically ozone with hydrogen peroxide in pilot scale, flow through tests. Specific objectives for this portion of the project are listed below.

1. Investigate the performance of ozone/H_2O_2 process for the removal of MTBE with flow through pilot scale setup
2. Investigate the performance of biological GAC column in removing MTBE degradation by-products from ozone/H_2O_2 process
3. Evaluate the effects of water source (groundwater and surface water) on ozone/H_2O_2 process in pilot scale

Materials & Methods

Pilot Scale Ozone System

Ozone generators installed at the La Verne Pilot Plant are capable of using either air or liquid oxygen as the feed gas for generating ozone at low concentrations. However, liquid oxygen was needed to generate up to 30 mg/L of applied ozone for some of the tests. Based on the initial tests, a single source of feed gas (liquid oxygen) with the fixed gas flow rate were chosen, so that the consistent setting can be applied to all tests. The applied ozone dosage was varied by changing the ozone gas concentration and keeping all the other parameters, such as gas flowrate, constant throughout the ozone pilot tests.

Figure 4.8 is a detailed schematic of the pilot-scale ozone setup, consisting of an ozone contactor column, an ozone reactor column, a sump pump, biological GAC column, and equalization tanks. For each test, 99.5-percent pure MTBE (Burdick & Jackson, Inc., Muskegon, Mich.), in a predetermined amount, was added directly to a tank and mixed with source water. Water was then pumped to the top of the ozone contactor from a 500-gal Nalgene® tank by a 1/3 hp sump pump (Sta-Rite, Delavan, Wis.) with a pumping flow of 1.9 gpm (2.0×10^{-5} m^3/s). Ozone

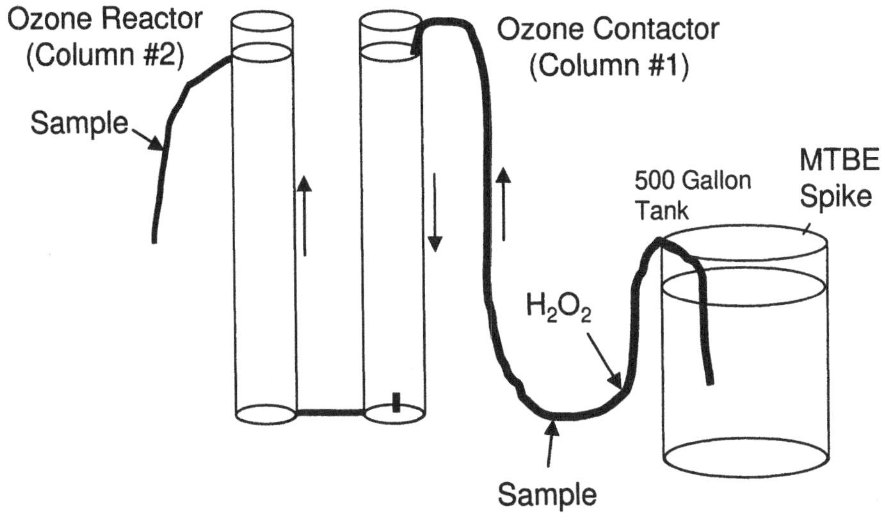

Figure 4.8 Schematic of the pilot scale ozone/H_2O_2 setup (not to scale)

gas was applied countercurrently in the column. The water flowed hydraulically upward to the ozone reactor column, and each ozone column was operated at a residence time of 12.5 min. Water levels were maintained at 16 ft (4.88 m) in the 6-in.-diameter columns. Ozone was bubbled through 4-in. (0.102-m)-diameter ceramic static diffusers to generate fine bubbles. Ozone transfer efficiencies for countercurrent operation was typically above 98 percent. The ozone generator was a tube-type, water-cooled unit operating on a low-frequency (60 Hz), variable-voltage power supply (model Labo-76, Praxair/Trailigaz Ozone Co., Cincinnati, Ohio). The capacity of the ozone generator was approximately 1 lb/day, with liquid oxygen as the feed gas. The target gas-phase concentration of ozone produced was up to 2.4 percent by weight to apply 30 mg/L of ozone with the conditions used. As mentioned previously, the gas phase ozone concentration was varied to control the applied ozone dose with fixed gas flow rate to minimize the variations in potential stripping effect by the bubbles. The ozone concentration in the feed gas was measured by a gaseous ozone monitor (model HC-12, PCI Ozone Corp., West Caldwell, NJ) and off-gas from the contactor was pumped through a thermocatalytic ozone destruction unit before discharged to the atmosphere. When used, hydrogen peroxide (3 percent by weight, USP grade) was injected upstream of the ozone contactor, in predetermined amounts, by the peristaltic a pump.

Pilot Scale Biological GAC System

The carbon used to pack the BAC column was virgin Filtrasorb 400 (Calgon Carbon, Pittsburgh, PA). The apparent density provided by the manufacturer for Filtrasorb 400 is 0.48 g/cm^3. Filtrasorb 400 has some MTBE sorption capacity, and the sorption isotherm is shown in Figure 4.9. However, it should be noted that the goal of the project was to use the advanced oxidation process to remove MTBE and to use (biological) GAC to remove MTBE degradation byproducts, although some MTBE sorption to GAC may take place.

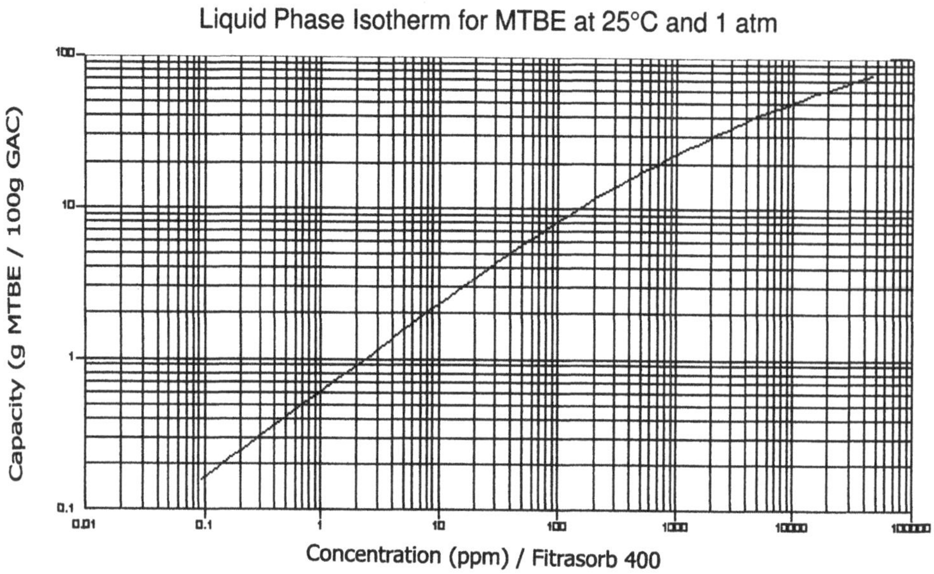

Figure 4.9 Calgon Carbon GAC (Filtrasorb 400) isotherm for MTBE used for biological columns

The BAC columns were equipped with dechlorinated water tank for backwashing, and an additional GAC tank was also installed in series with the biological columns in case the biological columns fail to remove all the byproducts. The GAC tank can ensure that all the byproducts are treated before discharging the water. The schematic of the BAC columns is shown in Figure 4.10. The flow rate to the biological filters was adjusted to achieve 20-minute retention time in the BAC filters. When not in use for testing, 100% raw SPW was pumped through the biological columns continuously and the filters were typically backwashed once every two days.

Sampling

For each pilot scale flow through test, influent and effluent samples were taken for analyses of MTBE and byproducts, aldehydes, ozone residual, hydrogen peroxide residual, and bromate. All samples were quenched as soon as the samples were taken. Third sample was also taken from the effluent of a BAC column, which was used as a post treatment for the MTBE degradation byproducts, to investigate whether the byproducts can be removed in the BAC columns.

Results & Discussion

Raw Water Quality

The blends of Colorado River Water (CRW) and State Project Water (SPW) was initially proposed as the source of the surface water for the pilot scale tests. However, because of the fluctuation in the background water quality characteristics and the percentage of SPW in the blends, SPW was chosen as a single source of raw water. SPW is contaminated with low level of MTBE due to the personal watercraft activities, and also it has high concentration of bromide

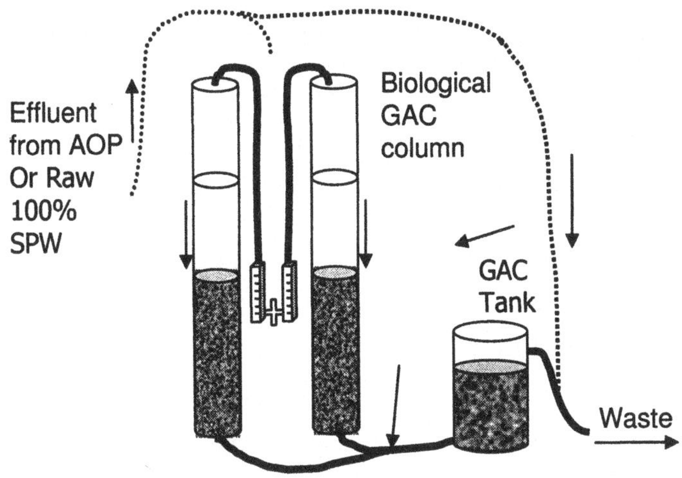

Figure 4.10 Schematic of pilot scale biological GAC system

(~ 200 μg/L), which would also provide valuable information regarding the potential bromate formation, so that mitigation strategy can be developed. The groundwater from the City of Santa Monica was used as the other source water for the pilot scale ozone test.

The following table (Table 4.4) provides the background water quality characteristics of the groundwater from the City of Santa Monica, CA and 100% State Project Water used for the pilot scale tests. The effect of difference in the levels of TOC, alkalinity, and nitrate are discussed in the following sections.

Volatilization Losses

For ozone/H_2O_2 testing, a series of control tests were performed to estimate the MTBE stripping by ozone gas bubbles in the ozone contactor. This was done with the a gas flow rate used for all the tests (without using ozone generator), to simulate the "stripping" effect by ozone bubbling during the high gas flow rate through the ozone columns.

High applied ozone dose was used for the pilot scale tests to reduce the MTBE concentration to below the secondary standard of 5 μg/L. Due to the limitation of the ozone gas concentration generated, high gas flow rate of up to 6 liters per minute (lpm) of ozone gas was used to achieve up to 30 mg/L of applied ozone concentration in the contactor. Because of the high bubbling velocity in the ozone contactor column, there was a concern regarding whether some MTBE may be stripped in the ozone contactor column. Number of control tests were performed with and without the hydrogen peroxide to test how much MTBE might be stripped during the contact with the bubbles.

Table 4.5 shows the results of the control tests with 100% SPW with fixed oxygen gas flow rate (6 lpm) and water flow rate (1.9 gpm or 7.2 lpm) for the gas to liquid ratio (G/L) of 0.83. The tests were done with liquid oxygen as the feed gas but without powering up the ozone generator. Thus, only oxygen gas was delivered to the ozone columns to simulate the bubbling in the ozone contactor column. For actual tests, the ozone generator was powered up to produce up

Table 4.4
Typical background water quality characteristics

Parameters	Unit	State Project Water Surface Water	Santa Monica Groundwater
TOC	(mg/L)	3.33	0.52
EC	(μmhos)	500	1319
Nitrate as NO_3	(mg/L)	1.98	21.5
Alkalinity	(mg/L)	82	326
Turbidity	(NTU)	1.7	0.14
Total Hardness	(mg/L)	114	538
pH		8.05	7.21
UV_{254}	(abs/cm)	0.087	0.01
Bromide	(mg/L)	0.2	0.18

Table 4.5
Concentration of MTBE in μg/L from ozone control tests

Test Conditions	Oxygen Only	Oxygen and H_2O_2	Oxygen Only	Oxygen and H_2O_2
Ozone Influent	40	40	145	150
Ozone Effluent	40	35	130	130
Percent Removed	0	12.5	10	13

Units in μg/L of MTBE, H_2O_2 concentration is 15 mg/L when used

ND – Not Detected, NA – Not Analyzed

to 2.4% of ozone gas. As indicated in the table, approximately 10% of MTBE was removed by stripping of MTBE in the ozone contactor columns by bubbles at this gas flow rate. This range of MTBE loss is comparable to the values observed using the similar setup in recirculation mode. The presence of 15 mg/L of hydrogen peroxide also did not contribute in the MTBE loss in the ozone columns. Thus, it is expected that the observed MTBE reduction in the ozone columns, when ozone gas is applied, can be attributed to the reaction of MTBE with either molecular ozone or hydroxyl radicals generated during AOP.

Separate control tests were done to find out the range of ozone residual in the ozone columns. Samples were taken at the effluent of the first column (ozone contactor), the effluent of the second column in series (ozone reactor), and the effluent of biological GAC column. Table 4.6 shows the ozone residuals for the test done with 30 mg/L of applied ozone without hydrogen peroxide spike. As expected, high residual ozone concentration of 7 mg/L was measured in the ozone effluent in the absence of hydrogen peroxide. The residual ozone concentration showed approximately 2 mg/L of ozone going in to the BAC column, which might have modified the surface of the GAC. In the case of PEROXONE process with hydrogen peroxide, low ozone residual was measured (data not shown) as ozone decomposes readily as it reacts with hydrogen peroxide.

Table 4.6
Ozone residual with 30 mg/L of applied ozone without hydrogen peroxide

Sample	O_3 Residual
Location	(mg/L)
Column 1 Influent	ND
Column 1 Effluent	7.06
Column 2 Effluent	2.03
BAC Effluent	0.03

Table 4.7
MTBE and byproducts for 15 mg/L ozone and 15 mg/L hydrogen peroxide

Sample Location	MTBE (μg/L)	TBA (μg/L)	TBF (μg/L)	Acetone (μg/L)	Acetaldehyde (μg/L)	Formaldehyde (μg/L)	Glyoxal (μg/L)	M-Glyoxal (μg/L)
Detection Limit	0.1	0.1	5	10	1	5	1-2	1-2
Influent	122.2	ND	ND	ND	ND	6	1	ND
Effluent	ND	15.9	ND	ND	3	33	14	12
BAC Effluent	ND	ND	ND	ND	ND	6	3	ND

MTBE and By-products Removal With Ozone/H_2O_2 in Surface Water

Initial ozone and hydrogen peroxide tests were done with 15 mg/L of applied ozone and 15 mg/L hydrogen peroxide with 100% SPW (surface water). MTBE was effectively removed during the PEROXONE process at this condition (Table 4.7), but the effluent TBA level was approximately 16 μg/L exceeding the action level of 12 μg/L proposed by CDHS. Elevated levels of acetaldehyde, glyoxal, and m-glyoxal (pyruvic aldehyde) were also observed from the ozone column effluent. These are not believed to be the degradation byproducts of MTBE, but rather degradation byproducts of natural organic matter when exposed to AOP. The levels of these compounds measured after the biological GAC columns, however, are below the detection limits except for formaldehyde and glyoxal, which are only slightly higher than the detection limits.

Although virgin carbon used to pack the GAC columns may have some adsorption capacity for the byproducts, generally an order of magnitude lower adsorption capacity compared with MTBE is expected for highly soluble, lower molecular weight compounds such as TBA and acetone. For example, at dissolved concentration of 100 μg/L, approximately 8 g of MTBE can be adsorbed to 100 g GAC, while approximately 0.7 ~ 0.8 g of TBA or acetone is adsorbed to the same amount of GAC. Figures 4.11 and 4.12 show the Filtrasorb 400 isotherm for TBA and acetone respectively for comparison with MTBE. Since the investigation of mechanism of biological GAC for removing MTBE and the by-products is beyond the scope of this project, no attempts have been made to try to differentiate chemical adsorption and biological degradation in

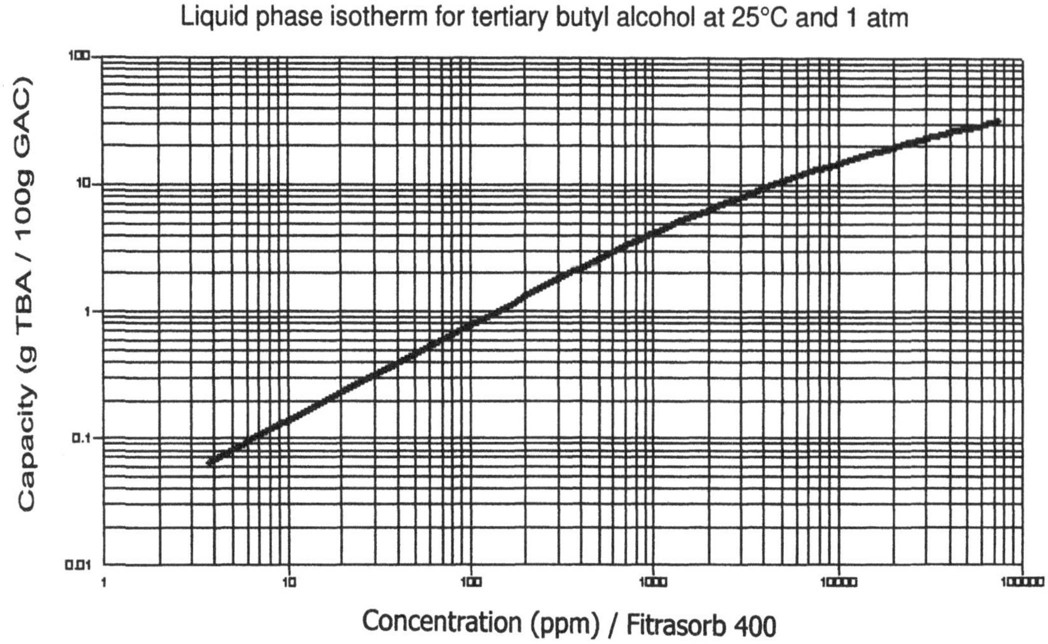

Figure 4.11 Calgon Carbon GAC (Filtrasorb 400) isotherm for TBA used for biological columns

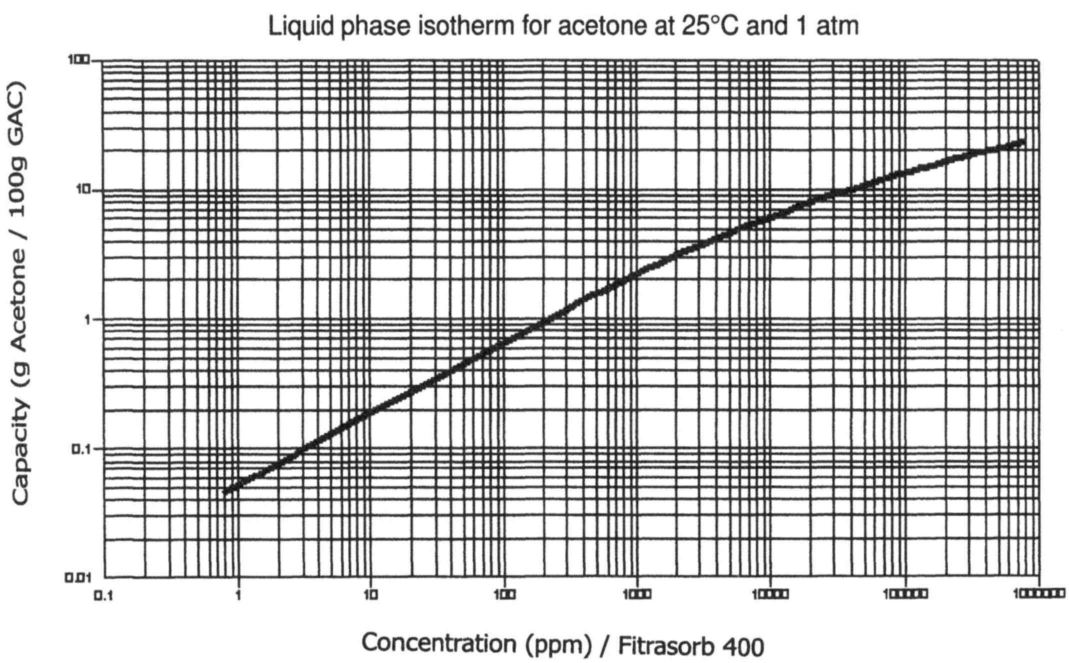

Figure 4.12 Calgon Carbon GAC (Filtrasorb 400) isotherm for acetone used for biological columns

Table 4.8
MTBE and byproducts for 30 mg/L ozone and 30 mg/L hydrogen peroxide

Sample Location	MTBE (μg/L)	TBA (μg/L)	TBF (μg/L)	Acetone (μg/L)	Acetaldehyde (μg/L)	Formaldehyde (μg/L)	Glyoxal (μg/L)	M-Glyoxal (μg/L)
Influent	113.4	ND	ND	ND	ND	7	2	ND
Effluent	ND	ND	ND	ND	2	28	11	10
BAC Effluent	ND	ND	ND	ND	ND	5	3	ND

Table 4.9
Low level of MTBE and byproducts for 15 mg/L ozone only

Sample Location	MTBE (μg/L)	TBA (μg/L)	TBF (μg/L)	Acetone (μg/L)	Acetaldehyde (μg/L)	Formaldehyde (μg/L)	Glyoxal (μg/L)	M-Glyoxal (μg/L)
Influent	81	ND	ND	ND	NA	NA	NA	NA
Effluent	ND	ND	ND	ND	3	26	11	7
BAC Effluent	ND	ND	ND	ND	ND	7	ND	ND

the GAC columns. Thus the reduction seen in biological GAC can be attributed to combination of chemical adsorption and biological activity.

When higher applied ozone of 30 mg/L was used with 30 mg/L of hydrogen peroxide (Table 4.8) similar removal efficiency was observed compared with the tests using 15 mg/L of O_3 and 15 mg/L of H_2O_2. One noticeable difference at higher applied ozone and hydrogen peroxide dose was non-detectable level of TBA in the ozone effluent sample. It may be possible that higher ozone dose applied contributed to higher production of hydroxyl radicals for more efficient oxidation of TBA during the ozone process with hydrogen perxoide. All other compounds were similar to the results shown for the lower applied ozone dose.

A series of control tests were done with ozone only to better understand the mechanics of MTBE degradation, and not necessarily to be used in real application as ozone in combination with hydrogen peroxide provides far more benefits in many aspects. The application of high ozone only can yield formation of bromate.

With the low levels of MTBE (~80 μg/L), no detectable amount of MTBE or primary degradation byproducts were measured. Some low levels of aldehydes were detected after the ozone process, but after the biological GAC, only formaldehyde was measured slightly above the detection limit (Table 4.9). Under the similar conditions with 30 mg/L of ozone only, comparable results were observed (Table 4.10).

With the initial MTBE level of approximately 350 μg/L, the applied ozone of 15 mg/L achieved 92% MTBE reduction (Table 4.11). TBA in the effluent measured was 56 μg/L, which exceeded the action level of 12 μg/L set by CDHS. After the BAC process, however, neither MTBE nor TBA was detected. As mentioned previously, the GAC used in the columns have some adsorption capacity for both MTBE and TBA, so it is possible that the small amount present in the effluent might have been removed by chemical adsorption in the columns. The levels and the trends for aldehydes were similar to the previous case with lower level of MTBE.

Table 4.10
Low level of MTBE and byproducts for 30 mg/L ozone only

Sample Location	MTBE (μg/L)	TBA (μg/L)	TBF (μg/L)	Acetone (μg/L)	Acetaldehyde (μg/L)	Formaldehyde (μg/L)	Glyoxal (μg/L)	M-Glyoxal (μg/L)
Influent	81	ND	ND	ND	NA	NA	NA	NA
Effluent	ND	ND	ND	ND	4	27	15	8
BAC Effluent	ND	ND	ND	ND	ND	5	1	ND

Table 4.11
Medium level of MTBE and byproducts for 15 mg/L ozone only

Sample Location	MTBE (μg/L)	TBA (μg/L)	TBF (μg/L)	Acetone (μg/L)	Acetaldehyde (μg/L)	Formaldehyde (μg/L)	Glyoxal (μg/L)	M-Glyoxal (μg/L)
Influent	345	ND	ND	ND	NA	NA	NA	NA
Effluent	29	56	ND	ND	4	35	12	9
BAC Effluent	ND	ND	ND	ND	ND	6	ND	ND

Table 4.12
Medium level of MTBE and byproducts for 30 mg/L ozone only

Sample Location	MTBE (μg/L)	TBA (μg/L)	TBF (μg/L)	Acetone (μg/L)	Acetaldehyde (μg/L)	Formaldehyde (μg/L)	Glyoxal (μg/L)	M-Glyoxal (μg/L)
Influent	432	ND	ND	ND	NA	NA	NA	NA
Effluent	ND	47	ND	ND	3	28	11	11
BAC Effluent	ND	ND	ND	ND	ND	5	1	ND

As the applied ozone dose was increased to 30 mg/L (ozone only), all MTBE present was removed below the detection limit during the ozone process as seen in Table 4.12. TBA, however, was detected at about 47 μg/L, which again was removed in the biological GAC column. Thus the combination of AOP and biological GAC have effectively removed both MTBE and the by-products.

Bromate Formation – Surface Water

The surface water used (SPW) for the initial tests contains approximately 200 μg/L of bromide. With lower concentration of ozone (15 mg/L) and hydrogen peroxide, approximately 5 μg/L of bromate was formed (Table 4.13) in the ozone effluent. When higher applied ozone (30 mg/L) and hydrogen peroxide were used, approximately five times more bromate was formed at 34 μg/L. As the ozone concentration approaches and exceeds the saturation level with 30 mg/L of applied ozone, more free ozone may be available to react with bromide to form more bromate. The higher level of bromate formation with higher level of applied ozone was also observed in the bench scale in recirculation mode.

Table 4.13
Effects of applied ozone dose on bromate formation for ozone/peroxide treatment

	15 mg/L Ozone and H_2O_2			30 mg/L Ozone and H_2O_2		
Sample Location	Bromate (μg/L)	O_3 Residual (mg/L)	H_2O_2 (mg/L)	Bromate (μg/L)	O_3 Residual (mg/L)	H_2O_2 (mg/L)
Column 1 Influent	ND	NA	20	ND	NA	24
Column 1 Effluent	NA	0.220	NA	NA	0.246	NA
Column 2 Effluent	5.2	0.146	12	34.3	0.061	11
BAC Effluent	ND	ND	ND	ND	ND	ND

ND = Non Detect, NA = Not Analyzed

Table 4.14
Effects of applied ozone dose on bromate formation for ozone-only treatment

	15 mg/L Ozone Only	30 mg/L Ozone Only
Sample Location	Bromate (μg/L)	Bromate (μg/L)
Column 1 Influent	ND	ND
Column 2 Effluent	317	317
BAC Effluent	20	66

However, no detectable amount of bromate was measured in the biological GAC effluent. Bromate can only be reduced biologically under anoxic condition (no or low dissolved oxygen), so the reduction in bromate seen is likely due to chemical adsorption to the GAC. Even with the limited adsorption capacity, the post treatment using BAC can assist in bromate control by adsorbing small amounts of bromate that leaves ozone column. The fluctuation in the system may leach out adsorbed bromate in the GAC back to the water, so the operation and check up of the system must be maintained.

As mentioned previously, the application of high ozone alone resulted in bromate levels exceeding 300 μg/L (Table 4.14), which is far more than the regulatory limit of 10 μg/L. Thus application of high ozone alone may be feasible to remove MTBE from bromide laden water, but impractical to adopt as a treatment method.

MTBE and By-products Removal With Ozone/H_2O_2 in Groundwater

Higher initial concentrations of MTBE were tested for Santa Monica groundwater, where the extent of contamination is higher than that of surface water used. With the initial MTBE level of 130 μg/L, 15 mg/L of applied ozone and 15 mg/L H_2O_2 removed approximately 96% of MTBE to the secondary standard level of 5 μg/L (Table 4.15). The low levels of MTBE degradation by-products were observed at the effluent of the ozone columns, but after the effluent was passed through the biological GAC column, no detectable amount of either MTBE or the by-products were found except for acetone and formaldehyde. As it was mentioned previously, it is

Table 4.15
Low level MTBE and byproducts for 15 mg/L O_3 and 15 mg/L H_2O_2

Sample Location	MTBE (μg/L)	TBA (μg/L)	TBF (μg/L)	Acetone (μg/L)	Acetaldehyde (μg/L)	Formaldehyde (μg/L)	Glyoxal (μg/L)	M-Glyoxal (μg/L)
Influent	132	8	NA	20	NA	NA	NA	NA
Effluent	5	10	NA	15	3	20	2	19
BAC Effluent	ND	ND	NA	30	ND	6	ND	ND

Table 4.16
Low level MTBE and byproducts for 15 mg/L O_3 and 15 mg/L H_2O_2 with TBA spike

Sample Location	MTBE (μg/L)	TBA (μg/L)	TBF (μg/L)	Acetone (μg/L)	Acetaldehyde (μg/L)	Formaldehyde (μg/L)	Glyoxal (μg/L)	M-Glyoxal (μg/L)
Influent	173	201	NA	11	NA	NA	NA	NA
Effluent	12	65	NA	57	1	29	2	54
BAC Effluent	ND	ND	NA	14	ND	7	ND	ND

Table 4.17
Low level MTBE and byproducts for 30 mg/L O_3 and 30 mg/L H_2O_2

Sample Location	MTBE (μg/L)	TBA (μg/L)	TBF (μg/L)	Acetone (μg/L)	Acetaldehyde (μg/L)	Formaldehyde (μg/L)	Glyoxal (μg/L)	M-Glyoxal (μg/L)
Influent	184	11	NA	12	NA	NA	NA	NA
Effluent	6	8	NA	40	4	18	2	23
BAC Effluent	ND	7	NA	9	ND	5	ND	ND

possible that combination of biological uptake and chemical adsorption could have contributed for removal of various types of constituents, although it was not clear how much each mechanism have contributed. Based on the previous data, no detectable amount of TBF was measured when the initial MTBE concentration was low, so TBF was not analyzed (NA) for the cases when the MTBE concentrations were below few hundred μg/L levels.

With the water that also has the initial TBA concentration of 200 μg/L under the same experimental conditions, similar trends in MTBE removal efficiency (93%) was observed. The presence of TBA did not hinder the MTBE degradation, and TBA itself was also reduced to non-detectable level after the BAC column with 15 mg/L of applied ozone and H_2O_2 (Table 4.16).

Since 15 mg/L was enough to achieve good MTBE removal efficiency, only slight improvement in removal efficiency (97%) was realized by applying higher ozone and H_2O_2 dose of 30 mg/L (Tables 4.17 and 4.18). With the initial TBA concentration of 200 μg/L, even the higher applied ozone/H_2O_2 dose of 30 mg/L in combination with BAC was not sufficient to remove TBA below the action level of 12 μg/L (Table 4.18). However, the presence of TBA did not interfere with the high MTBE removal efficiency (97%) under the conditions used.

Table 4.18
Low level MTBE and byproducts for 30 mg/L O_3 and 30 mg/L H_2O_2 with TBA spike

Sample Location	MTBE (µg/L)	TBA (µg/L)	TBF (µg/L)	Acetone (µg/L)	Acetaldehyde (µg/L)	Formaldehyde (µg/L)	Glyoxal (µg/L)	M-Glyoxal (µg/L)
Influent	183	204	NA	12	NA	NA	NA	NA
Effluent	5	114	NA	85	ND	27	2	73
BAC Effluent	ND	20	NA	13	ND	6	ND	ND

Table 4.19
High level MTBE and by-products for 15 mg/L O_3 and H_2O_2

Sample Location	MTBE (µg/L)	TBA (µg/L)	TBF (µg/L)	Acetone (µg/L)	Acetaldehyde (µg/L)	Formaldehyde (µg/L)	Glyoxal (µg/L)	M-Glyoxal (µg/L)
Influent	2037	ND	ND	82	NA	NA	NA	NA
Effluent	281	352	77	178	2	90	4	69
BAC Effluent	30	98	ND	93	ND	6	ND	1

Table 4.20
High level MTBE and by-products for 15 mg/L O_3 and H_2O_2 with TBA spike

Sample Location	MTBE (µg/L)	TBA (µg/L)	TBF (µg/L)	Acetone (µg/L)	Acetaldehyde (µg/L)	Formaldehyde (µg/L)	Glyoxal (µg/L)	M-Glyoxal (µg/L)
Influent	2378	400	ND	14	NA	NA	NA	NA
Effluent	326	466	45	198	2	100	3	86
BAC Effluent	25	57	ND	22	ND	5	ND	ND

The applied ozone dose of 15 mg/L with 15 mg/L H_2O_2 showed approximately 86% removal (either with or without initial TBA) when the initial MTBE concentrations were higher than 2,000 µg/L. The primary MTBE degradation byproducts (TBA, TBF, and acetone) were also much higher as seen in Tables 4.19 and 4.20 when compared with the levels for the tests done with initial MTBE concentrations of few hundred µg/L. Again the presence of initial TBA did not make any significant difference in terms of MTBE reduction or other by-products formation (Table 4.20). The biological GAC removed additional 10% of MTBE and the byproducts to yield the overall system removal efficiency of MTBE at greater than 98% (ozone in combination with biological GAC). All other by-products were also removed by BAC to some degree, but the residual levels were still higher than the regulatory standards for some compounds including MTBE and TBA.

The advantage of higher applied ozone and H_2O_2 dose of 30 mg/L is shown in Tables 4.21 and 4.22 where greater than 95% of initial MTBE was reduced by the ozone / H_2O_2 process. The small amount of MTBE remaining in the effluent was effectively reduced to below the detection limit in the biological GAC column. However, as MTBE breaks down to the primary degradation

Table 4.21
High level MTBE and byproducts for 30 mg/L O_3 and 30 mg/L H_2O_2

Sample Location	MTBE (μg/L)	TBA (μg/L)	TBF (μg/L)	Acetone (μg/L)	Acetaldehyde (μg/L)	Formaldehyde (μg/L)	Glyoxal (μg/L)	M-Glyoxal (μg/L)
Influent	1583	ND	ND	93	NA	NA	NA	NA
Effluent	63	162	ND	216	ND	58	3	110
BAC Effluent	ND	111	ND	ND	ND	5	ND	ND

Table 4.22
High level MTBE and byproducts for 30 mg/L O_3 and 30 mg/L H_2O_2 with TBA

Sample Location	MTBE (μg/L)	TBA (μg/L)	TBF (μg/L)	Acetone (μg/L)	Acetaldehyde (μg/L)	Formaldehyde (μg/L)	Glyoxal (μg/L)	M-Glyoxal (μg/L)
Influent	2081	396	ND	19	NA	NA	NA	NA
Effluent	99	311	ND	346	ND	84	3	170
BAC Effluent	ND	113	ND	76	ND	82	3	150

byproducts, TBA level increased from non-detect to 162 μg/L in the effluent. TBA in the effluent was not effectively removed in the BAC columns, and longer retention time in the BAC column may be necessary for the target removal efficiency. Typically, aldehydes have been effectively removed in the BAC column, so it is not clear why the aldehydes were not removed in the BAC column with TBA level of 200 μg/L (Table 4.22).

Bromate Formation – Groundwater

When ozone is applied in the presence of hydrogen peroxide, only small amount of bromate is formed during the process. Table 4.23 shows the bromate levels in the ozone effluent, which are lower than 10 μg/L when 15 mg/L of O_3 and H_2O_2 was applied. The BAC has some limited adsorption capacity for bromate, and during the ozone only test with SPW, the BAC columns have been exposed to the high levels of bromate formed during ozone only tests. The tests done with groundwater formed low level of bromate, which possibly created negative concentration gradient between the GAC and effluent. The elevated levels of bromate in the BAC effluent might be due to this phenomenon. However, the BAC columns were subjected to fresh water continuously, and it was suspected that any bromate adsorbed to GAC may have already leached out during non test period (typically 5 continuous days) where the water passing through GAC has no bromate. It is not clear how much bromate was adsorbed in the columns initially, and what the rate of bromate leaching from the biological GAC. It should be noted that the GAC has low adsorption capacity for bromate and the removal of bromate in GAC filters is hindered by the presence of dissolved oxygen. Bromate leaching was observed in the GAC columns when the columns were exposed to fluctuating levels of bromate in a previous AwwaRF study titled "Removal of Bromate and Perchlorate in Conventional Ozone/GAC Systems". An alternative explanation for the high bromate concentration in the GAC effluent is that the bromide entering

Table 4.23
Effects of applied ozone dose on bromate formation, bromate in μg/L

O_3 and H_2O_2 Condition (1:1)	MTBE (μg/L)	TBA (μg/L)	BrO_3^- in O_3 effluent (μg/L)	O_3 residual (mg/L)	H_2O_2 residual (mg/L)	BrO_3^- in BAC effluent (μg/L)	O_3 residual (mg/L)	H_2O_2 residual (mg/L)
15 mg/L	200	0	8	0	6	92	0	0
	200	200	4	0	6	24	0	0
	2,000	0	4	0	8	41	0	0
	2,000	200	5	0	12	20	0	0
30 mg/L	200	0	36	0	14	70	0	0
	200	200	29	0	14	25	0	0
	2,000	0	39	0	18	49	0	0
	2,000	200	48	0	20	19	0	0

GAC might have been oxidized at the surface of the GAC, which may have been exposed to various oxidants during the process. Further studies are needed to understand this issue more thoroughly.

Summary of Findings

Based on the stripping control study with oxygen bubbles, approximately 10% of MTBE was purged from the ozone column at the gas flow rate used. Overall approximately 90 ~ 100% of MTBE in the concentration range of few hundred μg/L were removed when ozone alone or ozone with hydrogen peroxide were applied. With the low concentration of MTBE, the type of source water did not significantly influence the removal efficiency of MTBE. For both source water, 92 ~ 100% removal of MTBE was consistently achieved for low initial MTBE concentrations. As the initial MTBE concentration increased to more than 2,000 μg/L, ozone only in the range of 15 ~ 30 mg/L did not provide enough conditions to effectively remove MTBE, and the MTBE removal efficiency decreased to 86%. The addition of hydrogen peroxide (1:1 ratio as ozone:H_2O_2) improved the removal efficiency to greater than 95% for higher concentration of MTBE in 2,000 μg/L range. The post treatment using biological GAC removed most of the residual MTBE and by-products in the ozone effluent. However, with MTBE concentration greater than 2,000 μg/L, even this combination did not provide enough removal efficiency for MTBE or by-products. Longer than 25 minute retention time in the ozone column and/or biological GAC column might be necessary to achieve the desired level of MTBE. In the absence of hydrogen peroxide, high levels of bromate were formed, well in excess of the regulatory limit. Addition of hydrogen peroxide significantly reduced the amount of bromate formed, but with 30 mg/L of applied ozone and hydrogen peroxide, the level of bromate formed in the effluent exceeded the regulatory limit. The biological GAC column enhanced overall MTBE and byproducts removal efficiency by providing some chemical adsorption and biological reduction, but the fluctuation in water quality or operating condition might provide circumstances where adsorbed species may leach out from the adsorbed GAC phase or modified GAC surface due to oxidant might provide conditions where some of the species might undergo reaction in the biological GAC column. The system should be designed to allow ozone in combination with hydrogen peroxide to provide sufficient removal of MTBE, and the biological GAC should serve as the means to remove the MTBE degradation by-products formed during the ozone process while providing additional treatment barrier for MTBE. However, the mechanism of bromate leaching or transformation in the biological GAC should be understood to implement this post treatment option.

IMPLICATIONS FOR FULL-SCALE USE

Process Design Criteria

Based on bench and pilot scale testing information presented in earlier sections it was inferred that an ozone and hydrogen peroxide dosages of 15 mg/L each and contact time of 25 minutes will be sufficient to lower the MTBE levels by 95% or higher. This is based on the assumption that the influent MTBE would be 200 µg/L. Pilot scale testing also showed that a loading rate of 2 gpm/sf and a retention time of 20 minutes in biological GAC will achieve further removal of MTBE (possibly to non-detect levels) and its by-products.

Process Schematics and Layout

Ozone/Hydrogen Peroxide System

A schematic of the ozone/hydrogen peroxide system is shown in Figure 4.13. Raw water supply is metered prior to pumping. Booster pumps provide the hydraulic head required for flow through the entire system. Following the pumps, ozone and hydrogen peroxide are added to the water at concentrations of 15 mg/L for each chemical. The design concentrations of ozone and hydrogen peroxide were determined from pilot testing. Hydrogen peroxide can either be fed through a venturi-type injector device which facilitates mixing or metered into the process stream and mixed downstream of the injection point via a static or dynamic mixing device. Ozone can be fed through a venturi-type injector prior to the contacting basin or through porous spargers within the contacting basin, however, venturi-type injector devices were preferred by most vendors.

Ozone generation for systems of 1 mgd and less use pressure swing absorption (PSA) to generate oxygen onsite. PSA requires additional feed gas equipment such as an air compressor, air chiller, and air dryer. For larger systems, liquid oxygen (LOX) is used and the bulk of the feed gas equipment is replaced with a LOX storage tank. The oxygen from either the PSA feed or the LOX feed is fed to the ozone generator and the resulting ozone gas is dissolved into the process stream as discussed above. Following the ozone/hydrogen peroxide mixing is a contacting basin that provides a total residence time of 25 minutes, determined in pilot testing. Ozone contactors must be covered and have systems for the collection of the ozone off-gas because ozone is toxic and must be kept within OSHA allowable limits. A negative pressure is maintained in the headspace of the contacting basin and the off-gas is fed to a thermal and catalytic ozone off-gas destruction device. The final unit process consists of biologically enhanced granular activated carbon (GAC) polishing to destroy any remaining MTBE and MTBE by-products. Allowing biological growth within the carbon beds enhances the ability of the process to destroy the remaining chemicals and extends the life of the carbon. Pilot testing of GAC resulted in a recommended hydraulic loading

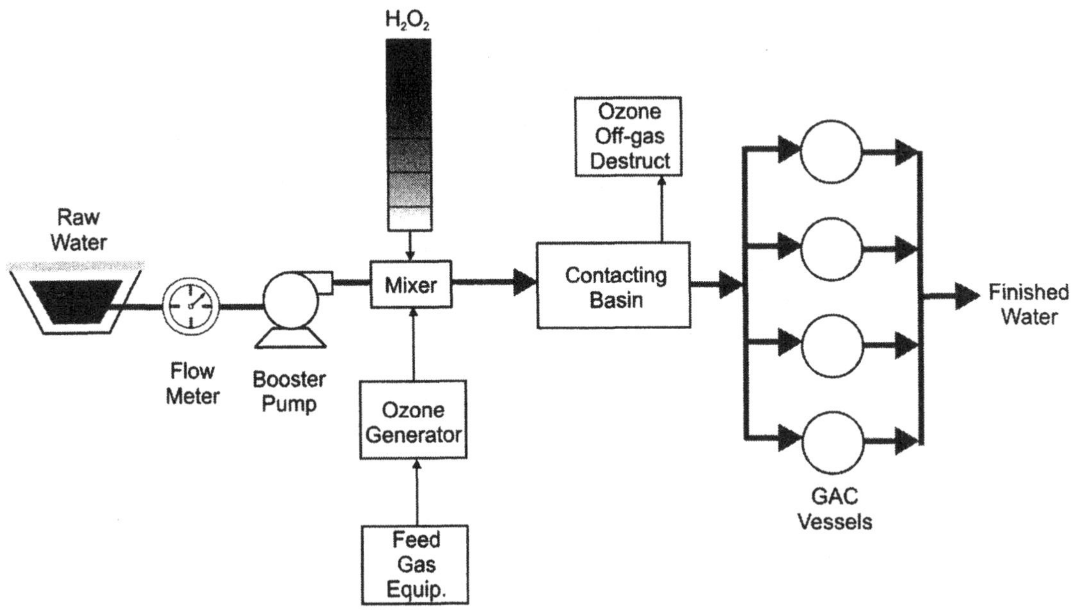

Figure 4.13 Ozone/peroxide schematic for 1 mgd system

of 2 gpm/sf and a contact time of 20 minutes. The design parameters require multiple pre-manufactured GAC vessels, depending on the flow rate through the system.

Based on the process schematic, potential site layouts were prepared for 1 mgd (Figure 4.14) and 10 mgd (Figure 4.15) treatment systems. Major equipment is shown on the site layout. MTBE contaminated water is pumped from the well pump base and boosted through the booster pump stations. From the booster pumps, the water flows through a mixing device where the ozone and hydrogen peroxide are added and mixed.

Alternatively, ozone can be added directly to the contacting basin via porous diffusers. Following mixing, the process water flows into the contacting basin. Contacting basins were assumed to be of on-site concrete construction with a length to width ratio of 2.5. Following the contacting basin, process water flows through a distribution header to the GAC vessels. Large pre-manufactured vessels are sold in pairs of 20,000 lb carbon capacity. The size of each vessel is 10 feet in diameter. To meet the hydraulic loading requirement, 4 GAC vessels are required for the 1 mgd system and 40 GAC vessels are required for the 10 mgd system. After determining the number of GAC vessels required to meet the hydraulic loading requirement, the volume of carbon required for each vessel was calculated to meet the required 20 minute contact time. Following the GAC polishing step, the treated water should meet all applicable MTBE and MTBE by-product goals.

Ozone and hydrogen peroxide delivery and storage systems are situated adjacent to the process water stream. For a 1 mgd system, an 8 foot diameter hydrogen peroxide tank will be required and a 10 foot diameter hydrogen peroxide tank will be required for a 10 mgd system. Ozone equipment for a 1 mgd system includes two ozone generators, two sets of PSA feed gas equipment, and at least one ozone off-gas destruct unit. Ozone equipment for a 10 mgd system includes three ozone generators, a LOX storage tank, and at least two ozone off-gas destruct units.

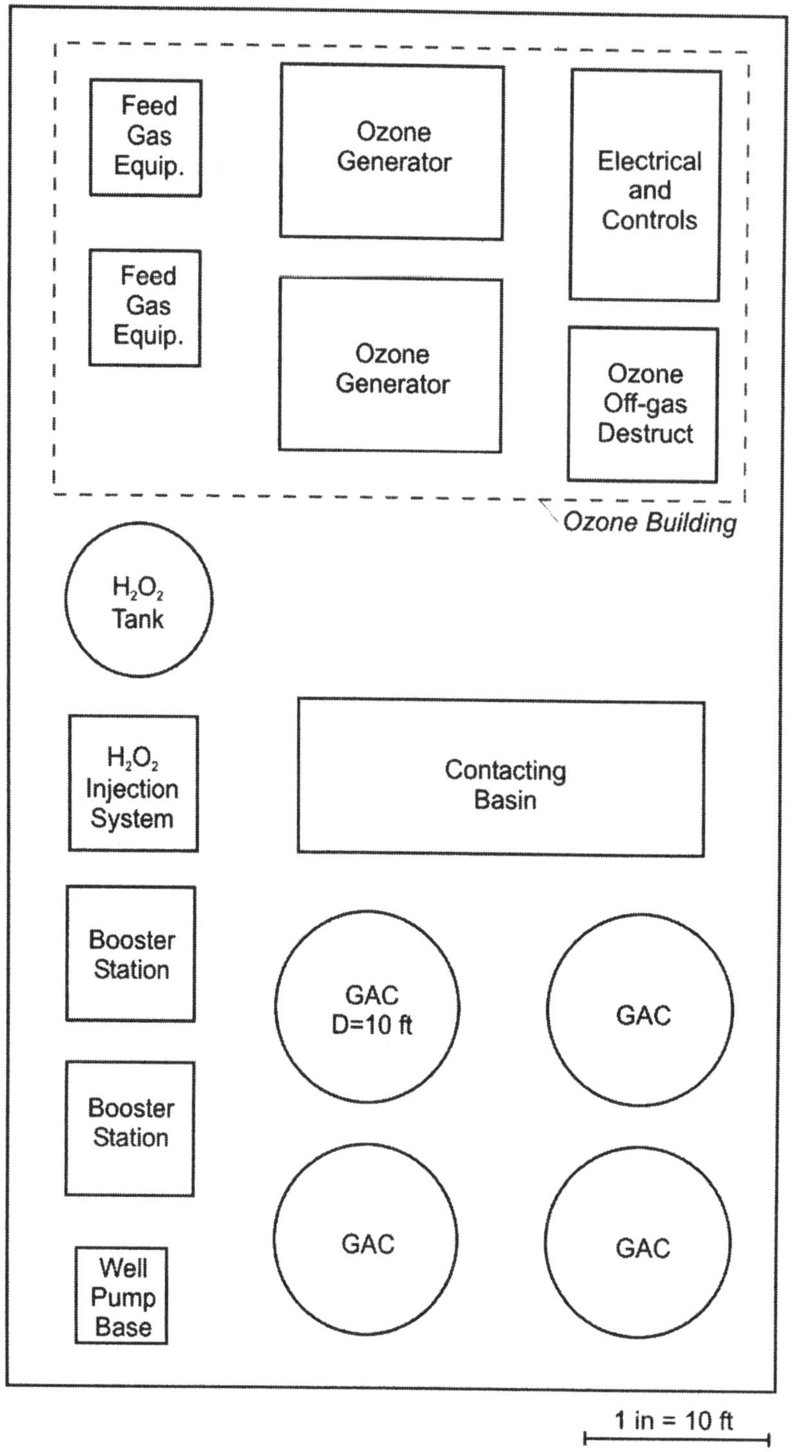

Figure 4.14 Ozone/peroxide layout for 1 mgd

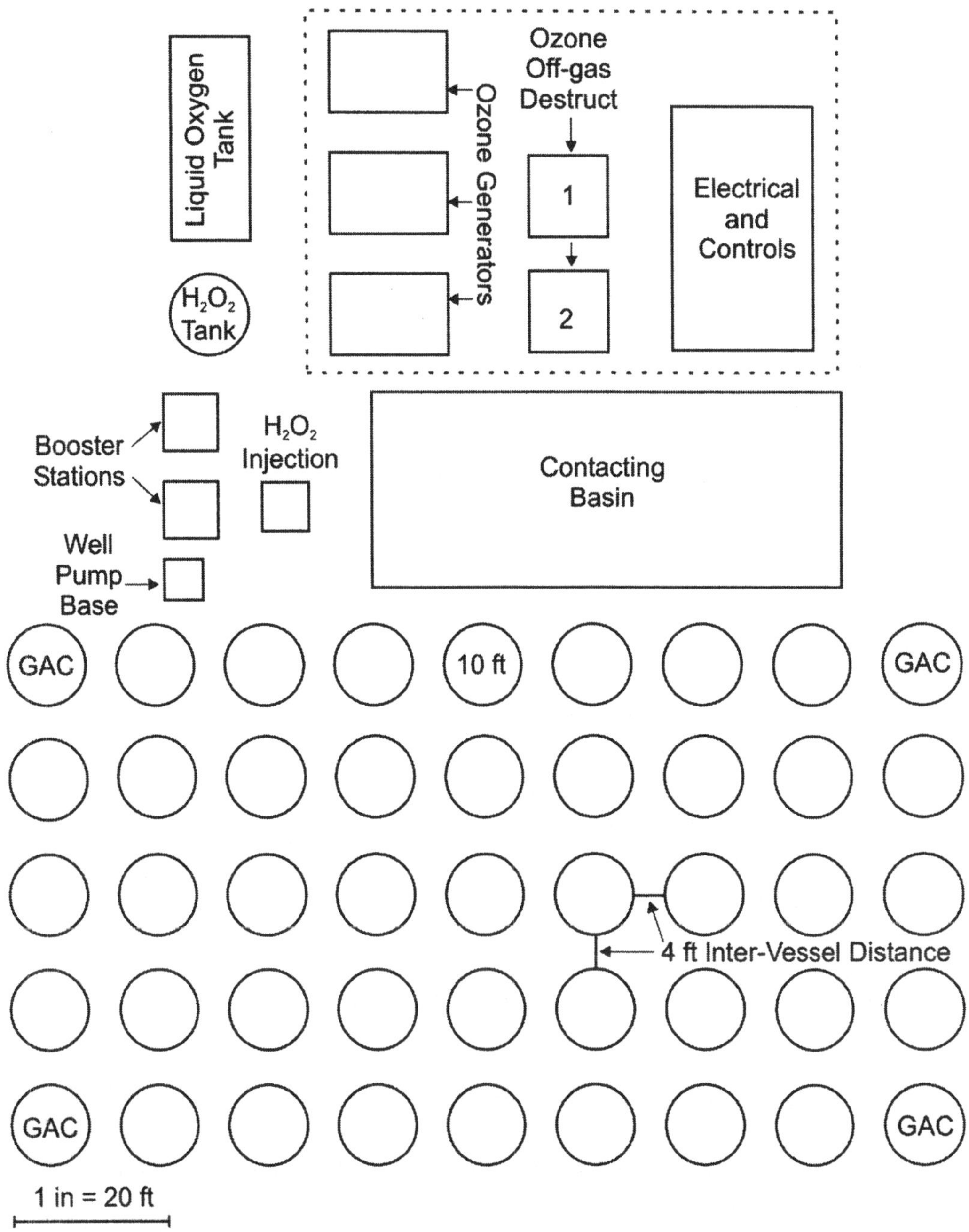

Figure 4.15 Ozone/peroxide layout for 10 mgd

Finally, an electrical control room was included to house electrical components and process control equipment. Approximate footprints of the required equipment are shown on the site layout.

Technology Costs

Cost spreadsheets were created for the following scenarios:

- Ozone/hydrogen peroxide
- Manufacturer A's proprietary ozone/hydrogen peroxide system
- Manufacturer B's proprietary ozone/hydrogen peroxide system and
- Biological activated carbon polishing treatment

Ozone/Hydrogen Peroxide

Shown in Table 4.24 are the costs for ozone/peroxide treatment that were estimated using the design assumptions identified above. Every row of the attached cost table is numbered and comments and assumptions for the costs are discussed below by row number.

Capital Cost Assumptions

(1) Design flow in million gallons/day (mgd). Four flow rates were chosen to represent the range of treatment sizes under consideration. The selected flow rates of 0.5, 1.0, 3.0, and 10.0 mgd are displayed across in the four columns.

(2) Conversion of mgd to gallons/minute (gpm) by multiplying with 694, the conversion factor.

(3) Ozone required for the design concentration of 15 mg/L is calculated using the following equation:

$$Ozone\left(\frac{lbs}{d}\right)=15\left(\frac{mg}{L}\right)\times 2.2\,x10^{-6}\left(\frac{lbs}{mg}\right)\times Q\left(\frac{MG}{d}\right)\times 3.785\left(\frac{gal}{L}\right)\times 10^{6}\left(\frac{gal}{Mgal}\right) \quad (4.1)$$

(4) Hydrogen peroxide (gallons/day) required for the design concentration of 15 mg/L are shown in this row. These values were calculated assuming 35% volume/volume H_2O_2.

(5) Volume of contacting basin (in gallons) required for 25-minute residence time is obtained by multiplying the flow rate in row 2 with 25.

(6) Ozone generating equipment cost quotes were obtained from the equipment manufacturers.

(7) Ozone feed gas equipment costs were provided by manufacturers. For the two lowest flow rates, oxygen is created onsite with pressure swing absorption. For the two highest flow rates, liquid oxygen was assumed to be use for ozone generation.

(8) Ozone diffuser system costs were obtained manufacturers. A venturi-type injection system was assumed to accomplish diffusion.

(9) Ozone off-gas destruction cost quotes were obtained from the equipment vendors.

(10) Tank cost for peroxide storage was obtained from tank suppliers.

Table 4.24
Capital, O&M, and total costs for ozone/peroxide treatment of MTBE (for sourced system)

Flow Rates					
Design Flow (mgd)	(1)	0.5	1	3	10
Design Flow (gpm)	(2)	347	694	2082	6940
Sizing Calculations					
Ozone Requirement (lbs/d)	(3)	63	125	375	1252
Hydrogen Peroxide Required (gal/d)	(4)	19	38	114	379
Contactor Volume (gal)	(5)	8675	17350	52050	173500
Capital Cost Summary					
Process Capital Cost					
Ozone Generating Equipment	(6)	$87,500	$140,000	$260,000	$460,000
Ozone Feed Gas Equipment	(7)	$62,500	$100,000	$130,000	$230,000
Ozone Diffuser System	(8)	$25,000	$40,000	$65,000	$115,000
Ozone Off-Gas Decomposer	(9)	$25,000	$40,000	$65,000	$115,000
Peroxide Storage Tank	(10)	$36,000	$36,000	$36,000	$40,000
Peroxide Injection System	(11)	$9,000	$9,000	$9,000	$10,000
Monitoring Instrumentation	(12)	$31,750	$46,750	$71,750	$122,500
Control Package	(13)	$31,750	$46,750	$71,750	$122,500
Pipes and Valves	(14)	$43,000	$58,000	$83,000	$135,000
Pumping	(15)	$60,203	$87,890	$169,078	$364,232
Contacting Basin	(16)	$34,788	$53,558	$106,129	$224,557
Process Cost Subtotal	(17)	$446,491	$657,948	$1,066,707	$1,938,789
Direct Capital Cost	(18)	$1,116,228	$1,644,871	$2,133,415	$3,877,579
Indirect Capital Cost					
Piloting	(19)	$10,000	$65,000	$65,000	$65,000
Permitting	(20)	$13,395	$19,738	$32,001	$58,164
Land	(21)	$13,395	$19,738	$32,001	$58,164
Operator Training	(22)	$1,796	$1,796	$1,796	$1,796
Housing	(23)	$63,074	$74,448	$119,944	$279,182
Indirect Cost Subtotal	(24)	$101,660	$180,721	$250,743	$462,306
Total Capital Cost	(25)	$1,217,888	$1,825,592	$2,384,158	$4,339,884
Annual O&M Summary					
Chemicals	(26)	$38,143	$76,285	$291,270	$964,513
Electricity	(27)	$25,792	$51,264	$89,904	$299,776
Labor	(28)	$18,221	$36,442	$54,662	$72,883
Total Annual O&M Cost	(29)	$82,156	$163,990	$435,836	$1,337,172
For a project period of 20 years with 7% interest					
Amortized Total Annual Cost	(30)	$197,116	$336,313	$660,884	$1,746,826
Cost/1000 Gallons Treated	(31)	$1.08	$0.92	$0.60	$0.48

(11) Peroxide injection system costs were obtained from vendors.
(12) Monitoring instrumentation costs for hydrogen peroxide equipment plus ozone equipment, provided by the above referenced vendors.
(13) Control package costs for hydrogen peroxide equipment plus ozone equipment, provided by the above referenced vendors.
(14) Piping and valves costs for hydrogen peroxide equipment plus ozone equipment, provided by the above referenced vendors.
(15) Pump capital and associated equipment costs are obtained using the Water/Wastewater Cost models (Culp, Wesner, Culp, 1990).
(16) Contacting basin costs are based on required contacting volume (row 5) multiplied by 1.15 for tank freeboard. These costs are derived from the concrete contactors cost worksheet developed for the EPA Stage-2/LT-2 Technology and Cost document (2002).
(17) Process cost sub-total is a sum of all the above-discussed process costs (rows 6 - 16).
(18) The direct capital cost is obtained by multiplying the total process cost (row 17) by a multiplier that would account for: site work, contractor's overhead and profit, contingencies, engineering and design, mobilization and bonding, legal and administrative, interest during construction, and installation. The process cost multiplier is 2.5 for flow rates equal to or below 1 mgd and 2.0 for flow rates above 1 mgd (NDWAC, 2001).
(19) Piloting costs were assumed to be similar to those calculated for the EPA Stage-2/LT-2 Technology and Cost document (2002). The piloting costs are assumed to be $10,000 for flow rates below 1 mgd and $65,000 for flow rates equal to and above 1 mgd. For large systems, the piloting costs were calculated to be $65,000 and for small systems it was assumed that limited piloting would be provided for by $10,000.
(20) Permitting costs were assumed to be 3% of the process cost subtotal (row 17), with a minimum of $2500 (EPA Stage-2/LT-2 Technology and Cost Document, 2002).
(21) Land costs were assumed to be 3% of the process cost subtotal (row 17).
(22) Operator training costs were based on 40 hours of supervisory training at a rate of $45/hour.
(23) The building size for housing was estimated from the footprints. Costs were developed from the RS Means Building Construction Cost (2002) data for warehouse/office space and equal to $62/sq.ft.
(24) Indirect cost sub-total consists of the sum of all indirect process costs (rows 19 - 23).
(25) Total capital cost is a sum of indirect capital cost (row 24) and direct capital cost (row 18).

O&M Cost Assumptions

(26) Chemical costs were based on unit costs for hydrogen peroxide at $5.50/gallon and liquid oxygen (where required) at $90/ton. These unit costs were obtained from vendors.
(27) Electricity costs were based on ozone equipment requirements plus main process pumping costs. Pumping costs were estimated by calculating energy required for pumping the desired flow rate under the assumptions of 50ft of head, 75% pump efficiency, and 85% motor efficiency. The total energy required was then multiplied by 1.1 to account for miscellaneous additional energy requirements and power was estimated at $0.08/kWh.

(28) Operational labor costs were estimated at the following amounts:

0.5 mgd system = 2 hours/day
1.0 mgd system = 4 hours/day
3.0 mgd system = 6 hours/day
10 mgd system = 8 hours/day

The labor rate was assumed at a typical technician rate of $25/hour.

(29) Annual O&M subtotal is a sum of all the itemized O&M costs (rows 26 - 28).

Total Cost Assumptions

(30) The total capital cost (row 25) was amortized over 20 years at a 7% interest rate and added to the annual O&M cost (row 29) to derive the amortized total annual cost.
(31) Cost per thousand gallons of treated water was calculated by dividing the amortized total annual cost (row 30) by the flow rate in thousands of gallons and the days in a year.

Manufacturer A's Packaged Ozone/Hydrogen Peroxide Treatment System Costs

Shown in Table 4.25 are the cost estimates that were developed for Manufacturer's A packaged ozone/peroxide treatment system. Listed below are some unique assumptions that are specific to Manufacturer A's system.

Capital Cost Assumptions

(6) Manufacturer A's system costs include all major capital equipment necessary for MTBE destruction.
(7) Control package costs are included in (6) for each Manufacturer A system unit, however, at 3 and 10 mgd flow rates, multiple systems are required and a master PLC was assumed for integral control of all the units.
(8) Piping and valves costs were assumed to be 10% of the Manufacturer A system cost. These items are required for connections with main pumping and connections between multiple Manufacturer A systems.
(9) Pumping costs for main plant pumping as calculated in the pumping costs worksheet.
(10) Contacting basins are included in (6).
(12) The direct capital cost is obtained by multiplying the total process cost (row 17) by a multiplier that would account for: site work, contractor's overhead and profit, contingencies, engineering and design, mobilization and bonding, legal and administrative, interest during construction, and installation. The process cost multiplier for packaged or pre-manufactured systems is 1.67 for all flow size systems.

O&M Cost Assumptions

(21) Electricity costs were based on equipment requirements provided by Manufacturer A system plus main process pumping costs as previously described.

Table 4.25
Capital, O&M, and total costs for ozone/peroxide treatment of MTBE using Manufacturer A's system

Flow Rates					
Design Flow (mgd)	(1)	0.5	1	3	10
Design Flow (gpm)	(2)	347	694	2082	6940
Sizing Calculations					
Ozone Requirement (lbs/d)	(3)	63	125	375	1252
Hydrogen Peroxide Required (gal/d)	(4)	19	38	114	379
Contactor Volume (gal)	(5)	8675	17350	52050	173500
Capital Cost Summary					
Process Capital Cost					
Manufacturer A's System	(6)	$296,000	$357,000	$970,000	$3,065,000
Control Package	(7)	$0	$0	$19,400	$61,300
Pipes and Valves	(8)	$29,600	$35,700	$97,000	$306,500
Pumping	(9)	$46,703	$74,390	$155,578	$349,232
Contacting Basin	(10)	$0	$0	$0	$0
Process Cost Subtotal	(11)	$372,303	$467,090	$1,241,978	$3,782,032
Direct Capital Cost	(12)	$621,746	$780,040	$2,074,104	$6,315,994
Indirect Capital Cost					
Piloting	(13)	$10,000	$65,000	$65,000	$65,000
Permitting	(14)	$11,169	$14,013	$37,259	$113,461
Land	(15)	$11,169	$14,013	$37,259	$113,461
Operator Training	(16)	$1,796	$1,796	$1,796	$1,796
Housing	(17)	$63,074	$74,448	$119,944	$279,182
Indirect Cost Subtotal	(18)	$97,208	$169,270	$261,260	$572,900
Total Capital Cost	(19)	$718,954	$949,310	$2,335,363	$6,888,894
Annual O&M Summary					
Chemicals	(20)	$38,143	$76,285	$291,270	$964,513
Electricity	(21)	$54,058	$78,052	$235,729	$788,321
Labor	(22)	$18,221	$36,442	$54,662	$72,883
Total Annual O&M Cost	(23)	$110,421	$190,778	$581,661	$1,825,717
For a project period of 20 years with 7% interest					
Amortized Total Annual Cost	(24)	$178,285	$280,386	$802,103	$2,475,980
Cost/1000 Gallons Treated	(25)	$0.98	$0.77	$0.73	$0.68

Manufacturer B's Packaged Ozone/Hydrogen Peroxide Treatment System Costs

Manufacturer B's system is a plug-flow reactor system. In this system, ozone and peroxide are added at multiple locations using injection ports. The amount of oxidant added decreases from the upstream (where the contaminants are at high concentrations) to the downstream end (wher the contaminants are at lower concentrations) within each reactor. Shown in Table 4.26 are the costs for Manufacturer B's ozone/peroxide treatment system. Listed below are some unique assumptions applicable to Manufacturer B's package treatment system.

Capital Cost Assumptions

(6) Manufacturer B system costs include all major capital equipment necessary for MTBE destruction.
(7) Control package costs are included in (6).
(8) Piping and valves costs were assumed to be 5% of the Manufacturer B system cost. These items are required for connections with main pumping.
(9) Pumping costs for main plant pumping as calculated in the pumping costs worksheet.
(12) Process cost multiplier is 1.67 for pre-manufactured systems. Multiplier accounts for the following items: site work, contractor's overhead and profit, contingencies, engineering and design, mobilization and bonding, legal and administrative, interest during construction, and installation. Process cost subtotal (row 11) × multiplier.
(17) Vendor B specified 500 s.f. for 0.5 and 1.0 mgd systems and 1000 s.f. for 3.0 and 10 mgd systems. In addition, it was assumed that an electrical and control room would be required and 150 s.f. was added to the 0.5 and 1.0 mgd systems and 540 s.f. for 3.0 and 10 mgd systems.

O&M Cost Assumptions

(21) Electricity costs were based on equipment requirements provided by Manufacturer B plus main process pumping costs as previously described.

Summary of Ozone/Hydrogen Peroxide Costs

Costs for ozone/hydrogen peroxide treatment of MTBE were developed by two comparable methods. In the sourced system, the cost of each major component of the system were solicited from vendors and added to get the total system cost. The other two MTBE treatment systems were costed from vendors that supply a complete system to accomplish MTBE destruction. In the proprietary vendor systems, less individual costing was necessary because the systems provided from the vendors had most of the components included. Table 4.27 shows cost per thousand gallons treated for four system capacities and three separate cost estimates. A comparison of the costs between the sourced system and proprietary vendor B shows the costs are within 8 percent of each other for the four capacities. Therefore, costs for the vendor B system are very similar to the sourced system. Vendor A costs are less expensive for small systems and transition to more expensive for large systems. For small systems, vendor A is less expensive on capital costs, which leads to the lesser costs shown in Table 4.27. However, vendor A has larger O&M costs for all of their systems and at the larger flow rates the greater O&M costs lead to greater

Table 4.26
Capital, O&M, and total costs for ozone/peroxide treatment of MTBE using Manufacturer B's system

Flow Rates					
Design Flow (mgd)	(1)	0.5	1	3	10
Design Flow (gpm)	(2)	347	694	2082	6940
Sizing Calculations					
Ozone Requirement (lbs/d)	(3)	63	125	375	1252
Hydrogen Peroxide Required (gal/d)	(4)	19	38	114	379
Contactor Volume (gal)	(5)	8675	17350	52050	173500
Capital Cost Summary					
Process Capital Cost					
Manufacturer B's System	(6)	$600,000	$800,000	$1,500,000	$3,200,000
Control Package	(7)	$0	$0	$0	$0
Pipes and Valves	(8)	$30,000	$40,000	$75,000	$160,000
Pumping	(9)	$46,703	$74,390	$155,578	$349,232
Contacting Basin	(10)	$34,788	$53,558	$106,129	$224,557
Process Cost Subtotal	(11)	$711,491	$967,948	$1,836,707	$3,933,789
Direct Capital Cost	(12)	$1,188,190	$1,616,474	$3,067,301	$6,569,428
Indirect Capital Cost					
Piloting	(13)	$10,000	$65,000	$65,000	$65,000
Permitting	(14)	$21,345	$29,038	$55,101	$118,014
Land	(15)	$21,345	$29,038	$55,101	$118,014
Operator Training	(16)	$1,796	$1,796	$1,796	$1,796
Housing	(17)	$40,326	$40,326	$95,542	$95,542
Indirect Cost Subtotal	(18)	$94,812	$165,199	$272,540	$398,365
Total Capital Cost	(19)	$1,283,002	$1,781,673	$3,339,842	$6,967,793
Annual O&M Summary					
Chemicals	(20)	$38,143	$76,285	$291,270	$964,513
Electricity	(21)	$15,976	$32,570	$96,199	$320,397
Labor	(22)	$18,221	$36,442	$54,662	$72,883
Total Annual O&M Cost	(23)	$72,340	$145,296	$442,132	$1,357,793
For a project period of 20 years with 7% interest					
Amortized Total Annual Cost	(24)	$193,446	$313,474	$757,389	$2,015,503
Cost/1000 Gallons Treated	(25)	$1.06	$0.86	$0.69	$0.55

Table 4.27
Total costs for ozone/peroxide treatment - itemized cost approach and packaged system supplier quotations

	Cost per thousand gallons treated ($/1,000 gal)			
	0.5 mgd system	1.0 mgd system	3.0 mgd system	10 mgd system
Sourced system using itemized cost approach	$1.08	$0.92	$0.60	$0.48
Manufacturer A's package treatment system	$0.98	$0.77	$0.73	$0.68
Manufacturer B's package treatment system	$1.06	$0.86	$0.69	$0.55

system costs. Overall vendor A is within 28% of vendor B and within 34% of the sourced system. This is reasonable agreement considering the conceptual nature of the cost development efforts.

Biological Activated Carbon Polishing Treatment

Shown in Table 4.28 are the costs for polishing treatment with biological activated carbon. Every row of the attached cost table is numbered and comments and assumptions for the costs are documented below by row number.

Capital Cost Assumptions

(1) Row 1 shows the design flow in million gallons/day (mgd). Four flow rates were chosen to represent the range of treatment sizes under consideration. The flow rates costed out are 0.5, 1.0, 3.0, and 10.0 mgd, displayed across four columns.
(2) Conversion of mgd to gallons/minute (gpm). mgd is multiplied by 694 to obtain the flow rate in gpm.
(3) Number of contactors required to meet the system sizing requirements. Contactors are sold only in pairs.
(4) Loading rate in gpm/sq.ft.

$$loading\ rate = \frac{Q}{A \times number\ of\ contactors} \tag{4.2}$$

Design loading rate was assumed as 2 gpm/sq.ft. A final loading rate of 2.2 gpm/sq.ft. was allowed in the cost analysis because standard sized vessels were used.

(5) Pounds of granular activated carbon (GAC) required per vessel with design empty bed contact time (EBCT) of 20 minutes is calculated using the equation:

$$GAC = \frac{EBCT \times Q \times GAC\ density}{\#\ of\ vessels} \tag{4.3}$$

GAC density = 29 lbs/ft^3 and Q is the design flow.

Table 4.28
Capital, O&M, and total costs for polishing treatment with biological activated carbon

Flow Rates					
Design Flow (mgd)	(1)	0.5	1	3	10
Design Flow (gpm)	(2)	347	694	2082	6940
Sizing Calculations					
Number of Contactors Required (Sold in Pairs)	(3)	2	4	12	40
Loading Rate (gpm/sq ft)	(4)	2.21	2.21	2.21	2.21
Pounds Carbon Required per Vessel @ 20 min EBCT	(5)	13453	13453	13453	13453
Capital Cost Summary					
Process Capital Cost					
Activated Carbon Vessels (Sold in Pairs)	(6)	$107,000	$214,000	$642,000	$2,140,000
Initial Carbon Purchase	(7)	$15,337	$30,673	$92,020	$306,733
Monitoring Instrumentation	(8)	$5,350	$10,700	$32,100	$107,000
Control Package	(9)	$0	$0	$0	$0
Pipes and Valves	(10)	$10,700	$21,400	$64,200	$214,000
Pumping	(11)	$0	$0	$0	$0
Process Cost Subtotal	(12)	$138,387	$276,773	$830,320	$2,767,733
Direct Capital Cost	(13)	$345,967	$691,933	$1,660,640	$5,535,466
Indirect Capital Cost					
Piloting	(14)	$10,000	$65,000	$65,000	$65,000
Permitting	(15)	$4,152	$8,303	$24,910	$83,032
Land	(16)	$4,152	$8,303	$24,910	$83,032
Operator Training	(17)	$0	$0	$0	$0
Concrete Slab for Carbon Vessels	(18)	$2,778	$5,556	$16,667	$55,556
Indirect Cost Subtotal	(19)	$21,081	$87,162	$131,486	$286,620
Total Capital Cost	(20)	$367,048	$779,095	$1,792,126	$5,822,086
Annual O&M Summary					
Activated Carbon	(21)	$2,293	$4,586	$13,757	$45,856
Electricity	(22)	$0	$0	$0	$0
Labor	(23)	$0	$0	$0	$0
Total Annual O&M Cost	(24)	$2,293	$4,586	$13,757	$45,856
For a project period of 20 years with 7% interest					
Amortized Total Annual Cost	(25)	$36,939	$78,127	$182,921	$595,420
Cost/1000 Gallons Treated	(26)	$0.20	$0.21	$0.17	$0.16

(6) Cost of carbon vessels was obtained from vendors for the desired sizes.
(7) Initial GAC purchase cost was assumed to be $0.57 per pound. This was based on vendor quotations.
(8) Monitoring instrumentation costs assumed at 5% of carbon vessel cost.
(9) Assumed control package purchased for upstream treatment system would have enough capacity for GAC system control, therefore no additional control package required.
(10) Piping and valves costs assumed to be 10% of carbon vessel cost (row 6).
(11) Assumed upstream pumps provide the necessary capacity for GAC system; therefore, no pumping costs associated with GAC system.
(12) Process cost subtotal consists of the sum of all process costs (rows 6 - 11).
(13) Process cost multiplier is 2.5 for flow rates equal to or below 1 mgd and 2.0 for flow rates above 1 mgd (NDWAC, 2001). Multiplier accounts for the following items: site work, contractor's overhead and profit, contingencies, engineering and design, mobilization and bonding, legal and administrative, interest during construction, and installation. Direct capital cost is obtained by multiplying the process cost subtotal with the appropriate multiplier.
(14) Piloting costs were assumed to be $10,000 for flow rates below 1 mgd and $65,000 for flow rates equal to and above 1 mgd. For large systems, the piloting costs were calculated to be around $65,000 and for small systems it was assumed that limited piloting would be provided for by $10,000 (EPA Stage-2/LT-2 Technology and Cost Document, 2002).
(15) Permitting costs were assumed to be similar to those calculated for the EPA Arsenic Technology and Cost document (2001). Permit costs equal 3% of the process cost subtotal (row 12), with a minimum of $2500.
(16) Land costs were assumed to be 3% of the process cost subtotal (row 12) (EPA Stage-2/LT-2 Technology and Cost Document, 2002).
(17) Operator training costs were included in the upstream process costs and it was assumed that no additional training hours are required.
(18) Concrete slabs are required for the carbon vessels. Slab size was calculated as 300 sq. ft. per 2-10 foot diameter vessel systems. Assumed slab thickness at 12 inches. Cost of concrete slabs is $250 per cubic yard based on best professional judgment.
(19) Indirect cost subtotal consists of the sum of all indirect process costs (rows 14 - 18).
(20) Total capital cost is a summation of indirect capital cost (row 19) and direct capital cost (row 13).

O&M Cost Assumptions

(21) Chemical costs for activated carbon replacement. Assumed carbon is replaced at years 5, 10 and 15. Value of carbon replacement is amortized over 20 years with 7% interest.
(22) Electricity costs for pumping were calculated in upstream processes. Assumed no additional electricity for carbon polishing.
(23) Labor costs were calculated in upstream processes. Assumed previous labor costs account for labor necessary for carbon polishing system.
(24) Annual O&M subtotal consists of the sum of all O&M costs (rows 21 - 23).

Total Cost Assumptions

(25) The total capital costs (row 20) were amortized over 20 years at a 7% interest rate and then annual O&M (row 24) was added to derive the amortized total annual cost.

(26) Cost per thousand gallons treated was calculated by dividing the amortized total annual cost (row 25) by the flow rate in thousands of gallons and the days in a year.

CHAPTER 5
REMOVAL OF MTBE WITH UV/H_2O_2 PROCESSES

BENCH SCALE TESTING

Introduction

The University of Poitiers, France has focused on the removal of MTBE with advanced oxidation processes using UV-radiation in combination with hydrogen peroxide (H_2O_2). The aim of the study was to determine the effect of different water quality parameters (Total organic carbon content, alkalinity, nitrate) on the removal of MTBE with UV/H_2O_2 processes and to follow the formation of by-products in different types of water. Bench scale testing on the removal of MTBE was performed by using a low pressure (LP) mercury lamp and two medium pressure (MP) mercury lamps in the presence of H_2O_2. In this report, the performance of the experiments, i.e., the UV-lamps, reactor, and the analytical methods that were used will be described and the results obtained on the removal of MTBE from four different field water samples at a varying initial concentration of hydrogen peroxide, varying MTBE concentration, and with or without any addition of tert-butyl alcohol (TBA) or nitrates will be reported. The decay of MTBE and the formation of major oxidations by-products (*tert*-butylformate (TBF), TBA, acetone, formaldehyde etc.) were followed and the results will be reported herein. Finally a comparison of the removal of MTBE from the four different field waters will be made and also of the efficacy of the UV-lamps used in the study. Only a selection of the results will be presented in the report but all the figures can be found in Appendix A.

Experimental Plan

The experiments were conducted with an MTBE spike of 200 μg/L (2.3 μM) in the presence of three different concentrations of H_2O_2, (2, 3.4, and 6.8 mg/L), 36 mg/L nitrate or 200 μg/L (2.7 μM) TBA. In addition, water samples with an initial MTBE concentration of 50 μg/L (0.57 μM) and 2000 μg/L (23 μM) were treated with LP or MP-UV/H_2O_2 keeping either the initial H_2O_2 concentration or the molar ratio between MTBE and H_2O_2 constant. Table 5.1 presents the different test matrixes that have been used in the experiments. The number of experiments on the removal of MTBE from Long Beach water had to be reduced due to an insufficient water volume. The Teflon shielded MP TQ 718 lamp has only been used in the tests on the removal of MTBE from the Santa Monica water and in two experiments on Long Beach and Miami Dade water at an initial MTBE:H_2O_2 ratio of 1:85.

For MTBE and by-product analytical methods refer to Appendix B: MTBE and By-products Analysis.

Materials & Methods

Irradiation Sources

The bench scale testing has been done with three different UV lamps: one low-pressure mercury Heraeus NNI 40/20 lamp (LP-UV) and two medium-pressure mercury lamps (Heraeus

Table 5.1
Test matrix for MTBE oxidation experiments

Test no	Source water				Irr. source	MTBE (μg/L)	MTBE: H_2O_2*	H_2O_2 (mg/L)	Other conditions
1	SM	SC	MD	-	LP-UV	200	0	-	-
2	SM	-	-	-	LP-UV	200	1:25	2	-
3	SM	SC	MD	LB	LP-UV	200	1:43	3.4	-
4	SM	SC	MD	LB	LP-UV	200	1:85	6.8	-
5	SM	-	-	-	LP-UV	200	1:43	3.4	36 mg/L NO_3^-
6	SM	-	-	-	LP-UV	200	1:43	3.4	200 μg/L TBA
7	-	SC	MD	LB	LP-UV	200	1:85	6.8	36 mg/L NO_3^-
8	-	SC	MD	LB	LP-UV	200	1:85	6.8	200 μg/L TBA
9	SM	SC	-	-	LP-UV	50		3.4	-
10	SM	SC	-	-	LP-UV	2000		3.4	-
11	-	-	MD	LB	LP-UV	50	1:85	1.7	
12	-	-	MD	LB	LP-UV	2000	1:85	65	
13	SM	-	-	-	TQ 718	200	0	-	-
14	SM	-	-	-	TQ 718	200	1:25	2	-
15	SM	-	-	-	TQ 718	200	1:43	3.4	-
16	SM	-	-	LB	TQ 718	200	1:85	6.8	-
17	SM	-	-	-	TQ 718	200	1:43	3.4	36 mg/L NO_3^-
18	SM	-	-	-	TQ 718	200	1:43	3.4	200 μg/L TBA
19	SM	-	-	-	TQ 718	50		3.4	-
20	SM	-	-	-	TQ 718	2000		3.4	-
21	SM	SC	MD	LB	TQ 150	200	0	0	
22	-	SC	MD	LB	TQ 150	200	1:45	3.4	
23	SM	SC	MD	LB	TQ 150	200	1:85	6.8	-
24	SM	SC	MD	LB	TQ 150	200	1:85	6.8	36 mg/L NO_3^-
25	-	SC	MD	LB	TQ 150	200	1:85	6.8	200 μg/L TBA
26	-	-	MD	-	TQ 150	2000		6.8	

SM: Santa Monica, MD: Miami Dade, LB: Long Beach, and SC: Suffolk County

* Molar ratio between MTBE and H_2O_2

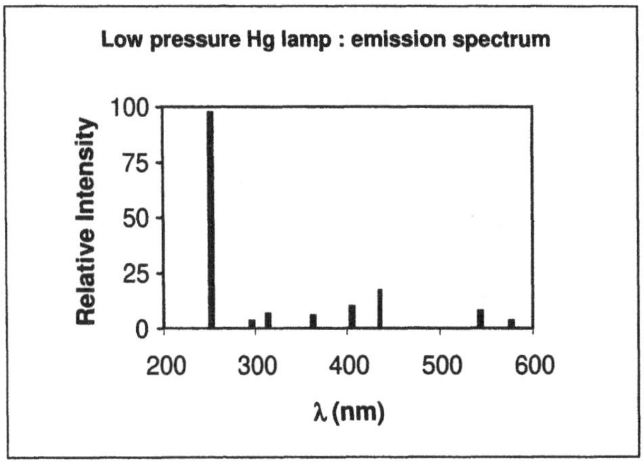

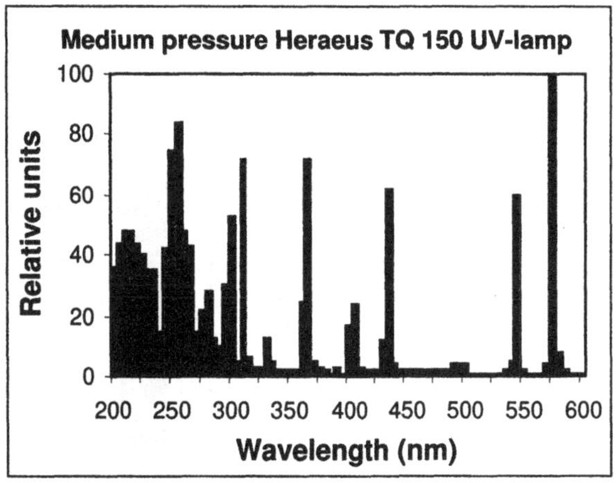

Figure 5.1 Emission spectra of the irradiation sources used in the study

TQ 718 and Heraeus TQ 150). Their emission spectra are shown in Figure 5.1. Whereas the low-pressure mercury lamp provides an almost monochromatic radiation at 254 nm, the medium-pressure lamps give a polychromatic emission.

MP-TQ 718 Lamp: Preliminary Experiments and Modifications

Our preliminary results with thc MP-TQ 718 lamp showed that MTBE was degraded by MP-UV irradiation lamp alone. After 30 min of irradiation, MTBE was completely decomposed in ultrapure water. By using a cooling system with running tap water (250 L/h) through the cooling jacket which surrounds the lamp and a cryostat (15°C) for the thermostated water in the Pyrex jacket, the temperature in the reactor was maintained at 20 ± 2×C. In order to retard the MTBE degradation the 39 cm cooling jacket was wrapped up in a double layer of Teflon tape as follows (from the top to the bottom): 13 cm- 5 cm - 5 cm - 5 cm - 5 cm – 4.5 cm and gaps of 3 mm were left between the Teflon bands.

A water solution of ~200 µg/L MTBE, TBF, TBA, and acetone was left in the reactor and analyzed after 12 h. The results (Figure 5.2) show that no noticeable stripping of the compounds or adsorption to the Teflon coated cooling jacket occurred in the reactor. In all the further experiments with the MP TQ 718 lamp, the Teflon covered cooling jacket was used, i.e. 96% of the surface was covered. This Teflon shielded TQ718 lamp was used in the studies on the removal of MTBE from Santa Monica field water sample and in a few tests on the removal of MTBE from Long Beach and Miami Dade water samples.

Comparison of the Irradiation Sources

Chemical actinometry was performed with the LP-UV lamp according to the procedure described by Nicole et al., (1999). The photonic flow was evaluated to be 1.1×10^{-5} $E.s^{-1}$. Since the MP-UV lamp emits a polychromatic radiation it is not possible to use a chemical actinometer to determine the photonic flow from the medium-pressure lamp. However, in order to make a comparison of the three lamps used, the degradation of hydrogen peroxide was followed as a function of irradiation time. Figure 5.3 compares the degradation of hydrogen peroxide with the

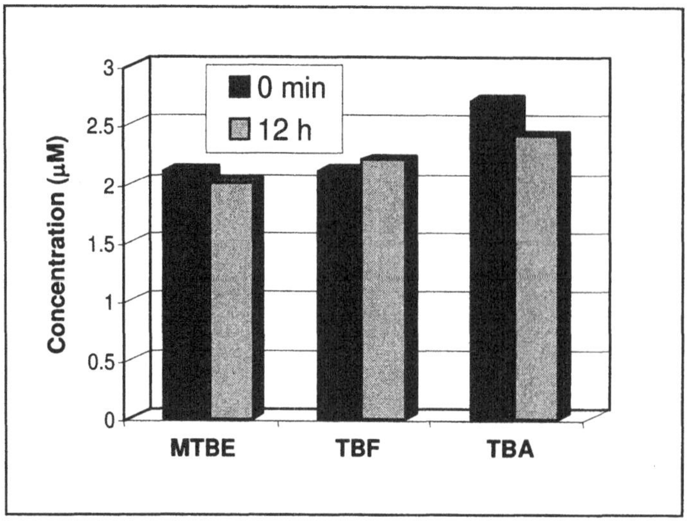

Figure 5.2 Stability of MTBE, TBF, and TBA after 12 h in the reactor in the absence of any UV-radiation

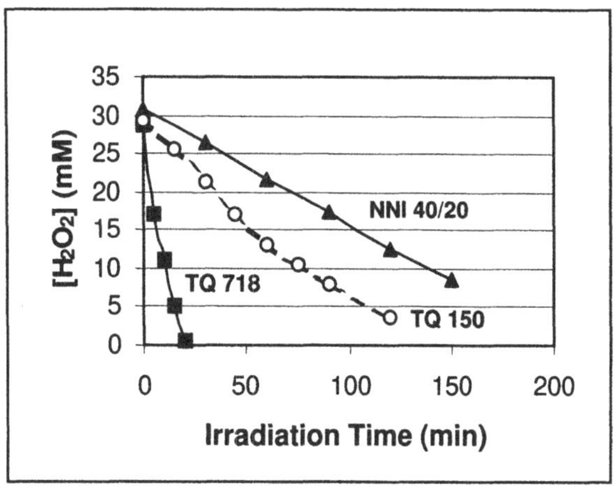

Figure 5.3 Decomposition of hydrogen peroxide in ultrapure water with three different UV-lamps (Heraeus NNI 40/20, TQ 150, and TQ718 with no shielding) as a function of irradiation time

UV-lamps used in this study. After 30 min of irradiation, 13% and 28% of the hydrogen peroxide was degraded by the LP-UV (NNI 40/20) and the MP-UV lamp (TQ 150), respectively. When the solution was irradiated with the more powerful MP-UV lamp (TQ 718) without any shielding, hydrogen peroxide was completely degraded after 30 min of irradiation. When the lamp was shielded with Teflon tape, 25% of the hydrogen peroxide was degraded after 30 min of irradiation. The degradation rate of H_2O_2 when using the shielded MP TQ 718 lamp was 5.5 µM/s compared to a rate of 25 µM/s with no shielding and 2.4 µM/s for the LP-UV lamp. The slower degradation of hydrogen peroxide with the TQ 150 MP-UV lamp indicates that this lamp can be used without any shielding of the cooling jacket around the lamp.

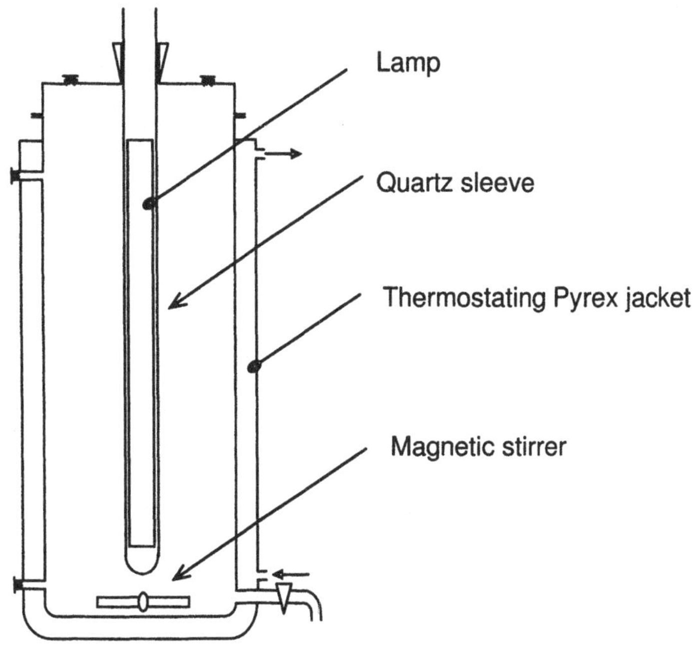

Figure 5.4 Schematic of the UV reactor

Reactor

Reactor design. The photoreactor is a cylindrical batch thermostated reactor (Figure 5.4) with an internal diameter of 145 mm The UV lamps are placed on the axis of the reactor in a quartz sleeve with a diameter of 28 mm for the LP-UV lamp and 50 mm for the MP-UV lamp. Since the working temperature of the MP-UV lamp is ~700 °C, a double walled sleeve made of quartz glass surrounds the MP-UV lamp and tap water circulates in the interstitial space (1 cm) to prevent warming-up of the solution in the reactor. The volume of the irradiated solution is 5 L, and the solution is continuously stirred during the irradiation. The UV-lamps are submersed in the solution during the treatment. Complete pictures of the reactors with their dimensions are presented in Figures A.1 and A.2 in Appendix A.

Agitation. In the experiments, the water solution in the reactor was stirred with a magnet during the irradiation. However, to evaluate the effectiveness of the magnetic stirring, the degradation of H_2O_2 was also followed by combining stirring and recirculation of the water in the reactor, i.e. a semi-batch reactor. The water samples were recirculated through the reactor at a rate of 2.3 L/min or 3.8 L/min. Figure 5.5 shows that H_2O_2 was slightly slower degraded with the MP-UV lamp at a higher flow rate (3.8 L/min), which can be explained by the fact that the solution stayed longer in the recirculation tubes in the absence of any irradiation. In the LP-UV experiments, on the other hand, the recirculation of the water ameliorated the mixing of the solution in the reactor and the decay rate of H_2O_2 increased slightly with an increased flow rate. However, due to the small difference between using recirculation or not and in order to keep similar experimental conditions with both lamps, the magnetic stirring only was used in all our further bench scale testing.

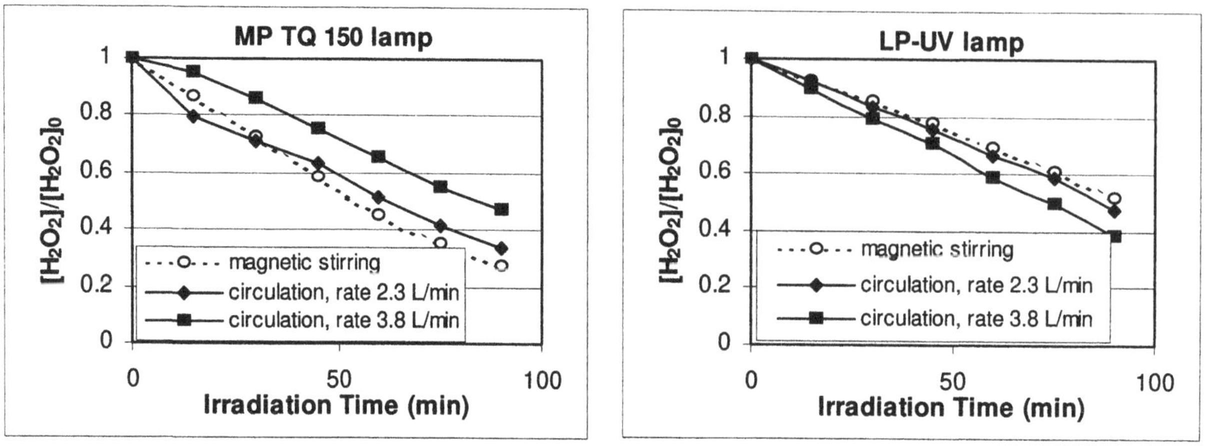

Figure 5.5 Decomposition of hydrogen peroxide with LP- and MP-UV treatment as a function of irradiation time. The water sample was stirred by a magnet or by a magnet combined with water recirculation.

Water Sources

In order to determine the effectiveness of UV/H_2O_2 processes on the removal of MTBE from water samples with different water quality parameters, the following different field water samples have been used:

1. Water from the city of Santa Monica with a low DOC content and a high alkalinity
2. Suffolk County water with a low DOC and a low alkalinity
3. Long Beach water with a high DOC and a high alkalinity
4. Miami Dade water with a high DOC and a low alkalinity

In order to better understand and to interpret the results, the initial water quality parameters were determined and Table 5.2 presents the characteristics of the studied water sources. The dissolved and total organic carbon contents were determined on a Shimadzu TOC 5000 analyzer. Figure 5.6 shows the UV-absorbance spectra of the field water samples between 200 and 400 nm and it can be seen that the water samples from Santa Monica and Suffolk County that contain a high amount of salts (e.g. nitrates), which absorb UV-radiation in the region where MTBE absorbs (~200 nm). On the other hand, Miami Dade and Long Beach waters, enriched in natural organic matter (NOM) as compared to the two other water sources exert higher UV-absorbance at 254 nm due to the presence of humic-type organics. The Long Beach water is also characterized by a higher SUVA (i.e., higher proportion of UV absorbing organics per DOC unit) as compared to the water from Miami Dade.

Results & Discussion

Stability of MTBE in the Reactor

Control samples of MTBE spiked ultrapure water showed that no degradation of MTBE occurred in the reactor in the absence of any UV/vis radiation even in the presence of H_2O_2 after

Table 5.2
Characteristics of the water sources

Parameters	Miami Dade	Long Beach	Santa Monica	Suffolk County
pH	7.9	8.1	8.3	7.6
DOC	4.3 mg/L	5.1 mg/L	0.7 mg/L	0.7 mg/L
A_{254} (cm^{-1})	0.07	0.21	0.005	0.004
SUVA*	2.0	4.0	0.7	0.6
HCO_3^-	61 mg/L	183 mg/L	399 mg/L	33 mg/L
NO_3^-	< 1 mg/L	-	21 mg/L	13 mg/L

* SUVA = A_{254}/DOC (m^{-1} L/mg C)

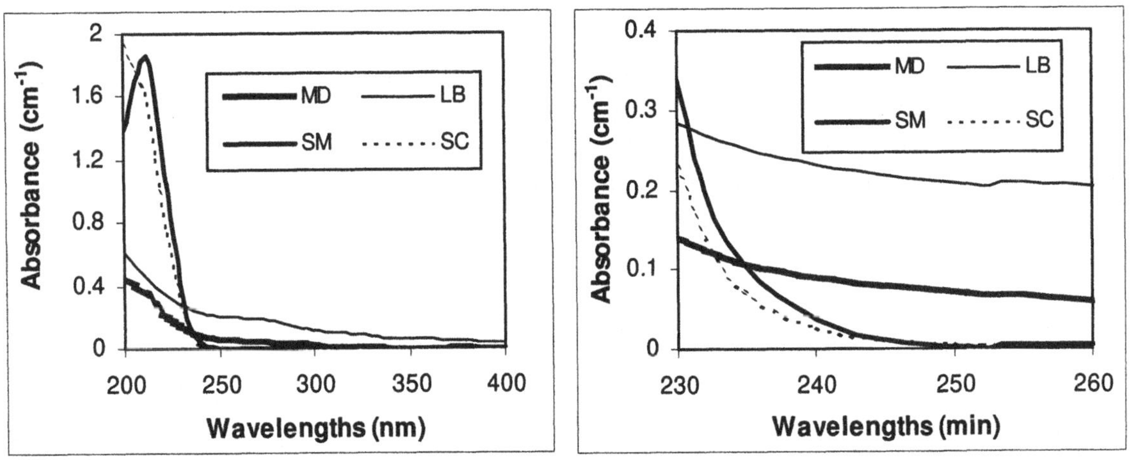

Figure 5.6 UV-absorbance spectra of Miami Dade (MD), Long Beach (LB), Santa Monica (SM), and Suffolk County (SC) water

2 hours of agitation. Hence, the agitation of the solution with a magnetic stirrer did not lead to any significant stripping of MTBE. No significant increase of the temperature was noticed at the conditions that maintained the MTBE concentration stable for a few hours.

Removal of MTBE in the Absence of H_2O_2

LP-UV-irradiation. Our results indicate that MTBE dissolved in ultrapure water was not significantly degraded when exposed to monochromatic LP UV-radiation (254 nm) without the addition of H_2O_2 (not shown). This shows that LP UV-radiation alone does not degrade MTBE, due to the fact that MTBE does not absorb UV-radiation at 254 nm. Water samples from Santa Monica and Suffolk County spiked with 200 μg/L MTBE were exposed to LP-UV irradiation without any addition of hydrogen peroxide. As shown in Figure 5.7, MTBE degradation was observed in both Suffolk County and Santa Monica waters that were exposed to LP-UV in the absence of hydrogen peroxide. After 120 min of irradiation, 6% and 56% of MTBE was degraded in the Santa Monica and Suffolk County water samples, respectively. The faster degradation of

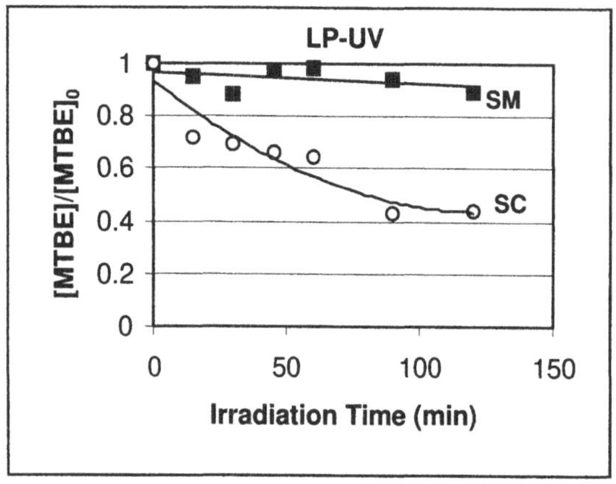

Figure 5.7 Removal of MTBE from Santa Monica (SM) and Suffolk County (SC) field water samples with LP-UV irradiation alone

MTBE in Suffolk County water compared to Santa Monica water may be attributed to the presence of higher concentration of bicarbonates in Santa Monica water. MTBE degradation in the absence of peroxide may be attributed to the production of hydroxyl radicals from UV light interaction with iron or nitrates in the water. Exact mechanism for this is not known. The suggested reactions for production of hydroxyl radicals from nitrates are shown below (Zepp 1988; Hoigné and Bader, 1976):

$$NO_3^- + UV \rightarrow NO_3^{-*} \rightarrow NO_2^- + O(^3P) \quad (5.1)$$

$$NO_2^- \rightarrow NO_2 + {}^{\bullet}O^- \quad (5.2)$$

$${}^{\bullet}O^- + H_2O \rightarrow {}^{\bullet}OH + OH^- \quad (5.3)$$

Complimentary experiments conducted with a 200 µg/L MTBE solution prepared at pH 8 in ultrapure water and spiked with 20 mg/L nitrate confirmed the hypothesis of a production of hydroxyl radicals during UV-photolysis of nitrates (Figure 5.8).

Because of the higher bicarbonate concentration in the Santa Monica water (399 mg/L as HCO_3^-) as compared to the one in Suffolk County (33 mg/L as HCO_3^-), the degradation of MTBE was much slower in the water from Santa Monica. Bicarbonates and carbonates are known to efficiently react with hydroxyl radicals ($k({}^{\bullet}OH/HCO_3^-) = 8.5 \times 10^6\ M^{-1}s^{-1}$; $k({}^{\bullet}OH/CO_3^{2-}) = 3.9 \times 10^8\ M^{-1}s^{-1}$ and thus retard the removal of an organic contaminant through radical pathways. Unfortunately the water volume of Miami Dade and Long Beach water samples was insufficient to perform the test with LP-UV irradiation in the absence of H_2O_2.

MP-UV irradiation. Our preliminary results with the medium pressure lamps showed that MTBE was degraded with MP-UV irradiation alone. After 30 min of irradiation, MTBE was completely degraded when using the Teflon shielded MP TQ 718 lamp compared to a 25% degradation with the less powerful MP TQ 150 lamp. Because the circulating water in the cooling jacket, which surrounds the MP-UV lamp, absorbs the part of UV-radiation that might generate hydroxyl radicals from direct photolysis of water, direct photolysis of MTBE is expected to be the major reaction pathway that can explain its degradation with MP-UV irradiation alone.

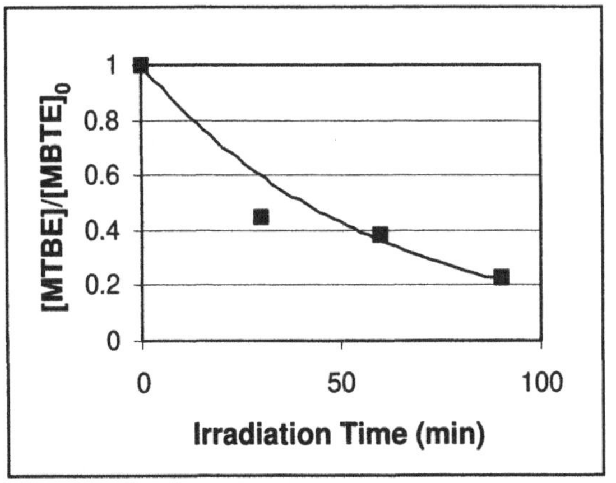

Figure 5.8 Degradation of 200 µg/L MTBE in ultrapure water with LP UV-irradiation alone in the presence of 20 mg/L nitrates

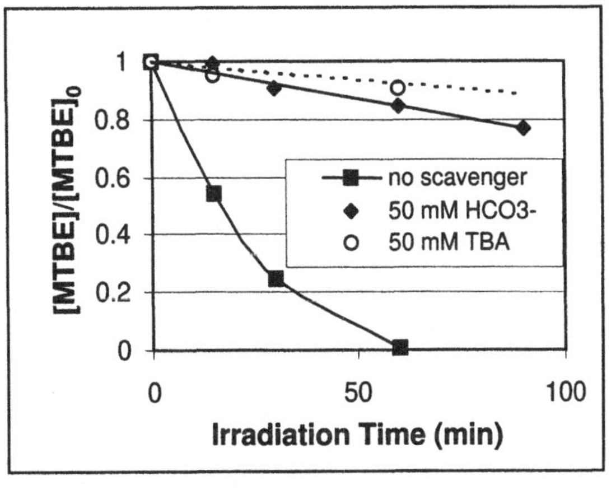

Figure 5.9 Relative concentration of MTBE in ultrapure water versus irradiation time when treated with the MP-UV (TQ 718) lamp in the presence of 50 mM TBA or 50 mM $NaHCO_3$

TBA and bicarbonates are known radical scavengers and in order to find out whether the MP-UV lamp caused a direct photolysis of MTBE or if hydroxyl radicals were generated in the treated water sample, a 200 µg/L solution of MTBE in ultrapure water was irradiated with the 96% Teflon shielded TQ 718 MP-UV lamp in the presence of 50 mM TBA or 50 mM $NaHCO_3$. According to our results (Figure 5.9), only 10% of the MTBE was degraded after 30 min of irradiation in the presence of a high concentration of bicarbonate compared to 75% with no addition of bicarbonates or TBA. In the presence of 50 mM TBA, no noticeable degradation of MTBE could be observed after 30 min of irradiation. A 50 mM bicarbonate solution shows an absorption maximum at 204 nm, i.e., in the domain where MTBE absorbs which indicates that bicarbonates may also inhibit the direct photolysis of MTBE by absorbing photons emitted by the MP-UV lamp. The absorption of TBA in this region is, however, very small and the reason why the removal of MTBE is retarded in the presence of TBA is somewhat unclear. One hypothesis is that MTBE absorbs radiation and gets in an excited state where after the excited MTBE molecule transfers its excess energy to TBA, which is present in large excess, and thereby avoiding its own degradation.

Experiments conducted with a 200 μg/L MTBE solution prepared in ultrapure water (pH 8, no bicarbonates) showed that the addition of 20 mg/L nitrates favoured the degradation of MTBE (Figure 5.10). In the presence of a high concentration of bicarbonates, as in the Santa Monica water, radical reaction pathways are expected to be scavenged. Thus, an UV-filter effect due to the presence of nitrates but also bicarbonates may explain the slower MTBE degradation rate observed in thc Santa Monica water as compared to the water from Suffolk County.

Figure 5.11 presents the removal of MTBE from the four different field water samples with the MP TQ 150 UV lamp in the absence of H_2O_2. MTBE was most efficiently removed from the Suffolk County field water with a low DOC content and a low alkalinity and most slowly removed from the highly colored Long Beach water with a high DOC content and alkalinity.

The DOC content in the Miami Dade water was approximately similar to the one in Long Beach water but the absorbance at 254 nm was much lower. The colored dissolved organic matter (DOM) in Long Beach water reduces the amount of photons that reaches the MTBE molecules and thus retards the removal of MTBE by acting as a filter. DOM is also known to act as a radical scavenger but UV-photolysis (254 nm) of DOM can also induce the production of radical species (Hoigné 1998). In such contest, the presence of a high a concentration of bicarbonates should scavenge the radical species generated at a very low concentration of DOM. The DOM is, however, more important as a scavenger than as a source of hydroxyl radicals. This can explain the fact that the MTBE decay was slower in the Miami Dade water as compared to the Suffolk County water.

Nitrates in solution may influence the MTBE degradation by two pathways. Nitrates may retard the degradation of MTBE due to their strong UV-absorption in the wavelength region of 190-200 nm where MTBE absorbs or enhance its degradation due to a production of radical species as discussed earlier. The degradation of MTBE in Santa Monica field water was relatively fast even though it contains a high concentration of bicarbonates that effectively scavenge hydroxyl radicals and consequently retards the degradation of MTBE. However, this water sample has a high nitrate concentration, which might explain the fast removal of MTBE. If the degradation rate of MTBE in Suffolk County and Santa Monica water samples are compared it can be seen that the removal of MTBE from the Santa Monica water is retarded due to the high alkalinity. The relatively fast degradation of MTBE in the Miami Dade water with a very high DOC content indicates that this water sample contains compounds that generate hydroxyl radicals during UV-photolysis.

Impact of Different Parameters on the Removal of MTBE

Effect of Initial H_2O_2 Concentration

Degradation of H_2O_2. To study the effect of the initial H_2O_2 concentration on the removal of MTBE from the field water samples with LP- and MP-UV/H_2O_2 processes, three different molar ratios between MTBE and H_2O_2 have been used for an initial MTBE concentration of 200 μg/L (2.3 μM): 1:25, 1:43, and 1:85, that gives an initial hydrogen peroxide concentration of 2, 3.4 and 6.8 mg/L (i.e., 60, 100, and 200 μM). Figure 5.12 presents the degradation of hydrogen peroxide with LP- and MP-UV irradiation in Miami Dade water at the three initial hydrogen peroxide concentrations used in the study. The amount of UV-radiation absorbed by H_2O_2

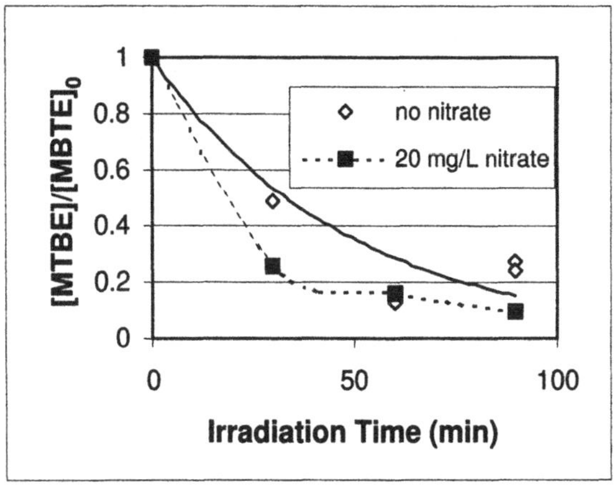

Figure 5.10 Degradation of 200 μg/L MTBE in ultrapure water with MP UV-irradiation alone in the presence of 20 mg/L nitrates

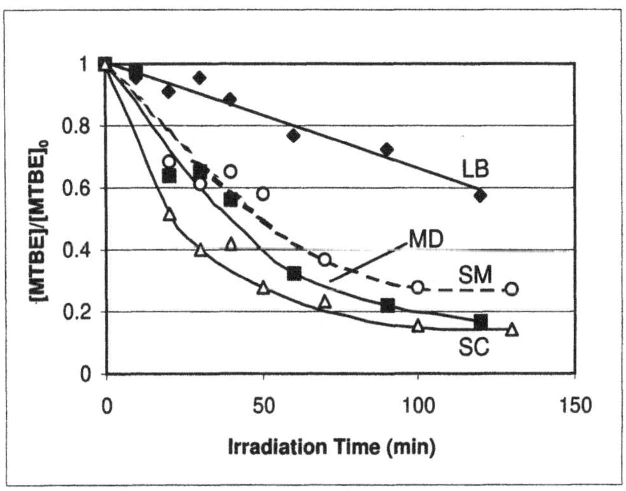

Figure 5.11 Removal of MTBE in Suffolk County (SC), Santa Monica (SM), Miami Dade (MD), and Long Beach (LB) field waters

increases with an increasing concentration of H_2O_2 and consequently the rate of the H_2O_2 disappearance increases. However, the relative rates of the H_2O_2 disappearance remain almost constant at the different concentrations of hydrogen peroxide used in our study Figure 5.13. The relative decrease in the H_2O_2 at an initial concentration of 2000 μM was slightly elevated. The decomposition of hydrogen peroxide was slightly faster when exposed to the MP-UV lamp compared to the LP-UV lamp.

Photolysis of H_2O_2 may be affected by suspended particles and other absorbing species in the aqueous solution being treated (Glaze 1989). Figure 5.14 presents the decomposition of H_2O_2 with LP- and MP-UV irradiation in the different field water samples used in this study. After 30 min of LP-UV irradiation, approximately 70% and 60% of the H_2O_2 was degraded in the Santa Monica and Suffolk County field water samples, respectively. In the field water samples with a higher DOC content, i.e. Long Beach and Miami Dade water samples, 35% and 45% of the initial H_2O_2 was degraded after 30 min of irradiation with the LP-UV lamp. In the MP-UV experiments,

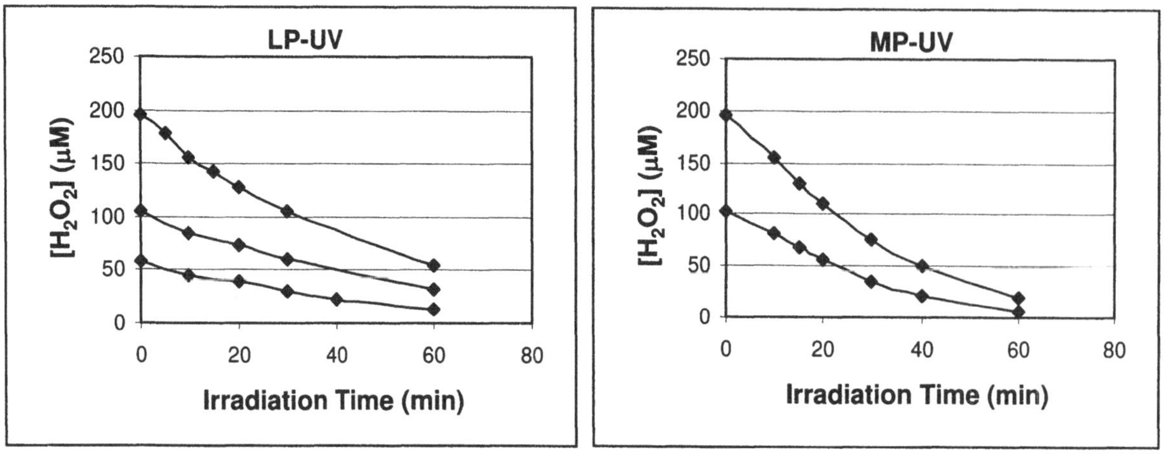

Figure 5.12 Decomposition of H_2O_2 with LP-UV irradiation or MP-UV irradiation as a function of time in Miami Dade water

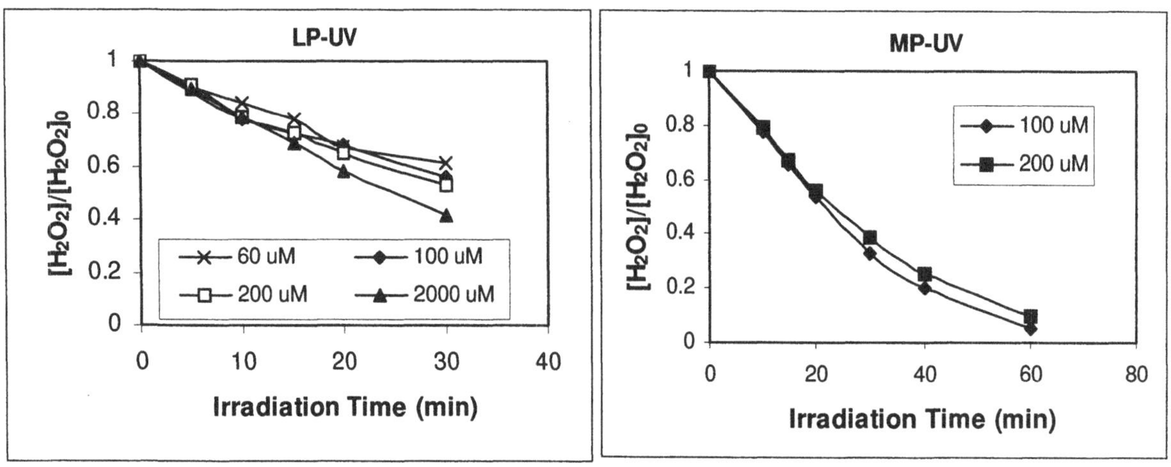

Figure 5.13 Relative decrease in H_2O_2 concentration when exposed to LP- or MP-UV irradiation

the decomposition rate of H_2O_2 was less affected by the background constituents in the water sample and approximately 65% of the H_2O_2 in Miami Dade, Santa Monica and Suffolk County water was degraded after 30 min of irradiation. The H_2O_2 decomposition rate was, however, retarded in the highly colored Long Beach water.

Degradation of MTBE. The removal of 200 μg/L (2.3 μM) MTBE was followed at three initial concentrations of H_2O_2 (60, 100, and 200 μM). Figure 5.15 presents an example of the decomposition of MTBE in water from Suffolk County with LP or MP-UV/H_2O_2 treatment at an initial H_2O_2 concentration of 0, 100 and 200 μM. In the Suffolk County water, the degradation rate of MTBE increased with an increasing concentration of hydrogen peroxide, due to the increased production of hydroxyl radicals at a higher initial H_2O_2 concentration. A similar trend was observed in the other field water samples (Figures A.3 and A.4 in Appendix A). However, no noticeable difference in the removal of MTBE from the Miami Dade water could be observed between an initial hydrogen peroxide concentration of 60 and 100 μM.

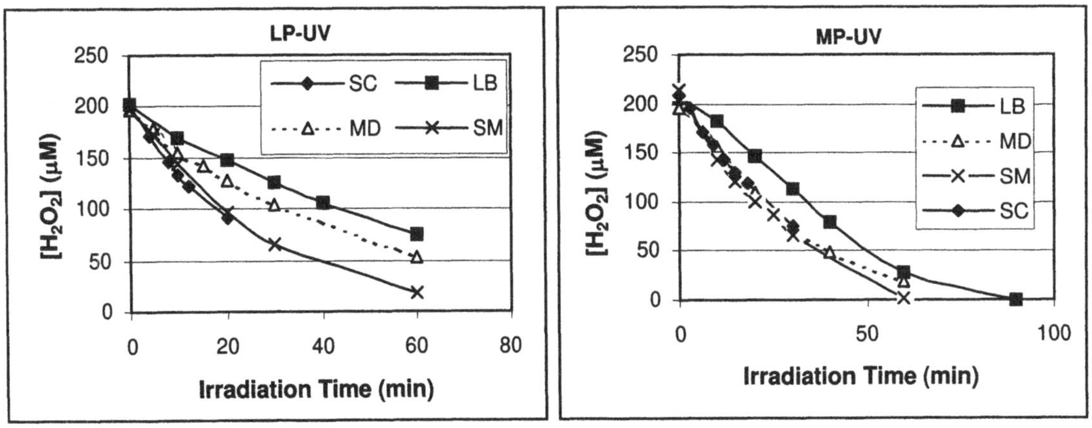

Figure 5.14 Decomposition of H_2O_2 versus irradiation time in Long Beach (LB), Miami Dade (MD), Santa Monica (SM), and Suffolk County (SC) water samples

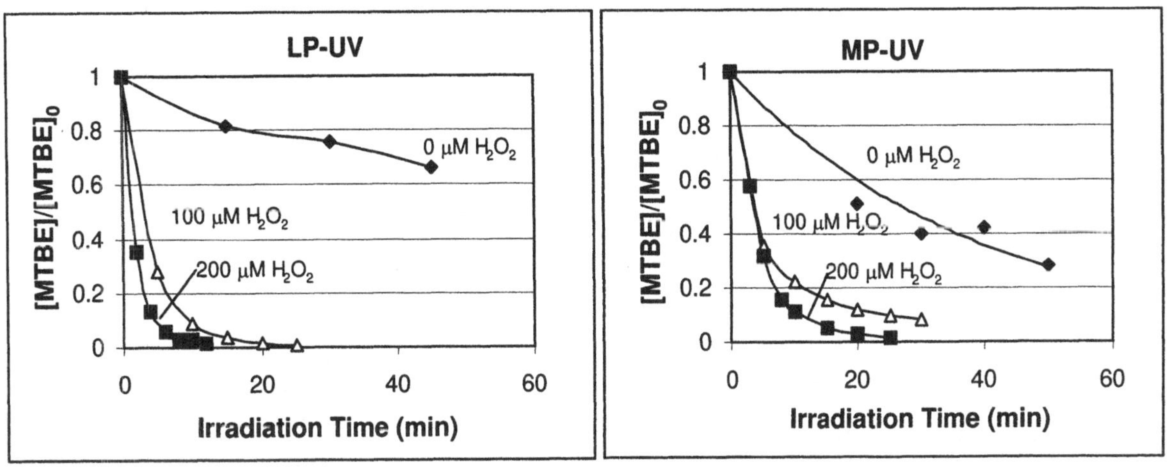

Figure 5.15 Normalized decay of MTBE with LP- and MP-UV irradiation at different initial concentrations of H_2O_2 in water from Suffolk County

To determine the pseudo-first order degradation rate constants, the degradation curves were plotted as natural logarithmic values as a function of irradiation time and fairly straight lines were obtained (Figure 5.14). The initial pseudo-first order degradation rate constants were then calculated and the constants obtained are presented in Table 5.3. According to these results, the MTBE degradation rate with LP- or MP-UV/H_2O_2 was highest in the Suffolk County water with a very low bicarbonate and DOC content followed by Miami Dade water with a high DOC content and a low alkalinity. The MTBE decay was most retarded in the Santa Monica and Long Beach water samples with a high alkalinity. The removal of MTBE was most retarded in the highly colored Long Beach water.

LP-UV irradiation. In Figure 5.16 two different types of waters can be distinguished: 1) the high-alkalinity Long Beach and Santa Monica water samples and 2) the low alkalinity Suffolk County and Miami Dade water samples. The degradation of MTBE was most retarded in

Table 5.3
Initial pseudo-first order rate constants for the degradation of MTBE with LP- and MP-UV irradiation at different initial concentrations of H_2O_2

	k_1' (s^{-1})			
$[H_2O_2]$ (μM)	Santa Monica	Suffolk County	Miami Dade	Long Beach
		LP-UV/H_2O_2		
60	0.3×10^{-3}	-	0.8×10^{-3}	-
100	0.4×10^{-3}	3.5×10^{-3}	0.7×10^{-3}	0.1×10^{-3}
200	0.6×10^{-3}	7.2×10^{-3}	1.7×10^{-3}	0.2×10^{-3}
		MP-UV/H_2O_2		
100	-	2.3×10^{-3}	0.4×10^{-3}	0.2×10^{-3}
200	1.0×10^{-3}	3.4×10^{-3}	0.8×10^{-3}	0.5×10^{-3}

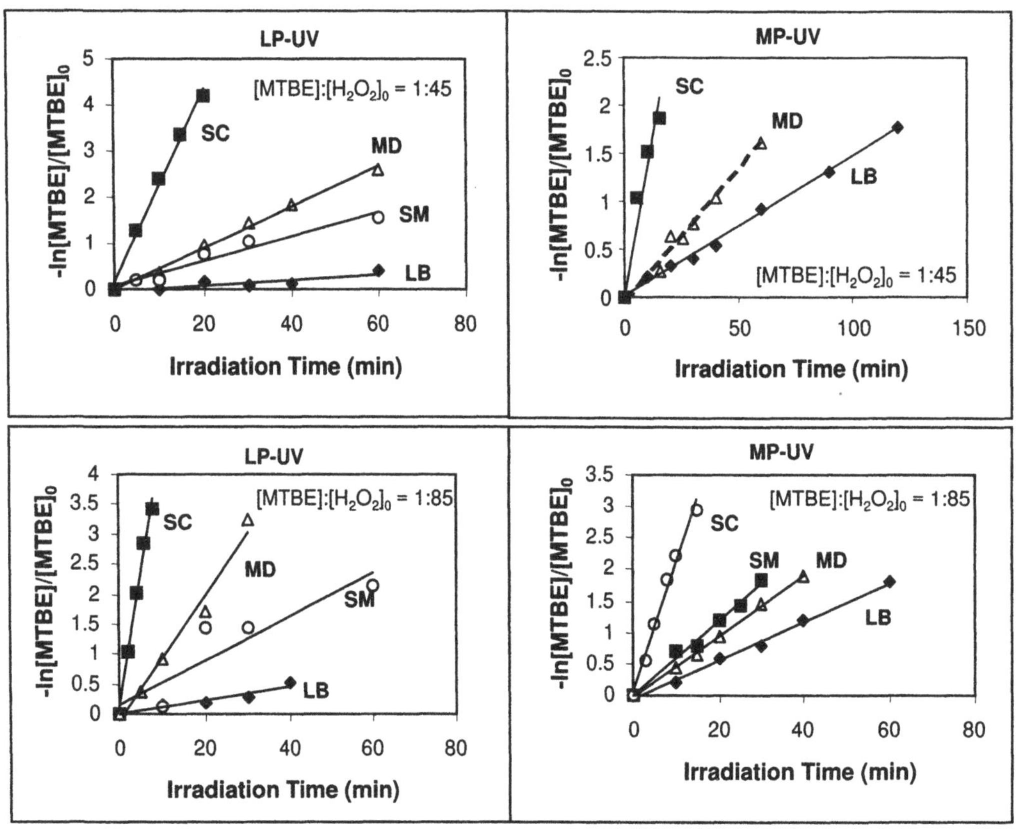

Figure 5.16 Pseudo-first order degradation of MTBE in Santa Monica (SM), Suffolk County (SC), Long Beach (LB), and Miami Dade (MD) water samples with LP-UV/H_2O_2 processes at different initial MTBE:H_2O_2 ratios

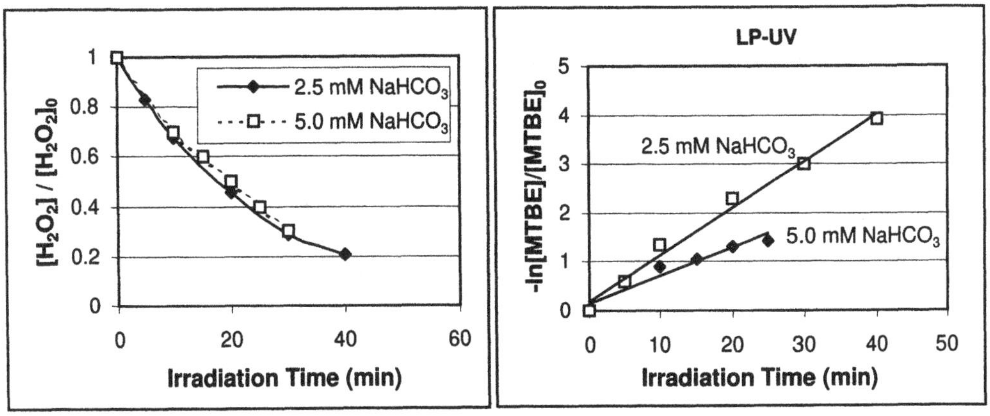

Figure 5.17 Decomposition of 100 μM H_2O_2 with LP-UV irradiation and the pseudo-first order degradation of 200 μg/L MTBE in ultrapure water with LP-UV/H_2O_2 at an initial H_2O_2 concentration of 100 μM and in the presence of 2.5 or 5.0 mM $NaHCO_3$

the high alkalinity water samples. For the two water types, a lower reaction rate constant is obtained at higher alkalinity and SUVA. Thus, the alkalinity and SUVA are the two major physical chemical parameters of the water samples that govern the efficiency of the UV-treatment for the removal of MTBE. The results in the LP-UV experiments indicate that MTBE is most slowly degraded in the highly colored Long Beach water, which can be explained with the so-called filter effect of the DOM, that is the DOM absorbs a part of the photons emitted by the UV-lamp and thereby retards the degradation of MTBE. The DOC content in the Miami Dade water is approximately similar to the one in the Long Beach water but the degradation of MTBE is much slower in the Miami Dade water due to the difference in UV-absorbance between the water samples. The absorbances at 254 nm in Miami Dade and Long Beach water samples were 0.07 cm^{-1} and 0.21 cm^{-1}, respectively, which means that during the LP-UV treatment of the Miami Dade water, the filter effect of the DOM was insignificant compared to the one in the Long Beach water. The fact that MTBE is more slowly removed from the water from Santa Monica than from Miami Dade can be explained by the higher concentration of bicarbonates in the Santa Monica water sample. The difference in the degradation rate of MTBE in the field waters was larger in the LP-UV experiments which, indicates that the water matrix plays a much larger roll in LP-UV processes than in MP-UV processes.

Experiments conducted with MTBE solutions prepared in ultrapure water confirmed that, in the concentration range studied, the amount of bicarbonates in the solution did not significantly affect the degradation rate of H_2O_2. As expected, the degradation rate of MTBE decreased with an increasing bicarbonate concentration (Figure 5.17). The obtained pseudo-first order degradation rate constants for MTBE in the presence of 2.5 or 5.0 mM $NaHCO_3$ were 1.6×10^{-3} s^{-1} and 0.9×10^{-1} s^{-1}, respectively.

Similar experiments were conducted with 200 μg/L MTBE solutions spiked with different amounts of aquatic fulvic acids in the presence of 5.0 mM $NaHCO_3$ and 100 μM H_2O_2 to create a molar ratio of 1:45 between MTBE and H_2O_2.The results show that the degradation of H_2O_2 and MTBE are retarded with an increasing FA concentration (Figure 5.18). The UV absorbances at 254 nm of the 4 and 8-mg/L FA solutions were 0.091 and 0.164 cm^{-1} respectively. Because the absorbance of the solution increases with an increasing FA concentration, the amount of radiation

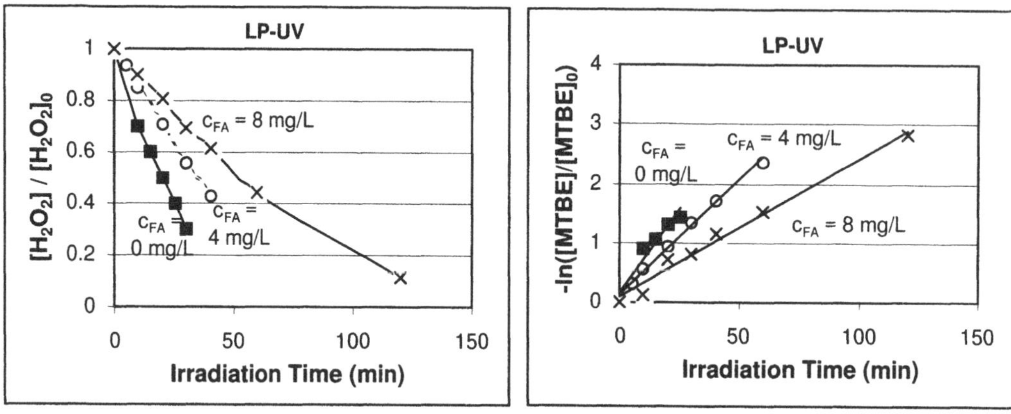

Figure 5.18 Decomposition of H_2O_2 with LP-UV irradiation at different concentrations of fulvic acid (FA) and pseudo-first order degradation of MTBE in LP-UV/H_2O_2 treated solutions of fulvic acid ($[MTBE]_0$ = 200 µg/L, HCO_3^- = 5.0 mM)

reaching the H_2O_2 decreases and the dissociation of H_2O_2 is consequently retarded. In turn, the decreased production of hydroxyl radicals retards the removal of MTBE. The pseudo- first order rate constants for the degradation of MTBE in the presence of 4 or 8 mg/L fulvic acid were 0.6×10^{-3} s^{-1} and 0.4×10^{-3} s^{-1} respectively.

Increasing the H_2O_2 concentration was found to increase the degradation rate of MTBE due to a higher production of $^{\bullet}OH$ radicals. In the concentration range studied, increasing the H_2O_2 concentration from 100 µM to 200 µM was found to increase the pseudo-first order degradation rate constant of MTBE by a factor of approximately 2. Figure 5.19 shows that the experiments conducted in ultrapure water gave similar results. The pseudo-first order degradation rate constants are 0.7×10^{-3} s^{-1}, 1.2×10^{-3} s^{-1}, and 1.7×10^{-3} s^{-1} for an initial MTBE: H_2O_2 molar ratio of 1:25, 1:45, and 1:85.

MP-UV irradiation. Similar results were obtained with the MP-UV lamp (TQ 150) as compared to the ones with the LP-UV lamp, i.e. MTBE was most efficiently removed from Suffolk County water followed by Miami Dade, Santa Monica and Long Beach water. Again a high alkalinity and SUVA in the water gave a slower removal rate of MTBE whereas the MTBE removal was fairly good in the low SUVA and low alkalinity waters. A major difference in the removal rate of MTBE in the LP- and MP-UV/H_2O_2 treated waters was that the physical chemical characteristics of the water seem to have less impact on the removal of MTBE with MP-UV/H_2O_2 than with LP-UV/H_2O_2. For example, the difference between the removal rate of MTBE in the Suffolk County and Long Beach water samples was less in the MP-UV/H_2O_2 treated water samples than in the LP-UV/H_2O_2 treated ones. The pseudo-first order degradation rate constants in the LP-UV treated Suffolk County and Long Beach waters were 7.2×10^{-3} s^{-1} and 0.2×10^{-3} s^{-1} respectively compared to 3.4×10^{-3} s^{-1} and 0.5×10^{-3} s^{-1} in the MP-UV/H_2O_2 treated water samples at an initial H_2O_2 concentration of 200 µM (Table 5.3). The degradation rates of MTBE in the MP-UV/H_2O_2 treated ($[H_2O_2]_0$ = 200 µM) Santa Monica and Miami Dade water samples were approximately similar. It can consequently be concluded that the alkalinity and NOM have a less impact on the MTBE removal with MP-UV than with LP-UV. The removal of MTBE was more efficient with MP-UV/H_2O_2 than with LP-UV/H_2O_2 in Long Beach and Santa

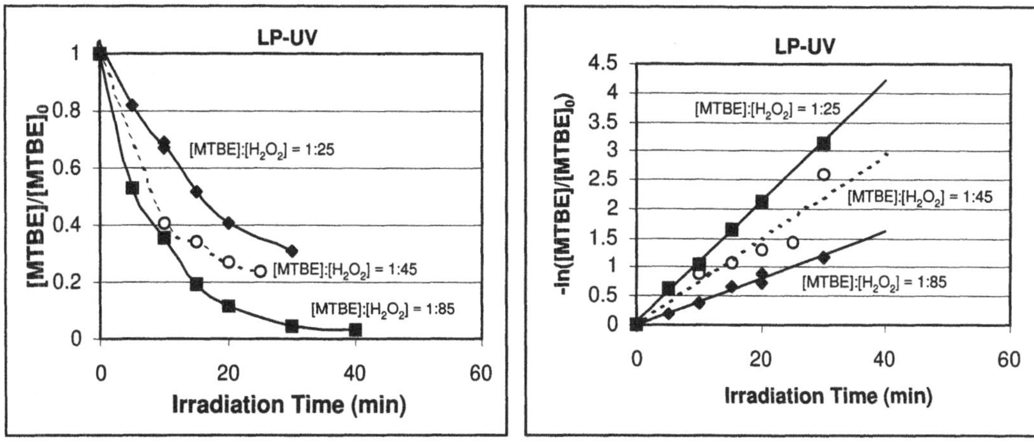

Figure 5.19 Removal of 200 µg/L MTBE from ultrapure water with LP-UV/H_2O_2 at different initial molar ratios between MTBE and H_2O_2 (c_{NaHCO_3} = 5.0 mM)

Monica water samples with a high alkalinity whereas the opposite was observed in Suffolk County and Miami Dade waters, i.e., MTBE was removed more efficiently with LP-UV/H_2O_2 than with MP-UV/H_2O_2.

Formation of By-Products

The evolution of the by-products TBF, TBA, formaldehyde and acetone has been followed in the LP- and MP-UV/H_2O_2 treated water samples. The results are presented as the concentration of MTBE and its by-products (in µg/L) versus irradiation time but also as histograms with the concentration expressed in µM. The results obtained in the LP-UV/H_2O_2 treated source waters at different initial MTBE/H_2O_2 molar ratios are presented in Figures 5.20 – 5.23. The corresponding results with in the MP-UV experiments are given in Figures 5.24 – 5.28.

LP-UV irradiation. In general, a maximum in the concentration of TBF and TBA appears as a function of irradiation time. Depending on the water samples and the concentration of H_2O_2 this maximum appears at different irradiation times. In general, it appears at shorter irradiation time when using higher H_2O_2 concentrations. In addition it can be noted that more TBF than TBA is formed and that the maximum concentration of TBA occurs at shorter irradiation times than the one of TBF. On the contrary, the concentration of formaldehyde and acetone increases with irradiation time and also with an increased initial H_2O_2 concentration. These compounds are consequently the major degradation products formed. The sum of the molar concentrations of the degradation products analyzed seems to decrease with irradiation time and also with an increased H_2O_2 concentration.

In the Suffolk County water where MTBE is rapidly degraded due to the low bicarbonate and NOM concentration, acetone is the major by-product formed and almost no TBA is formed. On the contrary, in the Santa Monica water with a high concentration of bicarbonates, no acetone was detected and TBF was the major by-product identified except in the experiment with a MTBE/H_2O_2 molar ratio of 1:85 where formaldehyde was the major by-product identified. In Miami Dade and Long Beach water samples with a high concentration of NOM, formaldehyde was the major degradation product. The total concentration of MTBE and by-products formed in

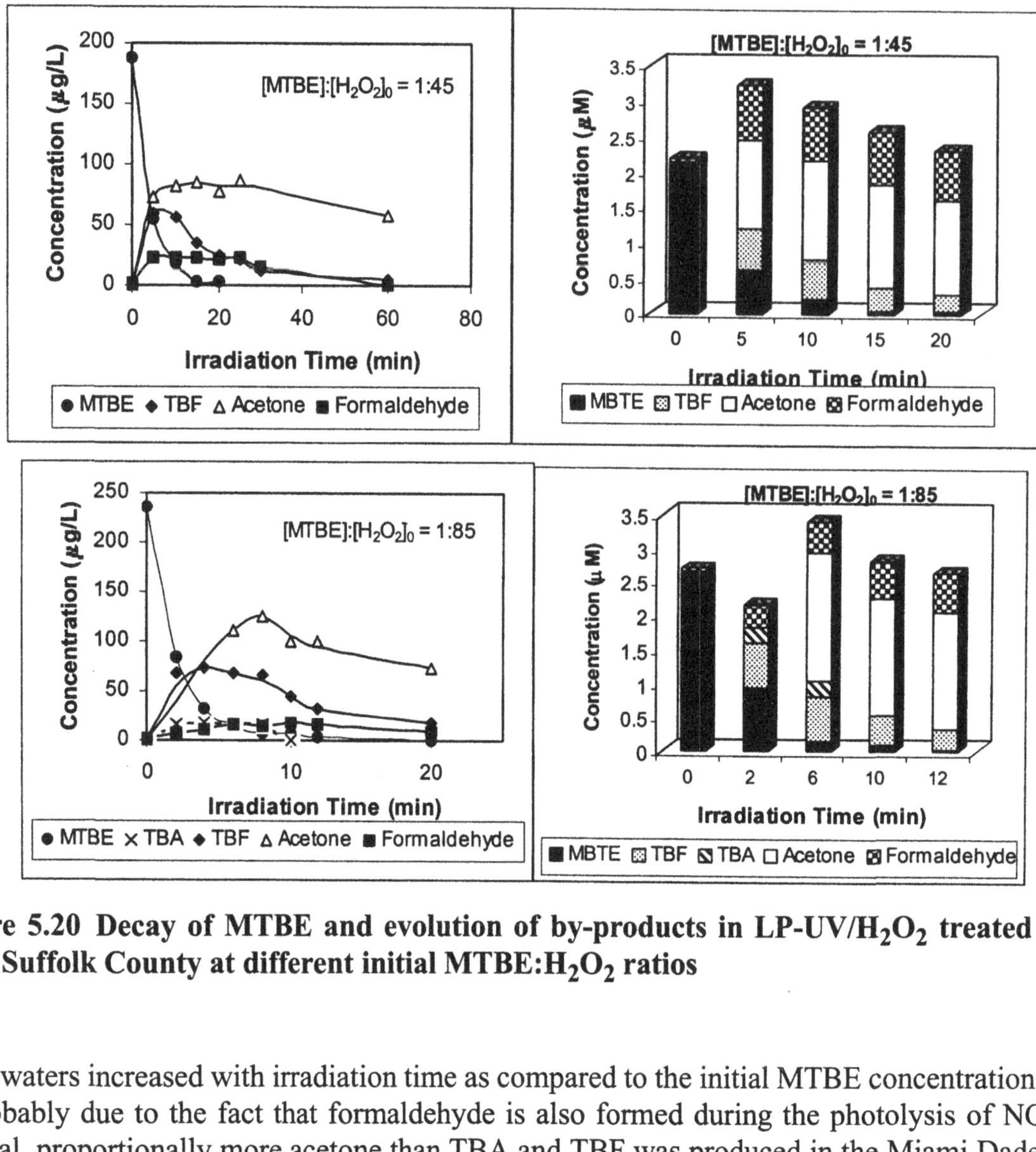

Figure 5.20 Decay of MTBE and evolution of by-products in LP-UV/H_2O_2 treated water from Suffolk County at different initial MTBE:H_2O_2 ratios

these waters increased with irradiation time as compared to the initial MTBE concentration which is probably due to the fact that formaldehyde is also formed during the photolysis of NOM. In general, proportionally more acetone than TBA and TBF was produced in the Miami Dade water with less bicarbonates than in the Long Beach water. This can also be observed at increasing irradiation time and H_2O_2 concentration.

MP-UV irradiation. Our results indicate that MTBE is degraded by MP-UV irradiation also in the absence of H_2O_2. In every water sample, formaldehyde was formed as a major degradation by-product and with prolonged irradiation times acetone was also produced. TBA was not always detected even if TBF was detected in the treated water sample. Previously Cater et al. 2000 has shown that these by-products were primary reaction products (i.e., TBF and acetone) and secondary reaction products (i.e., acetone and formaldehyde) when MTBE was treated with MP-UV/H_2O_2 through $^{\cdot}$OH radical pathways. Based on our results, it can be concluded that radical species are produced in the solution in the absence of H_2O_2. It is, however, somewhat surprising that similar by-products were found in the irradiated Suffolk County and Santa Monica water with different bicarbonate contents, 33 and 399 mg/L HCO_3^-, respectively. The large production of formaldehyde during the irradiation of Long Beach and Miami Dade waters may be due to the photolysis of UV absorbing moieties present in the NOM matrix.

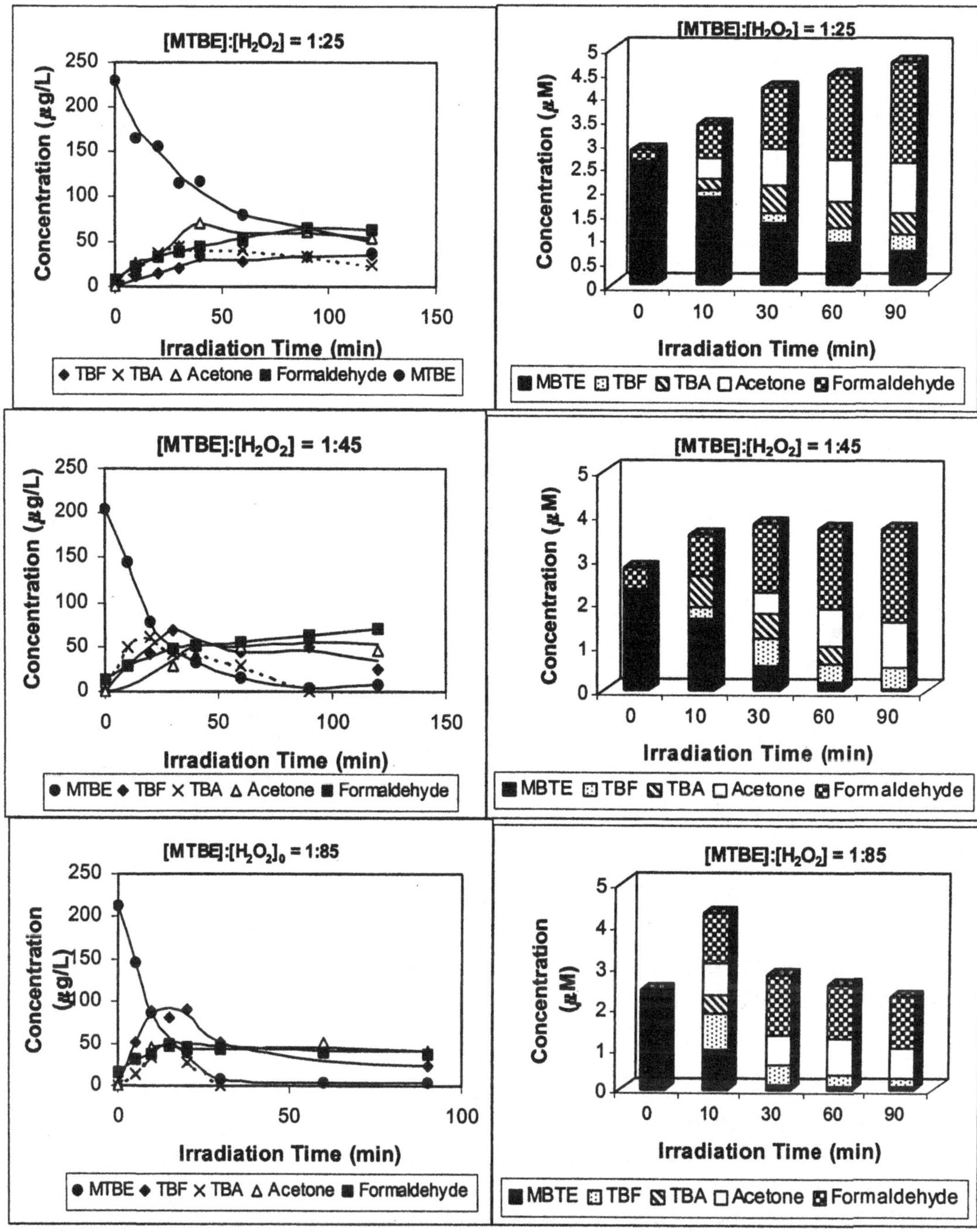

Figure 5.21 Decay of MTBE and evolution of TBF, TBA, acetone, and formaldehyde in LP-UV/H_2O_2 treated water from Miami Dade at different MTBE:H_2O_2 molar ratios

In the presence of H_2O_2, the general trends observed in the LP-UV experiments were also observed in the MP-UV experiments. However, significant differences can be observed in the relative abundance of the major degradation by-products. The most significant difference was found in the UV-treated Suffolk County water. In the water from Suffolk County, LP-UV/H_2O_2 degraded MTBE more efficiently than MP-UV/H_2O_2. For a similar MTBE removal and also whatever the irradiation time, less acetone but more formaldehyde, TBF and TBA was formed

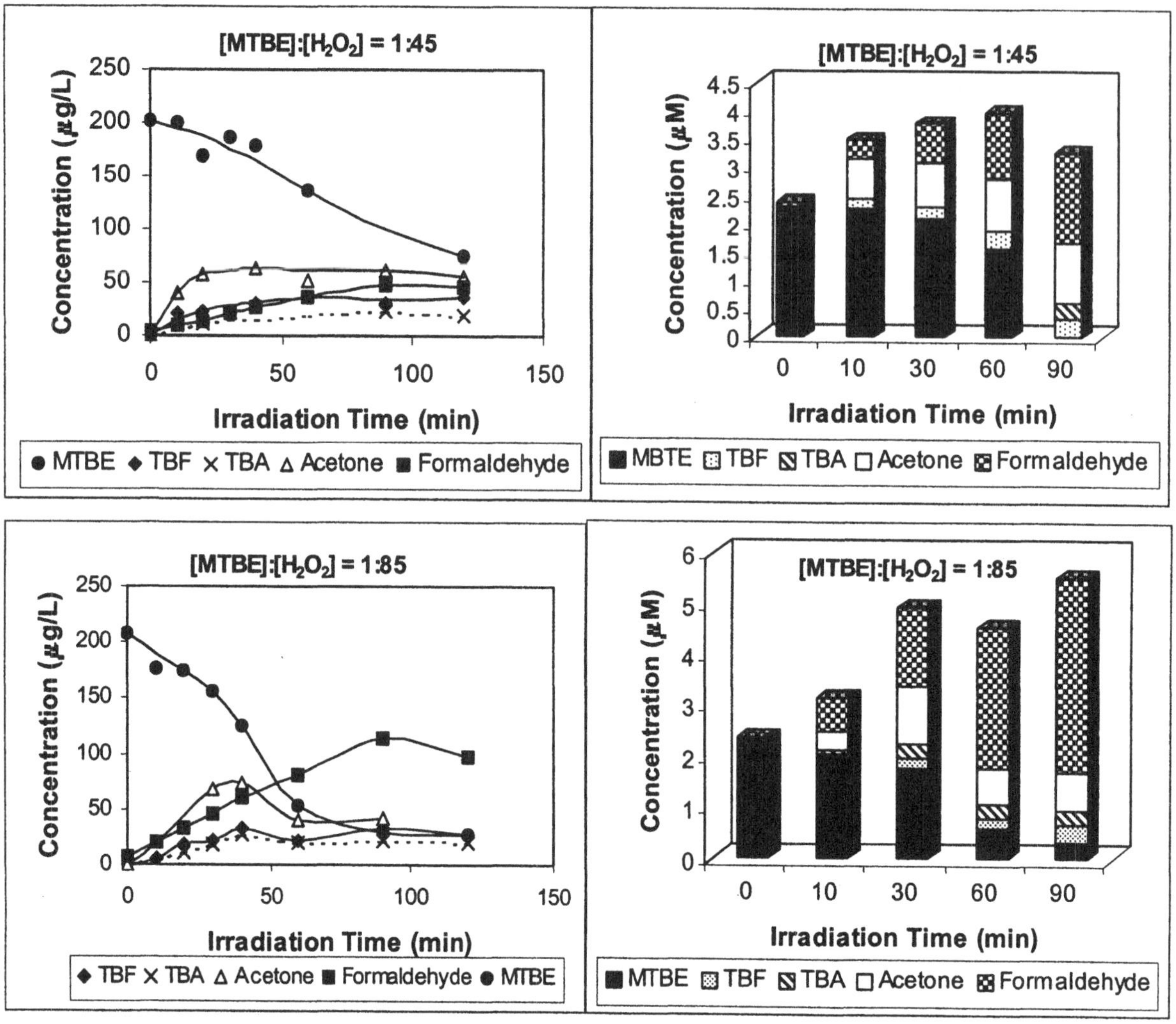

Figure 5.22 Decay of MTBE and evolution of by-products in LP-UV/H_2O_2 treated water from Long Beach at different MTBE:H_2O_2 molar ratios

when MTBE was degraded with MP-UV/H_2O_2. On the contrary, more acetone was detected in the MP-UV/H_2O_2 treated Miami Dade and Santa Monica water as compared to the corresponding LP-UV/H_2O_2 treated waters (acetone was not detected in the LP-UV treated Santa Monica water). As observed for the Suffolk County and Long Beach waters more formaldehyde is formed during MP-UV/H_2O_2 processes. In general, the production of by-products (the sum expressed in µM) increased as a function of irradiation time in Miami Dade and Long Beach water samples, which may indicate that the NOM may contribute to the production of these compounds when treated with MP-UV irradiation.

Impact of Initial MTBE Concentration

A high concentration of MTBE may retard the degradation of MTBE due to a larger competition for the hydroxyl radicals. Two different experiments were performed to study the impact of the initial MTBE concentration on the removal of MTBE with UV/H_2O_2 processes. The

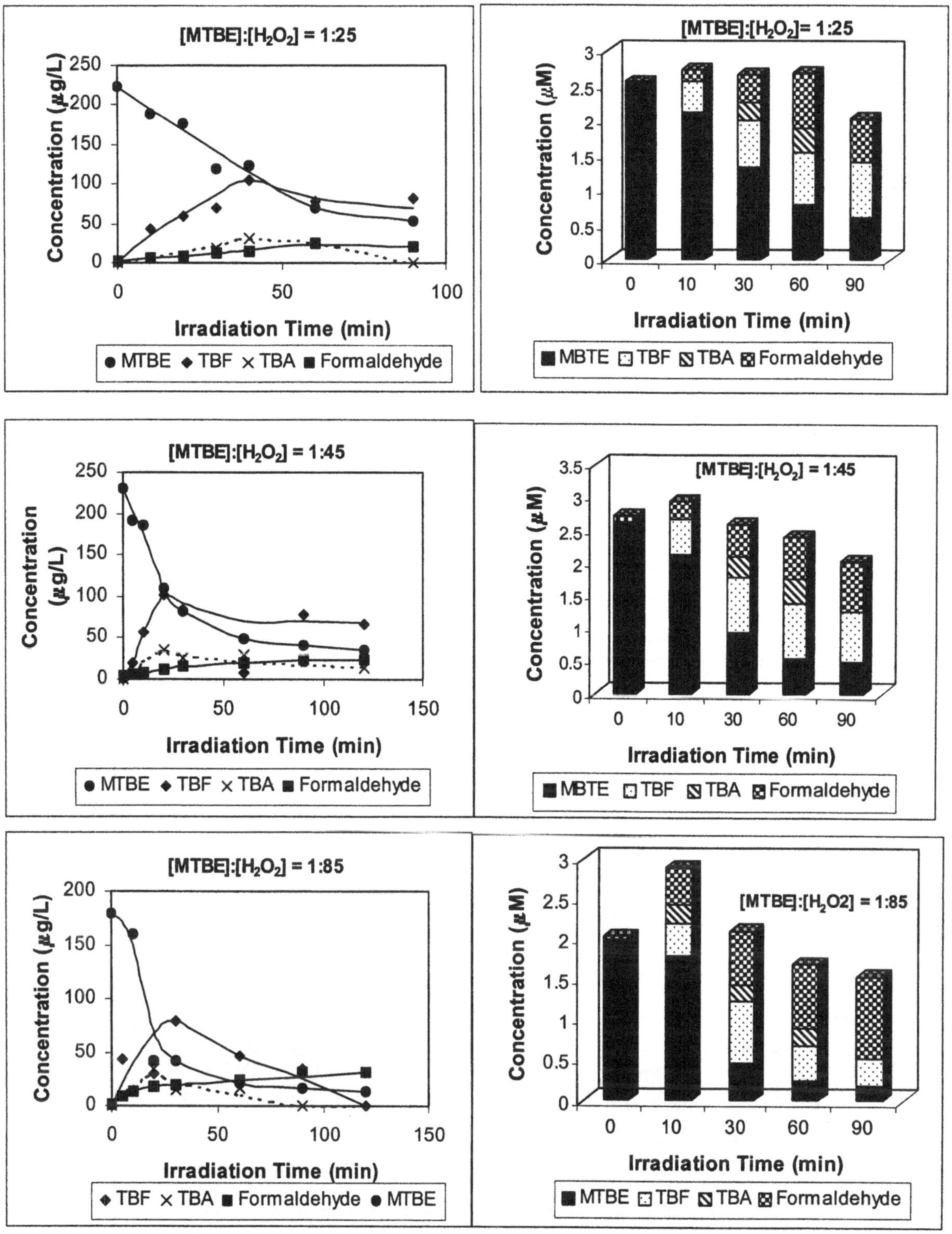

Figure 5.23 Decay of MTBE and evolution of TBF, TBA, and formaldehyde in Santa Monica water treated with LP-UV/H_2O_2 at different molar ratios between MTBE and H_2O_2

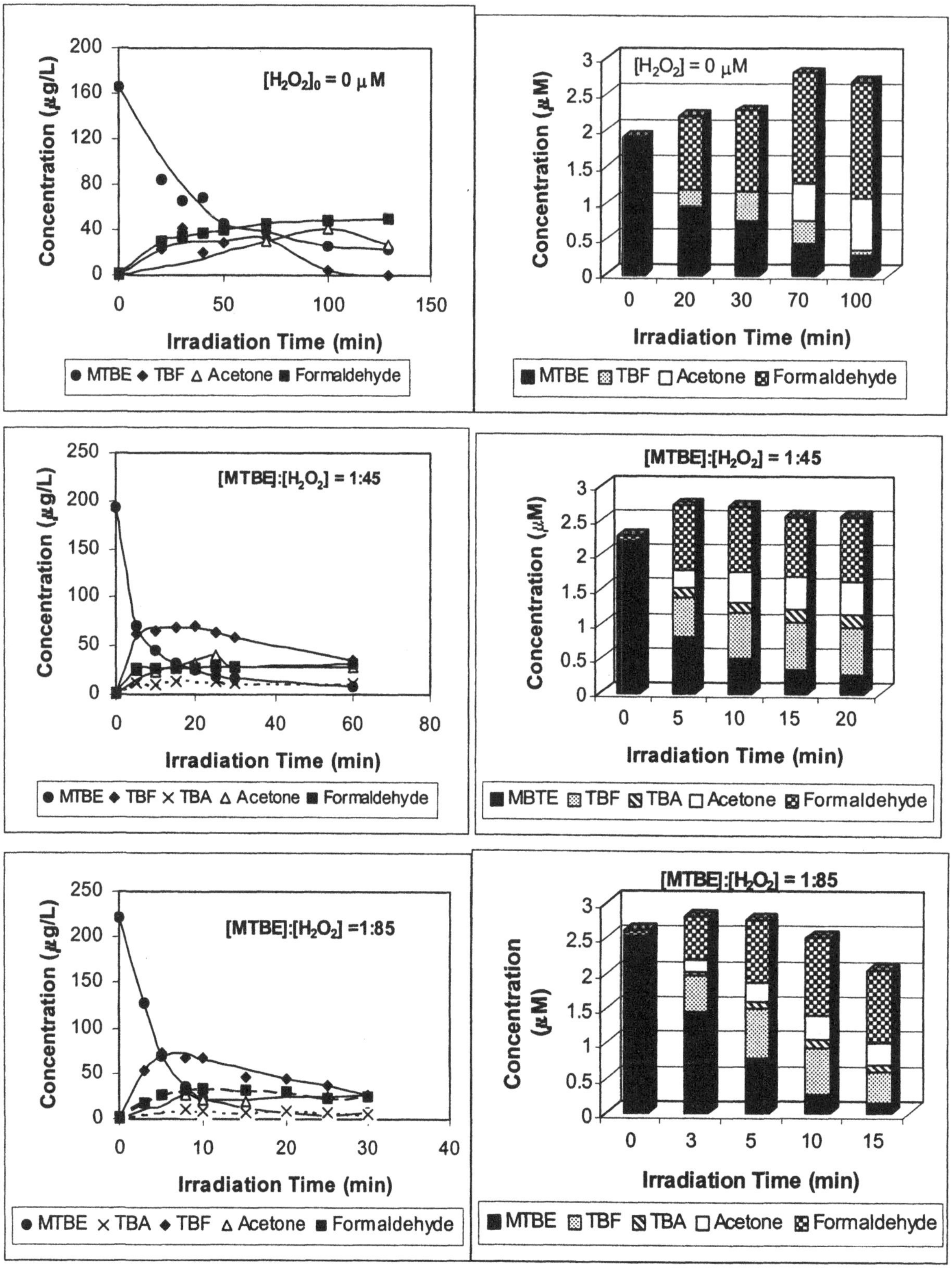

Figure 5.24 Removal of MTBE and formation of by-products in MP-UV/H_2O_2 treated water from Suffolk County at different H_2O_2 concentrations

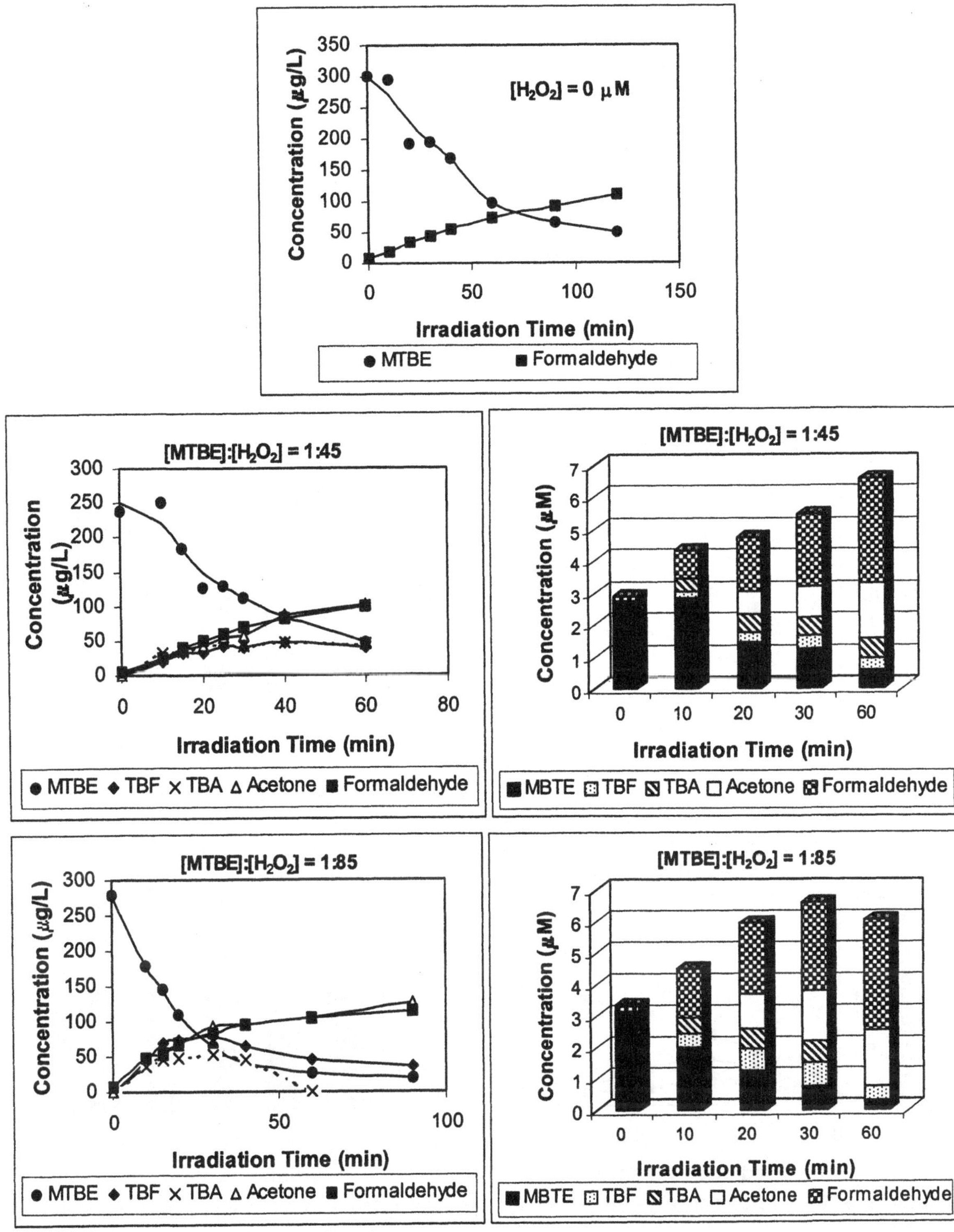

Figure 5.25 Decay of MTBE and evolution of by-products in MP-UV/H_2O_2 treated water from Miami Dade at different initial H_2O_2 concentrations

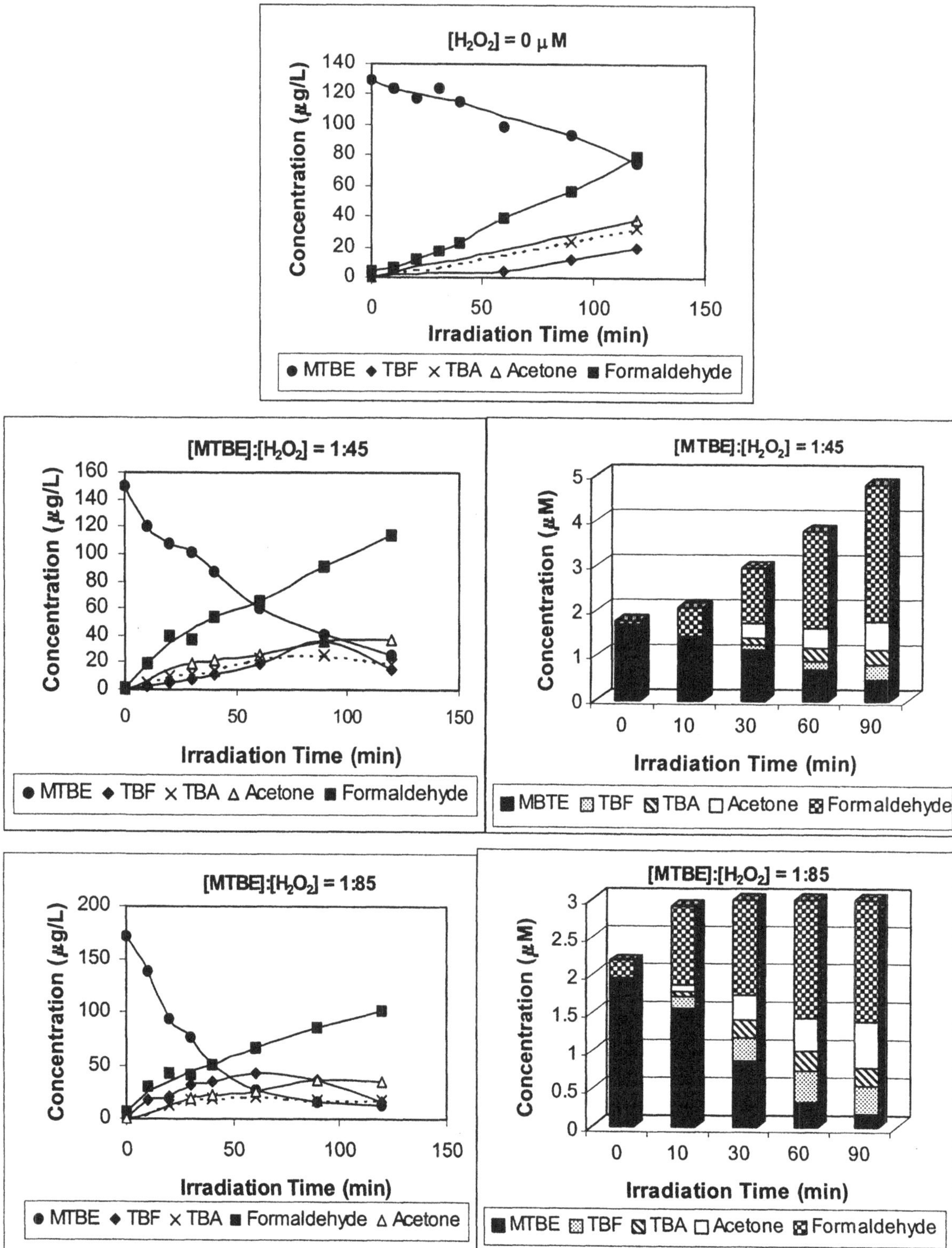

Figure 5.26 Degradation of MTBE and evolution of by-products in Long Beach water treated with MP-UV/H_2O_2 at different initial concentrations of H_2O_2

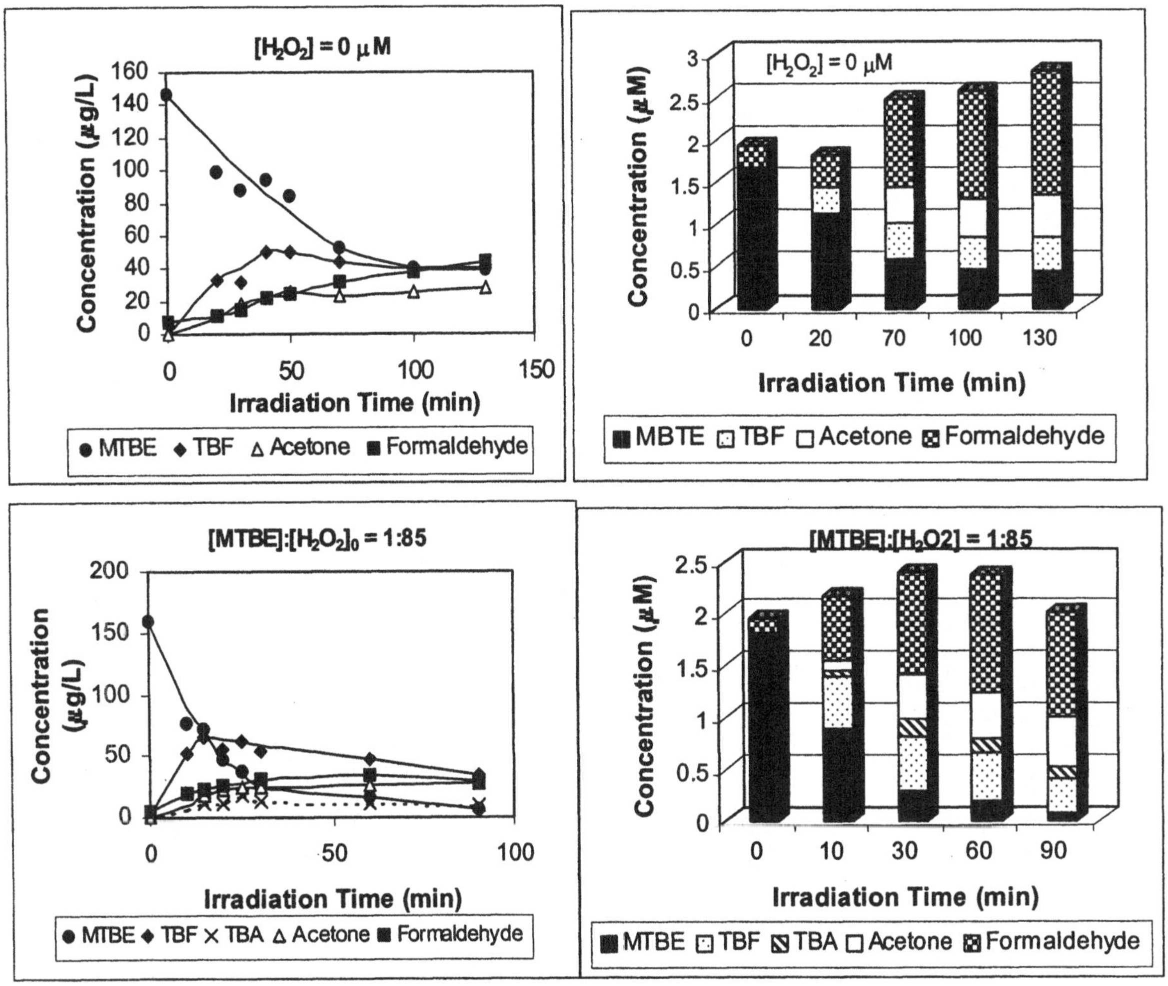

Figure 5.27 Degradation of MTBE and formation of by-products in MP-UV/H_2O_2 treated water from Santa Monica

testing in Santa Monica and Suffolk County water samples were done by keeping the initial hydrogen peroxide concentration constant and in the experiments with the Miami Dade and Long Beach water samples, the initial molar ratio between MTBE and H_2O_2 was kept constant.

Constant Initial Hydrogen Peroxide Concentration

The influence of the initial concentration of MTBE on the removal of MTBE in Santa Monica and Suffolk County field waters was investigated by keeping the initial H_2O_2 concentration constant. Our results indicate that an increased concentration of peroxide does not significantly influence the removal rate of MTBE with LP-UV/H_2O_2 in the Santa Monica water (Figure 5.23). The initial pseudo-first order rate constant obtained for the removal of 50, 200 or 2000 μg/L MTBE from the water of Santa Monica with LP-UV/H_2O_2 processes is approximately $k_1' = 0.6 \times 10^{-3}\ s^{-1}$.

An increasing initial concentration of MTBE generates a higher concentration of intermediates. These intermediates are also reactive towards hydroxyl radicals and may

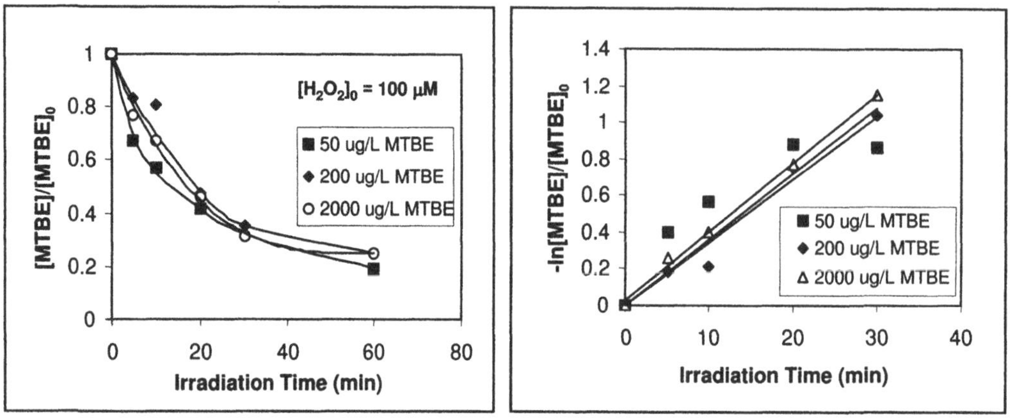

Figure 5.28 Effect of initial MTBE concentration on the removal of MTBE from Santa Monica water with LP-UV/H_2O_2 at an initial hydrogen peroxide concentration of 100 μM

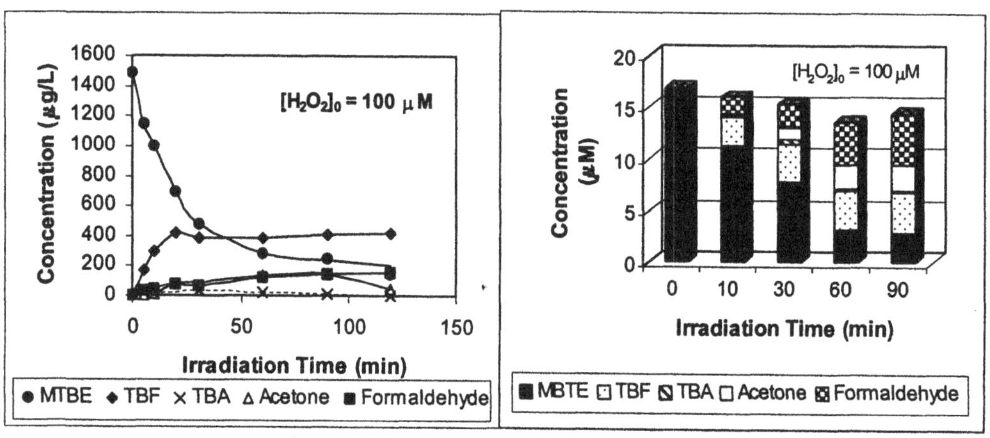

Figure 5.29 Decay of MTBE and evolution of by-products in LP-UV/H_2O_2 treated water from Santa Monica (the initial H_2O_2 concentration is 100 μM)

consequently compete with MTBE for the hydroxyl radicals and retard the removal of MTBE. In the water from Santa Monica, the very high concentration of bicarbonates will scavenge the hydroxyl radicals generated and the MTBE removal is therefore relatively slow.

Figure 5.29 shows the distribution of the major by-products formed as a function of irradiation time. An increasing concentration of MTBE generates a larger amount of by-products. The distribution of the by-products is, however, similar to the one at lower MTBE concentrations. When compared to the by-products generated in the UV/H_2O_2 treated Santa Monica water spiked with 200 μg/L MTBE, more acetone was formed in the Santa Monica water spiked with 2000 μg/L MTBE. Acetone formed in the 200 μg/L SM water was probably below the detection limits.

When the Santa Monica water was treated with the Teflon shielded MP TQ 718 UV lamp at an initial H_2O_2 concentration of 100 μM, a small difference in the degradation rate of MTBE was observed (Figure 5.30). The obtained pseudo-first order rate constant for the degradation of 200 and 2000 μg/L MTBE was $k' = 0.5 \times 10^{-3}$ s^{-1} is whereas the one for 50 μg/L MTBE was $k_1' =$

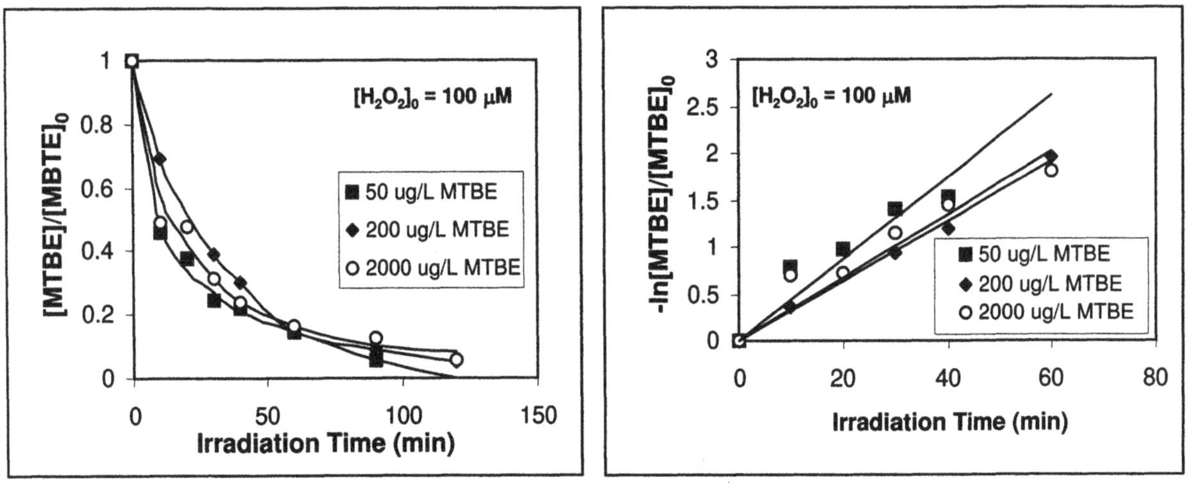

Figure 5.30 Effect of initial MTBE concentration on the removal of MTBE from Santa Monica water with MP-UV/H_2O_2 (Teflon shielded TQ 718 UV-lamp, the initial H_2O_2 concentration was 100 μM)

0.7×10^{-3} s^{-1}. Thus, the decay rate of MTBE with MP-UV/H_2O_2 (Teflon shielded TQ 718 UV lamp) was somewhat faster at a lower concentration of MTBE.

In the water from Suffolk County, the removal rate of MTBE with LP- and MP-UV/H_2O_2 (TQ 150 UV-lamp, $[H_2O_2]_0$ = 200 μM) was significantly retarded at an increased concentration of MTBE (Figure 5.31). The obtained initial pseudo-first order rate constant for the removal of 200 μg/L MTBE with LP-UV/H_2O_2 was $k' = 7.2 \times 10^{-3}$ s^{-1} compared to a rate constant of $k' = 3.6 \times 10^{-3}$ s^{-1} for a concentration of 2000 μg/L. The obtained pseudo-first order rate constants for the removal of MTBE with MP-UV/H_2O_2 were 3.4×10^{-3} s^{-1} and 1.8×10^{-3} s^{-1} for an initial MTBE concentration of 200 and 2000 μg/L, respectively. The degradation of MTBE in Suffolk County water was very fast due to the low concentration of bicarbonates and the relatively high H_2O_2 concentration. Thus, the by-products formed may affect the degradation of MTBE by scavenging the hydroxyl radicals generated.

Figure 5.32 presents the evolution of by-products formed in LP- and MP-UV/H_2O_2 treated water from Suffolk County as a function of irradiation time. The differences in the distribution of the by-products between LP- and MP-UV/H_2O_2 treatments are similar as discussed earlier with the experiments conducted at 200 μg/L MTBE. For a similar relative degradation of MTBE, a similar distribution of by-products was found in the experiments on the removal of MTBE at a concentration of 200 or 2000 μg/L.

A Constant Initial Molar Ratio Between MTBE and H_2O_2

The effectiveness of LP-UV/H_2O_2 treatment on the removal of MTBE from Miami Dade and Long Beach water at three different initial MTBE concentrations (50, 200 and 2000 μg/L) was studied by keeping the initial molar ratio between MTBE and H_2O_2 constant (1:85). The corresponding experiments with the MP TQ 150-UV lamp have not been done because of an insufficient water volume. According to our results, the removal of MTBE is accelerated at a higher initial concentration of MTBE due to the higher initial H_2O_2 concentration when keeping the molar ratio between MTBE and H_2O_2 constant (Figure 5.33).

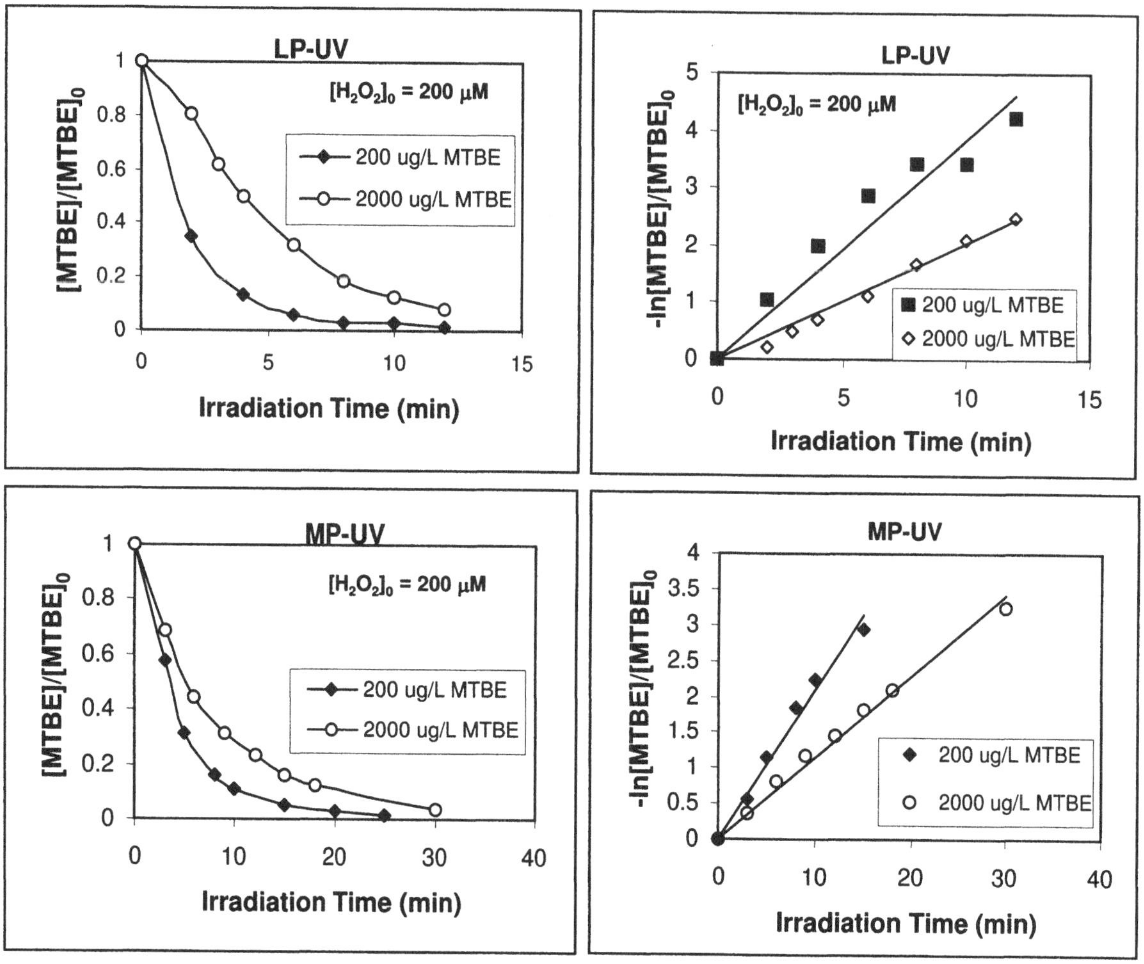

Figure 5.31 Removal of 200 and 2000 μg/L MTBE from Suffolk County water with LP- and MP-UV/H_2O_2 processes at an initial H_2O_2 concentration of 200 μM

If the degradation curves are plotted as natural logarithmic values of the relative decrease in MTBE concentration fairly straight lines are obtained and the initial pseudo-first order degradation rate constants can be calculated (Table 5.4). In these experiments, the MTBE:H_2O_2 ratio was kept constant and a larger degradation rate constant for MTBE is consequently obtained at a higher MTBE concentration.

Figures 5.34 and 5.35 show the degradation of 2000 μg/L MTBE and the distribution of the major degradation by-products as a function of irradiation time in the LP-UV/H_2O_2 treated Miami Dade and Long Beach water samples when keeping the initial MTBE/H_2O_2 molar ratio at 1:85.

For a similar irradiation time the production of secondary by-products (i.e., acetone and formaldehyde) is proportionally more important as compared to the experiment conducted with 200 μg/L MTBE at a similar molar ratio between MTBE and H_2O_2, which is due to the faster degradation of MTBE due to a higher production of hydroxyl radicals. The sum of the major by-products plus the MTBE residual expressed in μM rapidly decreased with increasing irradiation time showing that the organic compounds are completely mineralized or degraded into compounds not identified in this study at high H_2O_2 concentrations. Again a faster degradation of MTBE and its degradation products is observed in the Miami Dade water as compared to the one

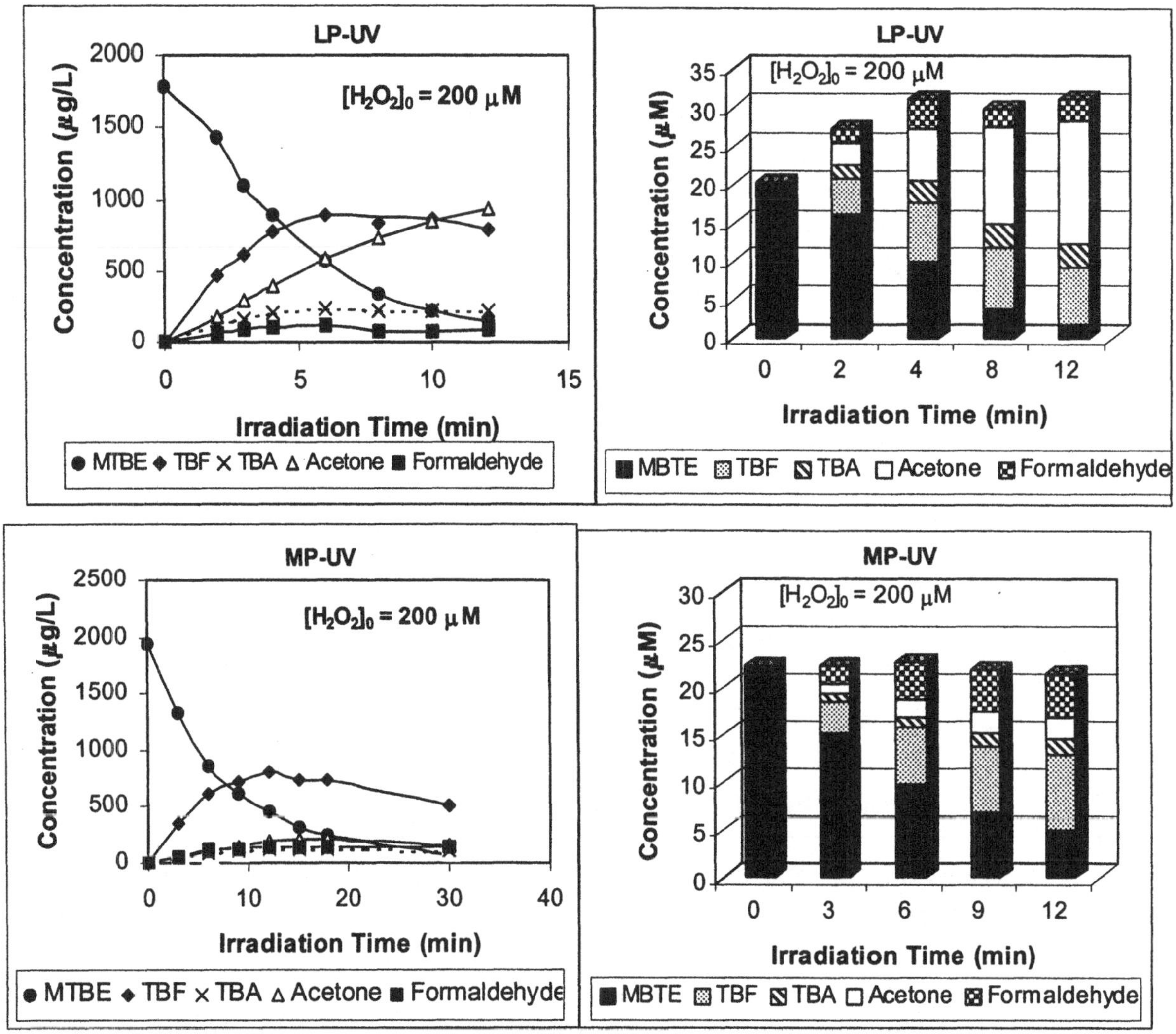

Figure 5.32 Degradation of MTBE and evolution of by-products in LP- and MP-UV/H_2O_2 treated water from Suffolk County (the initial H_2O_2 concentration: 200 μM)

in Long Beach water, of which the latter contains more bicarbonates and a higher content of UV-absorbing compounds.

Impact of Nitrates

As mentioned earlier, UV photolysis of nitrates may produce hydroxyl radicals. However, a highconcentration of nitrates may also retard the removal of MTBE by competing with MTBE for the hydroxyl radicals and by acting as a filter due to the strong absorption between 200 and 230 nm (Figure 5.36). Each of the four field water samples were spiked with 200 μg/L MTBE and treated with LP- or MP-UV irradiation in the presence of a 36-mg/L nitrate spike and a constant initial concentration of H_2O_2. According to our results there was no obvious difference in the degradation rate of MTBE in the Santa Monica and Suffolk County water samples with or without any addition of nitrates when treated with LP-UV/H_2O_2. However, at the end of the LP-UV/H_2O_2

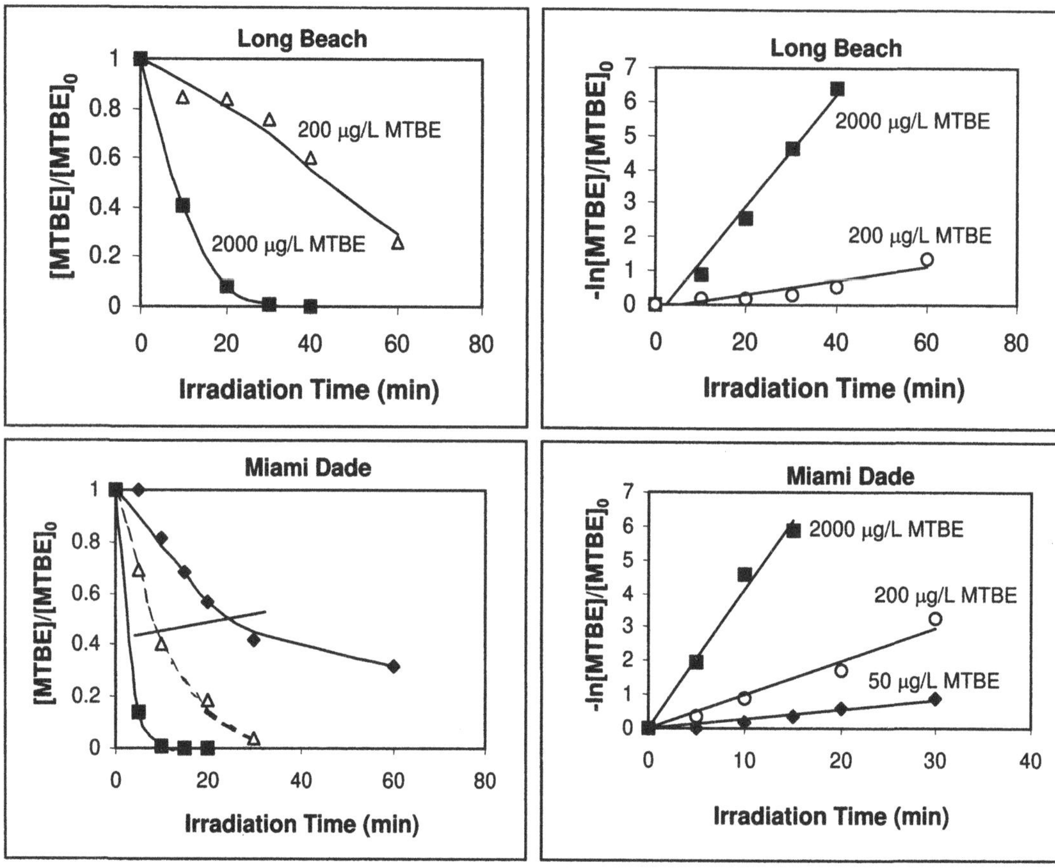

Figure 5.33 Removal of 50, 200, and 2000 μg/L MTBE from Miami Dade and Long Beach water samples with LP-UV/H_2O_2 with an initial molar ratio of 1:85 between MTBE and H_2O_2

Table 5.4
Initial pseudo-first order rate constants for the degradation of 50, 200, and 2000 μg/L MTBE in Miami Dade and Long Beach waters with LP-UV/H_2O_2. The initial molar ratio between MTBE and H_2O_2 was 1:85.

	k_1' (s^{-1})	
[MTBE] / (μg/L)	Miami Dade	Long Beach
50	0.5×10^{-3}	-
200	1.7×10^{-3}	0.2×10^{-3}
2000	6.8×10^{-3}	2.5×10^{-3}

treatment, the removal of MTBE from the high DOC Miami Dade and Long Beach waters was slightly retarded in the presence of nitrates.

Only minor differences were observed in the amount and in the distribution of degradation by-products formed during LP-UV/H_2O_2 treatment after addition of nitrates (see Figures A.5 and A.6 in Appendix A). In the 36-mg/L nitrate spiked water samples slightly less acetone was formed in the water from Suffolk County slightly and more formaldehyde and TBA were formed

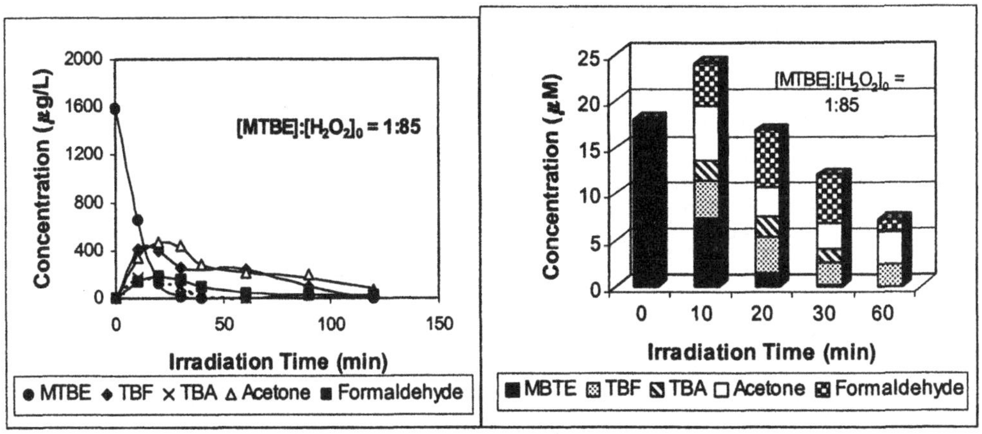

Figure 5.34 Degradation of MTBE and evolution of by-products in LP-UV/H_2O_2 treated Long Beach water at an initial molar ratio of 1:85 between MTBE and H_2O_2

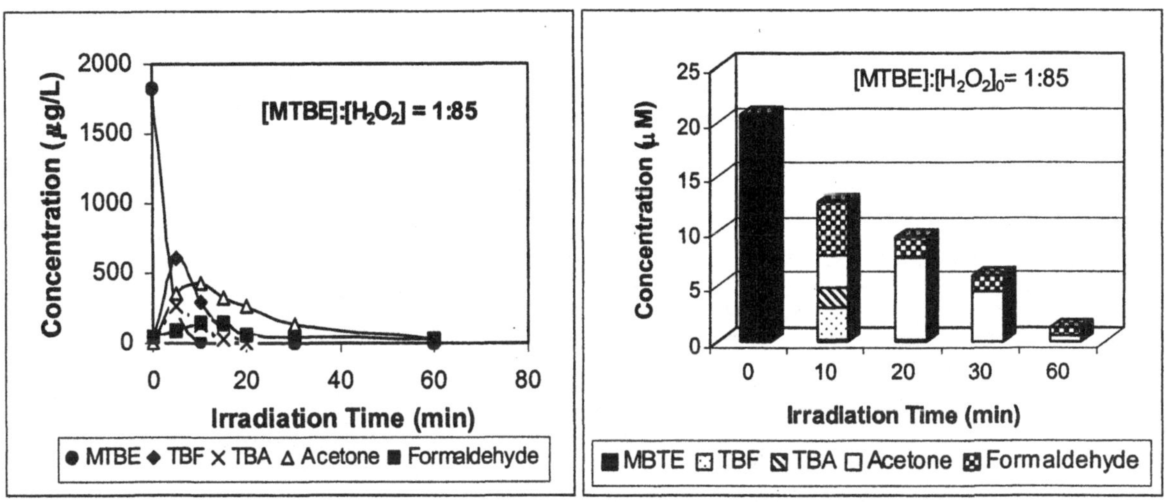

Figure 5.35 Degradation of MTBE and evolution of by-products in LP-UV/H_2O_2 treated Miami Dade water at an initial molar ratio of 1:85 between MTBE and H_2O_2

in the treated Miami Dade water compared to the production in the water samples without any nitrate spike. In the Long Beach water, less formaldehyde was formed probably due to the retarded degradation of MTBE. No TBA was detected in the Santa Monica water during the UV-treatment. As observed previously without the addition of nitrates, no acetone could be detected.

A similar trend was found when treating the four waters with MP-UV/H_2O_2 in the presence of 36 mg/L nitrates as compared to the original waters, showing the fastest MTBE removal in the low DOC and low alkalinity water from Suffolk County followed by Santa Monica, Miami Dade, and the highly colored Long Beach water (Figure 5.37).

If the degradation curves are plotted as natural logarithmic values of the relative decrease of MTBE fairly straight lines are obtained (Figure 5.38) and the initial pseudo-first order degradation rate constants can be calculated (Table 5.5). In each water sample, the removal of

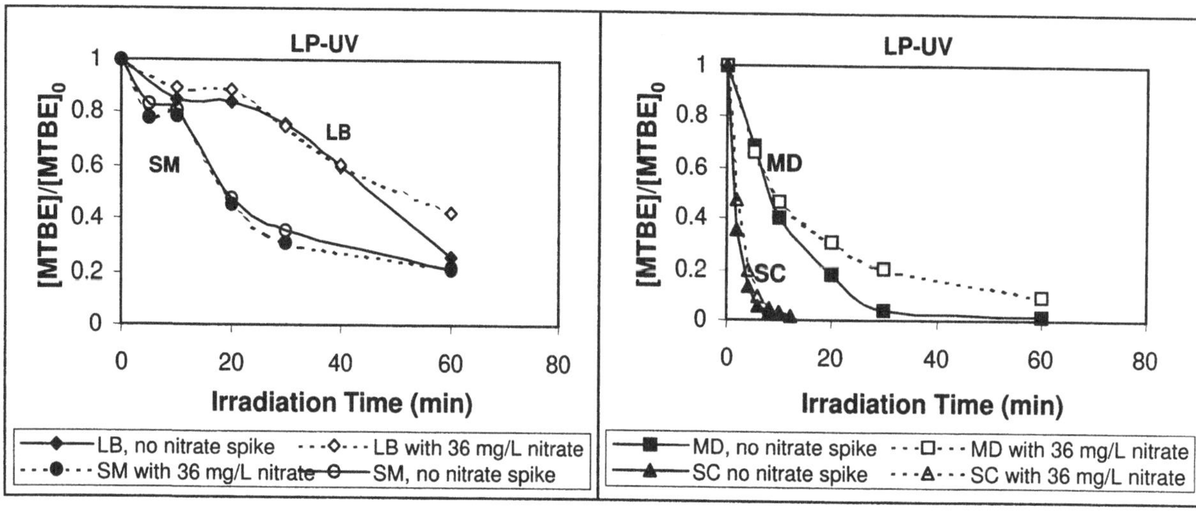

Figure 5.36 Influence of nitrates on the removal of MTBE from Long Beach (LB), Santa Monica (SM), Miami Dade (MD), and Suffolk County (SC) water samples

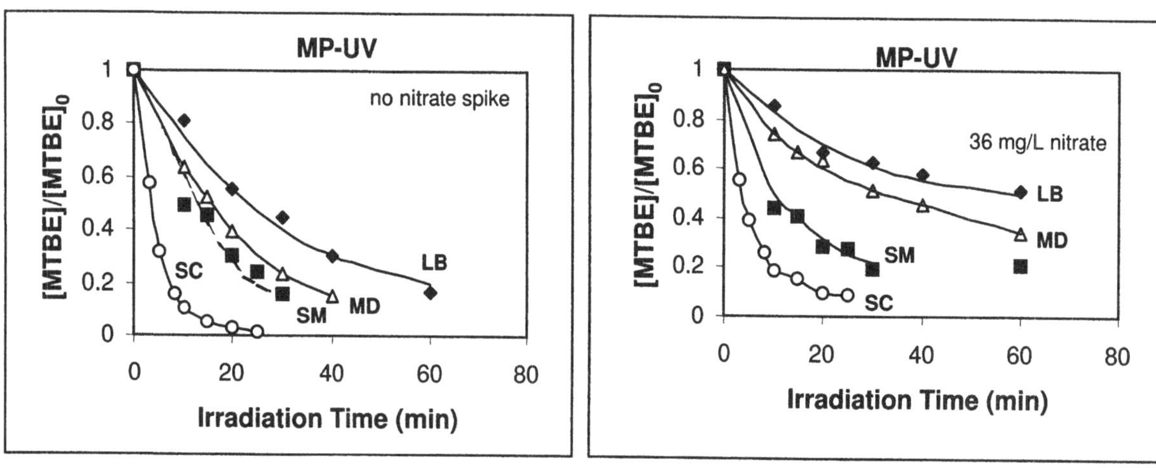

Figure 5.37 Removal of MTBE with MP-UV/H_2O_2 with or without any nitrate spike. The initial H_2O_2 concentration was 200 μM.

MTBE with MP-UV/H_2O_2 was retarded when the water samples were spiked with 36 mg/L nitrate. Before the nitrate spiking the Santa Monica water contained 21 mg/L nitrates and the additional spiking of the water sample did not noticeable affect the degradation rate of MTBE. The removal of MTBE was significantly retarded in the water samples from Miami Dade and Long Beach that did not contain any nitrates.

Concerning the formation of by-products, acetone and formaldehyde, it can be noted that the formation was delayed (see Figure A.7 in Appendix A). However, based on a similar degradation of MTBE the addition of nitrates in the water samples did not affect the amount and distribution of the major by-products.

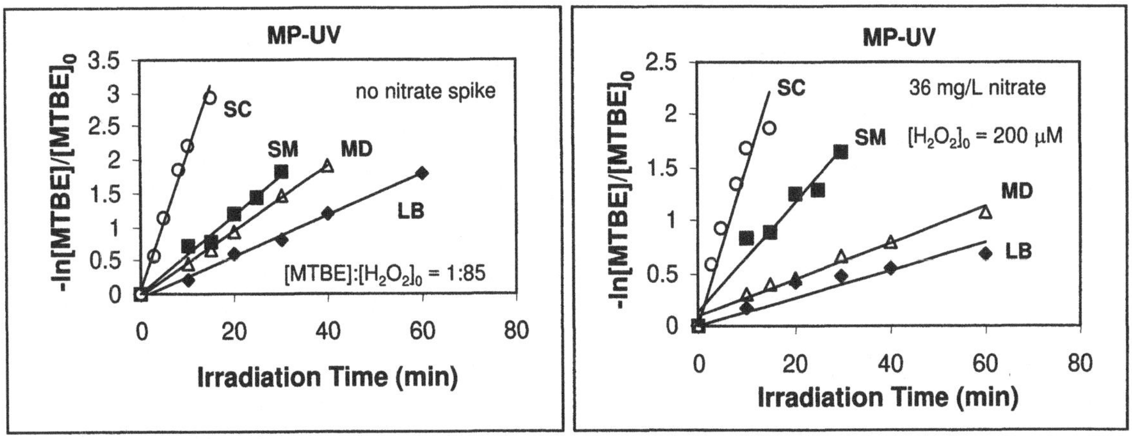

Figure 5.38 Pseudo first order degradation plots of the removal of MTBE from Santa Monica (SM), Suffolk County (SC), Miami Dade (MD), and Long Beach (LB) with MP-UV/ H_2O_2 treatment at an initial molar ratio of 1:85 between MTBE and H_2O_2

Table 5.5
Initial pseudo-first order rate constants for the degradation of MTBE in different water samples with MP-UV/H_2O_2 with or without any addition of nitrates. The initial molar ratio between MTBE and H_2O_2 was 1:85.

	k (s^{-1})	
Water sample	No nitrate spike	36 mg/L nitrate
Santa Monica (SM)	1.0×10^{-3}	0.9×10^{-3}
Suffolk County (SC)	3.4×10^{-3}	2.2×10^{-3}
Miami Dade (MD)	0.8×10^{-3}	0.3×10^{-3}
Long Beach (LB)	0.5×10^{-3}	0.2×10^{-3}

Impact of Tert-butyl Alcohol

TBA may retard the removal of MTBE with LP- and MP-UV/H_2O_2 processes because of its strong reactivity with hydroxyl radicals ($k = 6 \times 10^8\ M^{-1}\ s^{-1}$ as compared to $k_{\cdot OH/MTBE} = 1.6 \times 10^9\ M^{-1}s^{-1}$ According to Stefan et al., (2000) formaldehyde, acetone, and hydroxy iso-butyraldehyde are the major primary degradation intermediates of TBA. Figure 5.39 presents the results obtained when the field water samples were spiked with 200 μg/L MTBE and TBA and then treated with LP- or MP-UV/H_2O_2 in the presence of an initial hydrogen peroxide concentration of 200 μM.

The pseudo-first order degradation plots for MTBE in Suffolk County, Miami Dade, and Long Beach are presented in Figure 5.40.

According to our results, no obvious difference in the degradation rate of MTBE can be observed in the Santa Monica and Long Beach water samples in the presence or absence of TBA. However, the removal of MTBE in Suffolk County water with both LP- and MP-UV/H_2O_2 was

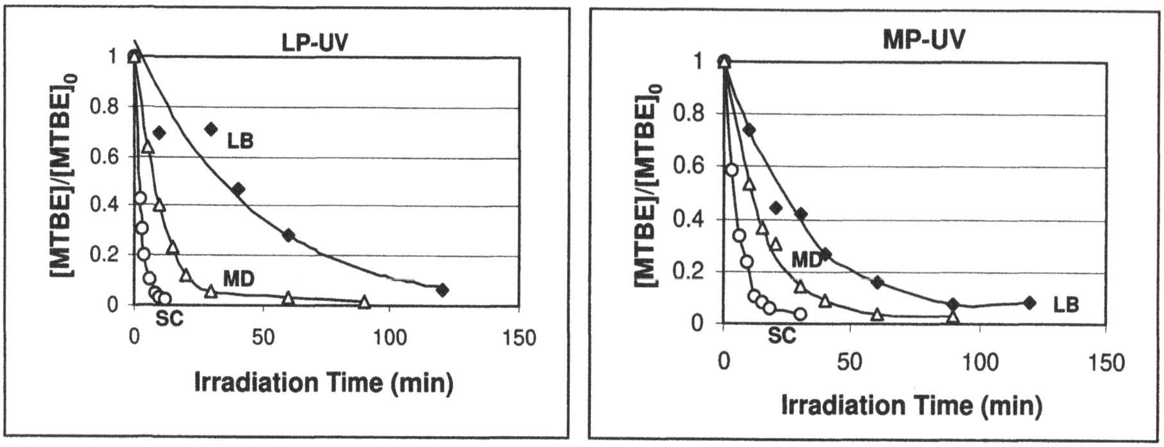

Figure 5.39 Normalized MTBE decay in Long Beach, Miami Dade, and Suffolk County waters with LP- and MP-UV/H_2O_2 in the presence of 200 μg/L TBA

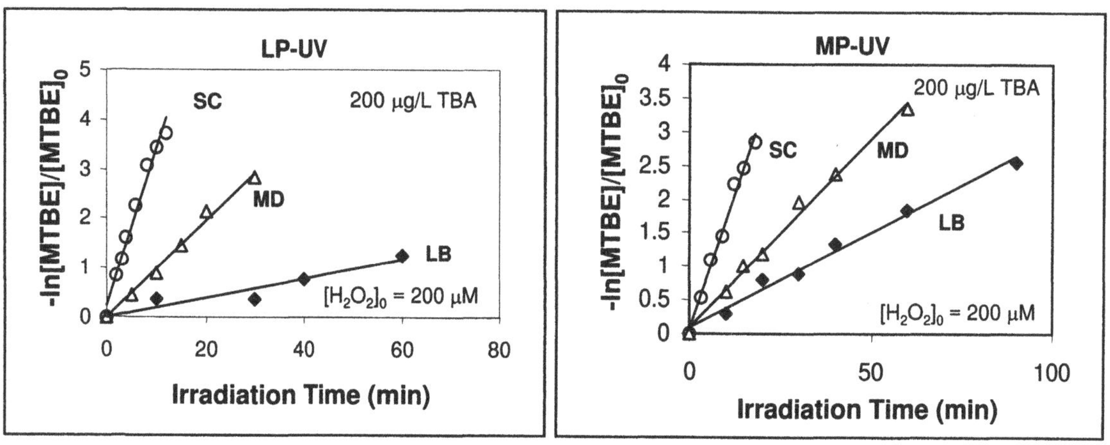

Figure 5.40 Pseudo-first order degradation of MTBE in Suffolk County (SC), Miami Dade (MD), and Long Beach (LB) water samples with LP- and MP-UV/H_2O_2 with an addition of 200 μg/L TBA

slightly retarded in the presence of TBA (Table 5.6). The physical chemical characteristics of the Suffolk County water, that is a low DOC content and a low alkalinity, promote the ˙OH radical degradation pathways of MTBE. Consequently, the presence of a high TBA concentration will consume proportionally a higher fraction of the ˙OH radicals available for the MTBE degradation.

The results of the major by-products analyzed are given in Figures A.8 and A.9 in Appendix A. The TBA concentration decreases as a function of irradiation time. As observed with MTBE, the fastest degradation of TBA occurred in the Suffolk County water followed by Miami Dade and Long Beach waters in which TBA was not considerably degraded even after 2 hours of LP- and MP-UV irradiation. The pseudo-first order rate constants determined for these three waters when treated with LP-UV/H_2O_2 are (no results for Santa Monica water): 2.1×10^{-3} s^{-1}, 0.6×10^{-3} s^{-1}, and 0.1×10^{-3} s^{-1} for Suffolk County, Miami Dade, and Long Beach waters,

Table 5.6
Initial pseudo-first order rate constants for the degradation of MTBE in different water samples with MP-UV/H_2O_2 with or without any addition of TBA. The initial molar ratio between MTBE and H_2O_2 was 1:85.

	k (s^{-1})	
Water sample	No TBA spike	200 μg/L TBA
	LP-UV	
Santa Monica (SM)	0.6×10^{-3} *	0.7×10^{-3} *
Suffolk County (SC)	7.2×10^{-3}	5.3×10^{-3}
Miami Dade (MD)	1.7×10^{-3}	1.6×10^{-3}
Long Beach (LB)	0.2×10^{-3}	0.3×10^{-3}
	MP-UV	
Suffolk County (SC)	3.4×10^{-3}	2.3×10^{-3}
Miami Dade (MD)	0.8×10^{-3}	0.9×10^{-3}
Long Beach (LB)	0.5×10^{-3}	0.4×10^{-3}

* $[H_2O_2]_0 = 100$ μM

respectively. The corresponding rate constants for the degradation of TBA with MP-UV/H_2O_2 are: 0.9×10^{-3} s^{-1}, 0.4×10^{-3} s^{-1}, and 0.1×10^{-3} s^{-1}. The addition of TBA in the source waters prior to UV-irradiation resulted in a more significant production of acetone and formaldehyde with both LP-and MP-UV irradiation as compared to the results obtained in the absence of TBA.

Summary of Findings

Figure 5.41 compares the removal of 200 μg/L MTBE with LP-UV/H_2O_2 or MP-UV/H_2O_2 in from Santa Monica, Suffolk County, Miami Dade and Long Beach water samples at an initial molar ratio of 1:85 between MTBE and H_2O_2. Both LP- and MP-UV/H_2O_2 (TQ 150) removed MTBE most efficiently from the Suffolk County water sample with a low DOC content and also a low alkalinity. With both LP- and MP-UV irradiation, the most retarded degradation of MTBE can be observed in the water from Long Beach characterized by a high DOC content, high SUVA, and a high alkalinity. Intermediate decomposition rates were determined for the Miami Dade and Santa Monica waters.

Table 5.7 indicates that the pseudo-first order rate constants determined for the Long Beach and Santa Monica waters were higher with the MP-UV/H_2O_2 treatment as compared to the LP-UV/H_2O_2 treatment using the same initial molar ratio between MTBE and H_2O_2. The opposite trend was found with the two other water samples. It is interesting to point out that the Long Beach and Santa Monica waters were enriched in bicarbonates as compared to the two other water sources.

According to our results, MTBE was degraded with MP-UV irradiation in all water samples also in the absence of H_2O_2. The reaction rates obtained for the different waters followed the same order as found in the presence of H_2O_2, that is SC>SM>MD>LB. Since the degradation of MTBE with MP UV irradiation generated similar major by-products in the presence or absence

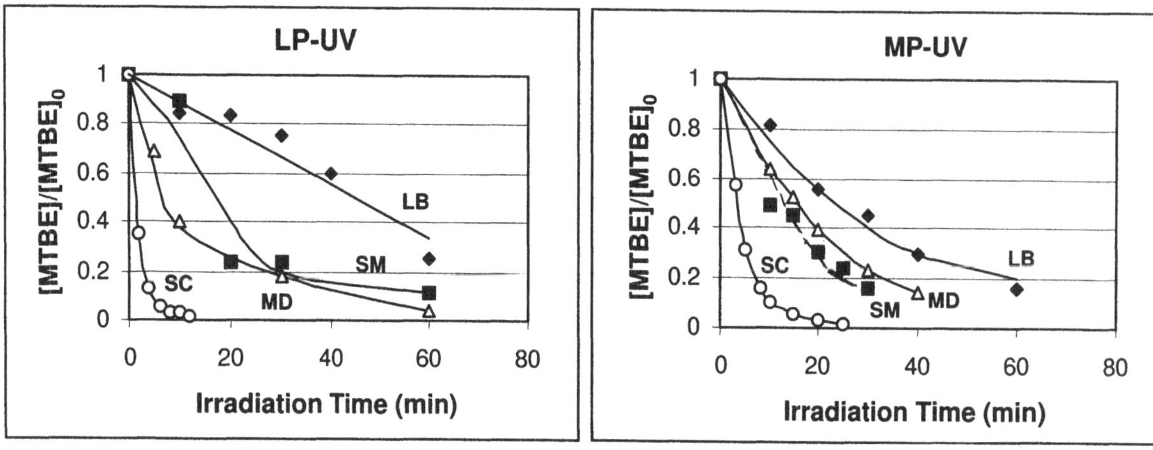

Figure 5.41 Decay of MTBE in Santa Monica (SM), Suffolk County (SC), Long Beach (LB), and Miami Dade (MD) water samples treated with LP- or MP-UV/H_2O_2. The initial molar ratio between MTBE and H_2O_2 was 1:85.

Table 5.7

Pseudo-first order rate constants for the removal of MTBE with different irradiation sources at an initial H_2O_2 concentration of 200 μM ($[MTBE]/[H_2O_2]_0$ = 1:85)

	k_1' (s^{-1})	
Water sample	LP-UV/H_2O_2	MP-UV/H_2O_2 (TQ 150)
Santa Monica	0.6×10^{-3}	1.0×10^{-3}
Miami Dade	1.7×10^{-3}	0.8×10^{-3}
Long Beach	0.2×10^{-3}	0.5×10^{-3}
Suffolk County	7.2×10^{-3}	3.4×10^{-3}

of H_2O_2, it is still unclear if the degradation of MTBE in the absence of H_2O_2 is due to direct photolysis and/or radical pathways. It can be noticed that MTBE was rapidly degraded in the Suffolk County water with LP-UV irradiation even without any hydrogen peroxide added. This can be explained by the presence of nitrates that may generate hydroxyl radicals during photolysis.

In the presence of hydrogen peroxide, similar major by-products (TBF, TBA, formaldehyde, and acetone) were produced in the LP- and MP-UV treated waters. These compounds may require further biological treatment (e.g., biological Activated Carbon) to guarantee the biological stability of the treated water in the network. The major findings regarding the formation of by-products are the following:

- TBF is formed in larger amounts than TBA
- Acetone and formaldehyde are the major secondary by-products
- A higher concentration of H_2O_2 degrades the degradation products more efficiently.
- More formaldehyde is formed in the high DOC water samples

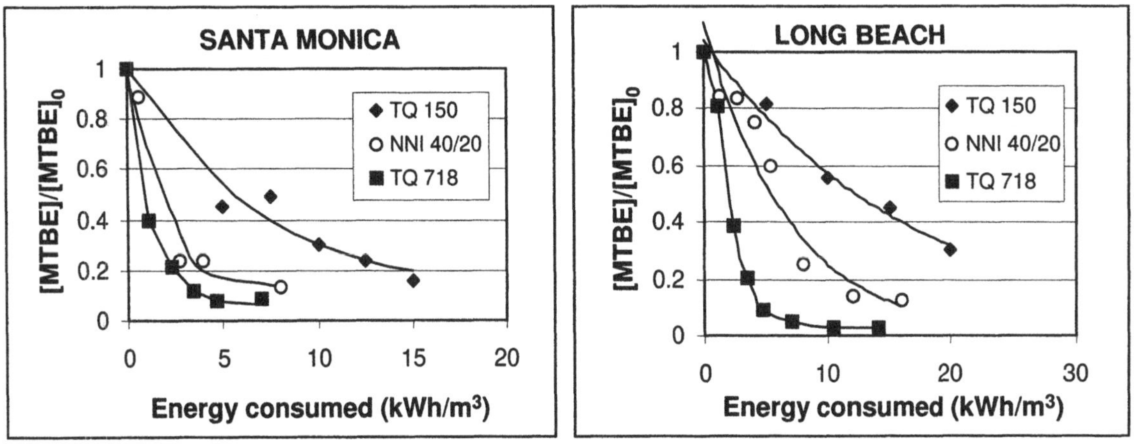

Figure 5.42 Removal of MTBE from Santa Monica, Long Beach, and Miami Dade field waters with different UV-lamps at an initial H_2O_2 concentration of 200 µM (NNI 40/20: LP-UV lamp)

- LP-UV/H_2O_2 treatment of the Suffolk County water formed fairly high amounts of acetone
- More formaldehyde was formed in the MP-.UV/H_2O_2 treated water samples

Also, the experiments conducted with the two waters containing UV-absorbing organic constituents (LB and MD) showed that the MP-UV treatment was more efficient in reducing the UV-absorbance of the water than the LP-UV radiation. These observations may be of some interest regarding the final disinfection since the production of disinfection by-products is known to be correlated with UV or SUVA, the higher the SUVA, the higher the formation of DBPs.

In the bench scale testing, three different UV-lamps have been used: one LP-UV lamp and two MP-UV lamps of different electrical power. To compare the efficiency of the different UV lamps on the removal of MTBE from the field water samples, the results are presented as the relative concentration of MTBE as a function of energy consumed by the UV-lamp. Figure 5.42 presents the results obtained on the removal of 200 µg/L MTBE from Santa Monica and Long Beach water as a function of energy consumed by the UV lamp used. The initial hydrogen peroxide concentration was 200 µM.

The appropriate figure of merit is electrical energy per order (E_{EO}), defined as the number of kilowatt hours of electrical energy required to reduce the concentration of a pollutant by 1 order of magnitude (90%) in 1 m^3 of contaminated water. The E_{EO} (kWh/m^3/order) can be calculated from the following equation for a batch type reactor:

$$E_{EO} = \frac{P \times t \times 1000}{V \times 60 \times \log(C_i / C_f)} \quad (5.4)$$

Where P is the power input (kW) from the wall to drive the UV-lamps, t is the irradiation time (min), V is the volume of the reactor, C_i and C_f are the initial and the final concentrations. Table 5.8 presents the electrical energy per order (E_{EO}) values for the degradation of MTBE in the different field waters.

Table 5.8

Electrical energy per order for the removal of MTBE from Santa Monica, Suffolk County, Miami Dade, and Long Beach water samples with LP- and MP-UV/H_2O_2 processes

[MTBE] (µg/L)	$[H_2O_2]_0$ (mg/L)	Other constituents	E_{EO} $Kwh/m^3/order$ Santa Monica	Suffolk County	Miami Dade	Long Beach
			LP-UV			
200	0	-	512	51.2	-	-
200	2	-	17.1	-	6.4	-
200	3.4	-	12.8	1.5	7.3	51.2
200	6.8	-	8.5	0.7	3.0	25.6
200	6.8	36 mg/L NO_3	8.5*	0.9	3.9	25.6
200	6.8	200 µg/L TBA	7.3*	1.0	3.2	17.1
50	3.4	-	8.5	-	-	-
2000	6.8	-	8.5*	-	-	
50		-	-	-	10.2	17.1
2000	68	-	-	-	0.7	1.8
			MP-UV (TQ 150)			
200	0	-	96	48	64	274
200	3.4	-	-	8.3	48	96
200	6.8	-	19.2	5.8	24	38
200	6.8	36 mg/L NO_3	21.3	8.0	64	96
200	6.8	200 µg/L TBA	-	7.1	21.3	38.4
2000	6.8	-	-	10.6	-	-
			MP-UV (TQ 718)			
200	0	-	12	-	-	-
200	2	-	9	-	-	-
200	3.4	-	7	-	-	-
200	6.8	-	3	-	2	4
200	3.4	36 mg/L NO_3	6	-	-	-
200	0	200 µg/L TBA	12	-	-	-
50	3.4	-	6	-	-	-
2000	3.4	-	6	-	-	-

* $[H_2O_2]_0 = 3.4$ mg/L

Based on the E_{EO} data presented in Table 5.8 it can be noted that MTBE is more efficiently removed with the Teflon shielded TQ 718 MP UV lamp than with the LP-UV lamp. The Teflon shielded TQ 718 MP- UV lamp is, however, only slightly more efficient when irradiating the Santa Monica and Miami Dade waters but much more efficient when applied to the water from Long Beach which is enriched in UV-absorbing organic moieties. One explanation may be that the MP-UV radiation decolorizes the water more efficiently than the LP-UV radiation and consequently the filter effect of the NOM decreases faster in the MP-UV treated Long Beach water and the removal of MTBE is more efficient with the MP-UV lamp than with the LP-UV lamp. It is, however, important to keep in mind that the calculated value of the energy consumed by the TQ 718 lamp is corrected by the fact that only 4% of the surface of the lamp was used. Both the MP-TQ 718 and the LP-UV lamp are more efficient in removing MTBE than the MP-TQ 150 lamp. One reason to the relatively slow degradation of MTBE with the MP-TQ 150 lamp may be the short arc length of the UV-lamp, i.e. the pathway for the photons to reach the MTBE molecules is longer than when using the two other lamps.

Our results show that the removal of MTBE from the water samples studied is improved with an increasing initial hydrogen peroxide concentration due to a larger production of hydroxyl radicals. In addition MP-UV/H_2O_2 treatment with the TQ 718 lamp removes MTBE faster than LP-UV/H_2O_2 treatment. The presence of nitrates or TBA did not significantly influence the decomposition rate of MTBE with LP-UV/H_2O_2 in Miami Dade and Long Beach field waters. The DOM and bicarbonates in the water samples retard the degradation of MTBE. The major intermediates identified were TBF followed by TBA and the major degradation by-products formed were acetone and formaldehyde. More formaldehyde was formed in water samples with a higher DOC content and in MP-UV/H_2O_2 treated water samples. Small amounts of acetaldehyde were also detected but no obvious trend in the evolution of acetaldehyde could be observed. However, no other low-molecular-weight aldehydes were detected.

PILOT SCALE TESTING

Introduction

The generation of hydroxyl radicals from the photo-dissociation of H_2O_2 is a well characterized process. The fundamental kinetic parameters describing this process are the quantum yield of the photo-dissociation reaction and the wavelength dependent extinction coefficient of H_2O_2. The rate of reaction is equal to the rate of UV absorption by H_2O_2 times the quantum yield. The rate of UV absorption by H_2O_2 depends on the emission from the UV source (i.e., lamp) and the competition for this incident UV with other constituents in the water. The efficiency of the overall process must combine these factors with the electrical efficiency of the UV lamp for the generation of UV photons and the effective pathlength of the photoreactor.

LP mercury lamps are characterized by relatively long lifetimes (e.g., 10,000 hours) and about 90% of the UV output at 254 nm with an electrical efficiency greater than 30%. MP mercury lamps have shorter lifetimes (e.g., 3000 – 5000 hours) and emit over a broad wavelength range but at only about 13% electrical efficiency in the UV range.

Most organic compounds, including MTBE, and H_2O_2 absorb low wavelength UV more strongly than higher wavelength UV. This characteristic is also true for most of the compounds that are present in the water matrix and compete for UV photons with H_2O_2 and MTBE (e.g., organics, nitrate). Therefore, for many 'real' water samples the absorption by the water matrix of the low wavelength UV photons generated by the medium pressure lamp is very significant. The absorption spectrum of the nitrate ion exemplifies this effect well. Nitrate absorbs strongly at wavelengths below 240 nm but almost not at all at 254 nm. This competition for these low wavelength photons combined with the lower electrical efficiency of the medium pressure lamp can result in a significant fraction of wasted energy.

The dependence of the process efficiency on reactor pathlength is relatively complex. This is especially true for medium pressure lamps due to the wavelength dependence of light absorption. In general, however, as the pathlength increases, more of the incident photons will be absorbed within the reactor before being lost through absorption at the reactor walls. This effect is especially true for the 254 nm emission line of the low pressure lamp. Since higher quality water has a relatively high transmittance at this wavelength, long pathlength reactors are required to capture the photons and thus make efficient use of the input energy. In addition, since a higher fraction of these photons are absorbed by H_2O_2, as opposed to the background matrix, the additional photons absorbed result in increased reaction rate and thus higher reactor efficiencies. The lower wavelengths emitted by the medium pressure lamps are absorbed much more strongly by the water matrix. Therefore, increasing reactor pathlength has a less significant impact on reactor performance at the lower wavelengths than at higher wavelengths.

The UV/peroxide pilot testing was performed on Santa Monica groundwater and Lake Huron surface water. This testing was conducted at Ontario Clean Water Agency's surface water treatment plant at Lake Huron in Grand Bend, Ontario, Canada. Approximately, 6,000 gallons of groundwater from Santa Monica was shipped to Grand Bend for this testing. This water was sent in a clean tanker, the tanker that was used for transporting bottled water. At Grand Bend, LP-UV and MP-UV testing was performed employing the reactors that are being provided by the Trojan Technologies, London, Ontario, Canada. Influent and effluent samples from this testing were sent to the same labs that were used by the other team members (MWD-SC and UC, Boulder) for consistency in analysis and data reporting. Detailed description of UV reactors, methods, results and discussion follows.

Experimental Plan

The UV/peroxide pilot tests were performed on Santa Monica groundwater and Lake Huron surface water. These tests were conducted at Ontario Clean Water Agency's surface water treatment plant at Lake Huron in Grand Bend, Ontario, Canada. Approximately, 6,000 gallons of groundwater from Santa Monica was shipped to Grand Bend for this testing. This water was sent in a clean tanker, the tanker that was used for transporting bottled water. At Grand Bend, LP-UV and MP-UV testing was performed employing the reactors that were provided by Trojan Technologies Inc., London, Ontario, Canada. Influent and effluent samples from this testing were sent to EPA certified laboratories for analysis.

The pilot tests were conducted using an orthogonal matrix of experiments. The baseline experimental conditions for this orthogonal matrix are:

- MTBE spike 1.2 μM or 100 μg/L
- Peroxide to MTBE 1 : 200, molar ratio
- pH Ambient (7.5)
- Nitrate Ambient (21.5 mg/L)

Table 5.9
MP-UV/peroxide pilot testing experimental matrix

Experiment #	Lamp	MTBE (μg/L)	MTBE: Peroxide	pH	Nitrate (mg/L)	TBA (μg/L)	Bromide (mg/L)
1 Baseline	MP	100	1 : 200	Ambient	Ambient	--	Ambient
2	MP	2,900	1 : 15	Ambient	Ambient	--	Ambient
3	MP	100	1 : 300	Ambient	Ambient	--	Ambient
4	MP	100	1 : 50	Ambient	Ambient	--	Ambient
5	MP	100	1 : 200	6.5	Ambient	--	Ambient
6	MP	100	1 : 200	8.5	Ambient	--	Ambient
7	MP	100	1 : 200	Ambient	36	--	Ambient
8	MP	100	1 : 200	Ambient	Ambient	200	Ambient
9	MP	100	1 : 200	Ambient	Ambient	Ambient	0.5
10	MP	100	1 : 200	Ambient	36	200	0.5
11	MP	2,900	1 : 15	Ambient	36	200	0.5
12	MP	100	--	Ambient	Ambient	--	Ambient
13	--	100	1 : 200	Ambient	Ambient	--	Ambient
14	MP	100	1 : 200	Adjusted to Ambient	Half of Ambient	--	Half of Ambient

- TBA Ambient (0 μg/L)
- Bromide Ambient (0.25 mg/L)

The MP-UV/peroxide pilot test matrix for is shown in Table 5.9. The proposed experimental matrix for MP-UV/peroxide evaluation had 14 experiments. Experiment 1 is the baseline experiment with water of ambient quality spiked with 100 μg/L of MTBE. In experiment 2, the MTBE will be spiked at an order of magnitude higher concentration (2,000 μg/L) than in the baseline water (100 μg/L). Experiments 3 and 4 will investigate the effect of varying peroxide dose on MTBE and its by-products oxidation. The MTBE to peroxide ratios that are being studied match those that were used in the bench-scale testing by the University of Poiters, France. Experiments 5 and 6 will investigate the effect of feed water pH at one unit higher than the ambient (7.5) and one unit lower than the ambient. In experiments 7, 8 and 9 the raw water will be respectively spiked with nitrate, TBA and bromide. Experiments 10 and 11 will investigate the extreme water quality conditions, i.e., with a higher concentration of radical scavengers and interfering compounds. In experiments 10 and 11 the feed water will be spiked simultaneously with nitrate, TBA and bromide to investigate the efficiency of UV/peroxide oxidation treatment under poor water quality conditions. The effect of applying MP-UV alone and peroxide alone will be evaluated in experiments 12 and 13. These experiments will establish the oxidation, if any, of MTBE and its by-products by UV photolysis and chemical oxidation from peroxide addition. In

Table 5.10
LP-UV/peroxide pilot testing experimental matrix

Experiment #	Lamp	MTBE (μg/L)	MTBE: Peroxide	pH	Nitrate (mg/L)	TBA (μg/L)	Bromide (mg/L)
15 Baseline	LP	100	1 : 200	Ambient	Ambient	--	Ambient
16	LP	2,900	1 : 15	Ambient	Ambient	--	Ambient
17	LP	100	1 : 300	Ambient	Ambient	--	Ambient
18	LP	100	1 : 50	Ambient	Ambient	--	Ambient
19	LP	100	1 : 200	6.5	Ambient	--	Ambient
20	LP	100	1 : 200	8.5	Ambient	--	Ambient
21	LP	100	1 : 200	Ambient	36	--	Ambient
22	LP	100	1 : 200	Ambient	Ambient	200	Ambient
23	LP	100	1 : 200	Ambient	Ambient	Ambient	0.5
24	LP	100	1 : 200	Ambient	36	200	0.5
25	LP	2,900	1 : 15	Ambient	36	200	0.5
26	LP	100	--	Ambient	Ambient	--	Ambient
27	--	100	1 : 200	Ambient	Ambient	--	Ambient
28	LP	100	1 : 200	Adjusted to Ambient	Half of Ambient	--	Half of Ambient

experiment 14, groundwater will be mixed with pure water (with minimal or no interfering compounds) in one to one volume ratio to yield a water with half the ambient concentrations for nitrate, bromide and alkalinity. Summarized in Table 5.10 is the experimental matrix for the LP-UV/peroxide system. Experiments 15 through 28 are structured similar to 1 through 14, except, in these experiments the performance of LP lamp oxidation will be evaluated. Dose-response curves for MTBE destruction and influence of water quality on MTBE oxidation was accessed using the data from these experiments.

Electrical energy dose-response curves for MTBE destruction and influence of water quality on MTBE oxidation was accessed using the data from these experiments. The average water quality of Santa Monica groundwater and Lake Huron surface water are summarized in Table 5.11.

Materials and Methods

MP-UV and LP-UV Pilot Reactors

The pilot UV reactors used in this study are commercial reactors. The MP lamp system (UV Swift 2L12, Trojan Technologies) had two medium pressure lamps that were arranged perpendicular to the flow. These MP lamps were rated for 2.8 kW power per lamp driven from ballast. The MP lamps were operated at 30% of the full power, i.e., 0.84 kW per lamp. The reactor volume of the MP-UV, Swift 2L12 is 56 liters (L). A picture of the MP lamp system used in this

Table 5.11
Santa Monica groundwater and Lake Huron surface water quality

Parameter	Santa Monica Water	Lake Huron Water
Alkalinity, mg/L as $CaCO_3$	325	82.5
Dissolved organic carbon, mg/L	0.45	1.4
Hardness, mg/L	540	95.0
pH	8.3	8.3
Total dissolved solids, mg/L	750	140
Turbidity, NTUs	0.14	0.91
Nitrate as NO_3, mg/L	21.5	2.4

study is shown in Figure 5.43. The LP lamp system (UV Logic 3AS20, Trojan) had three low-pressure lamps arranged parallel to the flow direction. The LP lamps were rated for 0.16 kW per lamp driven from ballast. During pilot testing, the LP lamps were operated at full power. The reactor volume of LP-UV Logic 3AS20 is 38 L. Picture of the LP lamp system is shown in Figure 5.44. Shown in Figures 5.45 and 5.46 are the irradiance versus wavelength and the intensity versus wavelength profiles for the LP and MP lamps, respectively.

MP-UV and LP-UV Test Methods

Both the MP and LP UV reactors were fed water from a storage tank of approximately 500-gallon capacity. The feed water was re-circulated through the UV reactors to achieve the conditions of a completely-stir tank reactor. The recirculation mode of testing simulates a single-pass system through a series of UV reactors, as in full-scale facilities. Batch mode, closed loop recirculation tests were performed with 900 liters (or 238 gallons) of water per test. The UV/peroxide tests were performed by re-circulating the feed water spiked with MTBE and other constituents (like nitrate in some experiments) at 2,650 liters/minute or 700 gallons/minute to yield a retention time of 1 second or less per cycle. Schematic of UV/peroxide pilot testing is shown in Figure 5.47. Since one storage tank was used to supply water to both the MP and LP units, these tests were staggered, i.e., MP test was followed by an LP test and so on.

Each test was run for 40 minutes to achieve the necessary MTBE destruction. For each test, influent and effluent samples were collected for MTBE and its by-products analysis. Sampling for MTBE would involve collecting water in triplicates of 2 mL vials. These vials were frozen to prevent hydrolyses of TBF. The frozen vials were shipped in coolers to the Oregon Graduate Institute (OGI) for MTBE and its by-products analysis. Some select samples were sent to the Montgomery Watson laboratories for aldehydes, acetone, nitrate/nitrite and bromate analysis.

In MP and LP-UV testing, control runs were done with lamps being off. No significant MTBE removal was observed in these experiments indicating that the test system used kept MTBE losses by volatilization to a minimum. Since 900 L of water was used in each experiment, the temperature change from lamp and pump heating was also minimal (3-4 °C) over the duration of an experiment.

Figure 5.43 Picture of the MP lamp system (UV Swift, Trojan)

Figure 5.44 Picture of the LP lamp system (UV Logic, Trojan)

Theory

The kinetic model describing the rate of oxidation of a micropollutant by UV/H_2O_2 is taken from the work of previous researchers, developed for the monochromatic irradiation at 253.7 nm. The validity of this model is dependent upon the following assumptions and simplifications:

- The decomposition of MTBE is only due to the reaction with H_2O_2 derived hydroxyl radicals. The possible reactions between MTBE and other species have been assumed to be negligible.
- The concentration of hydrogen peroxide is assumed to be constant during the course of the reaction.
- The UV absorbance of the water is assumed to be constant during the course of the reaction.

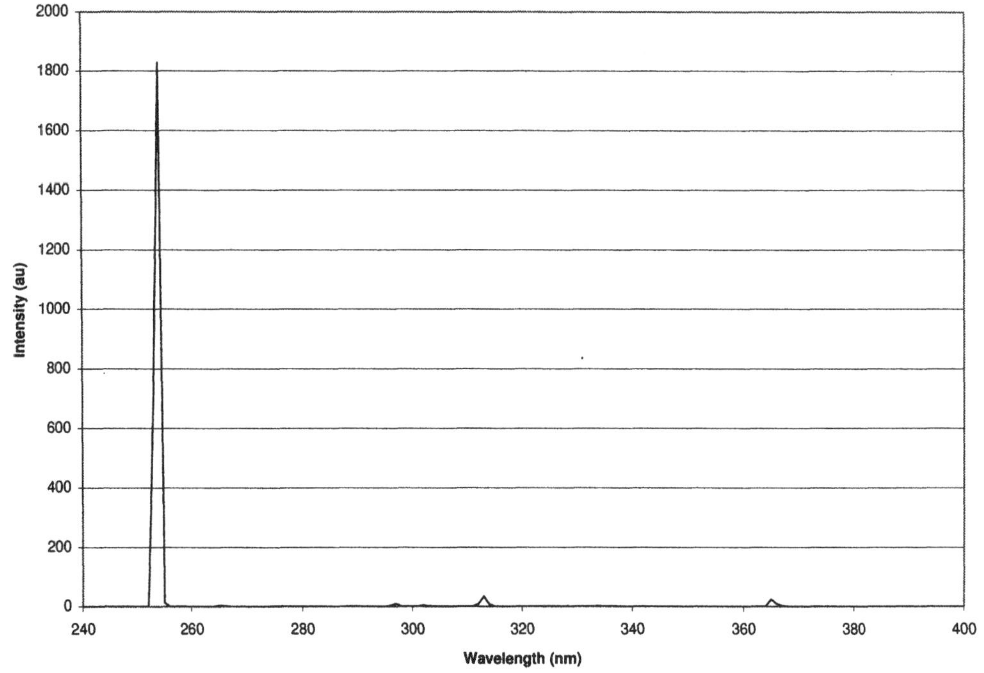

Figure 5.45 Intensity versus wavelength for the LP-UV lamp used in pilot testing

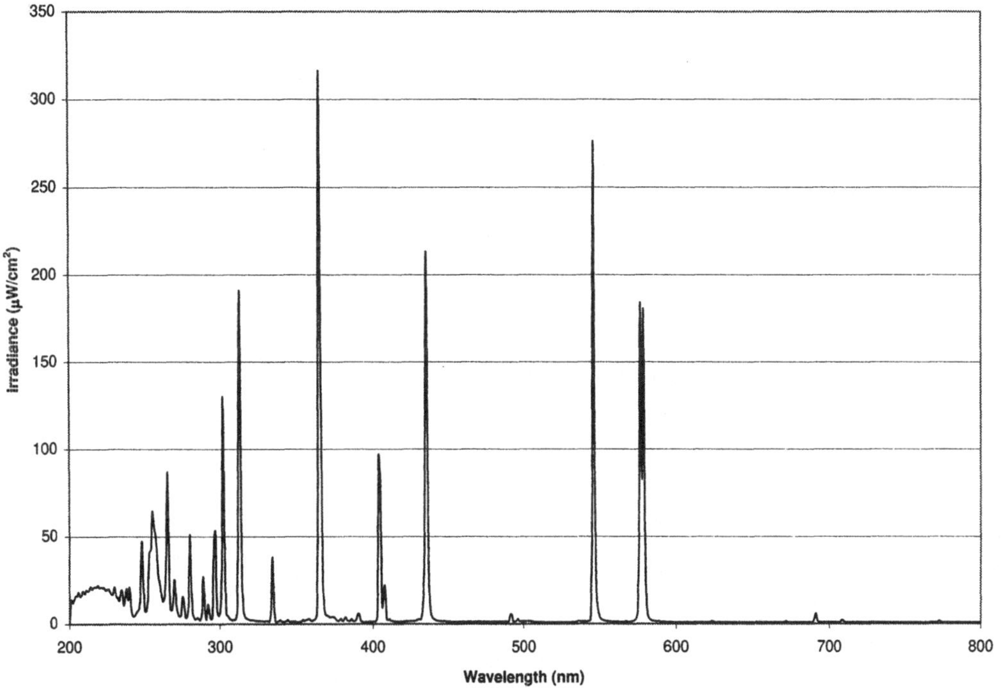

Figure 5.46 Irradiance versus wavelength for the MP-UV lamp used in pilot testing

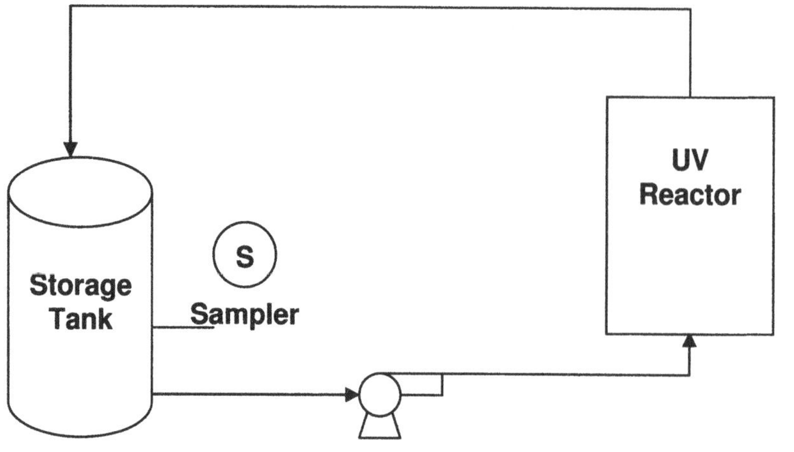

Figure 5.47 Schematic of UV reactor operation

The overall kinetic equation describing the photo-oxidative reactions of MTBE is:

$$\frac{-d[MTBE]}{dt} = k_{MTBE}[MTBE][\bullet OH]_{SS} = k_{MTBE}\left[MTBE\right]\frac{2\phi_{H_2O_2}N_{0\lambda}F_{s\lambda}F_{H_2O_2\lambda}}{V\left(k_{H_2O_2}\left[H_2O_2\right]+\sum k_s\left[S\right]\right)} \quad (5.5)$$

where:

k_{MTBE} is the rate constant for the reaction of MTBE with hydroxyl radicals
[MTBE] is the concentration of MTBE
$[\cdot OH]_{SS}$ is the steady state concentration of hydroxyl radicals
ϕ_{H2O2} is the quantum yield for photolysis of peroxide
$N_{0\lambda}$ is the incident photon flux
V is the reactor volume and
$\Sigma k_s[S]$ is the kinetic competition term due to all hydroxyl radical scavengers.

and $F_{S\lambda}$ and $F_{H2O2\lambda}$ are the fraction of incident light absorbed by the system (i.e., reactor fluid) and the fraction of that absorbed light that is absorbed by hydrogen peroxide respectively. The equations describing $F_{S\lambda}$ and $F_{H2O2\lambda}$ are:

$$F_{S\lambda} = \{1-10^{-Al}\} \quad (5.6)$$

and

$$F_{H_2O_2\lambda} = \frac{\varepsilon_\lambda[H_2O_2]}{A_\lambda} \quad (5.7)$$

where,

A is the water absorbance at the given wavelength (λ)
l is the optical path length of the reactor and
ε is molar absorption coefficient for the compound (peroxide).

The overall kinetic equation describing the photolytic and photo-oxidative reaction of MTBE is therefore:

$$-\frac{d[MTBE]}{dt}=\frac{2k_{MTBE}\phi_{H_2O_2}N_0(1-10^{-AL})\varepsilon_{H_2O_2}[H_2O_2][MTBE]}{VA\left(k_{H_2O_2}[H_2O_2]+\sum k_s[S]\right)} \quad (5.8)$$

and since

$$\frac{-d[MTBE]}{dt}=k[MTBE] \quad (5.9)$$

describes the pseudo first-order disappearance of MTBE, then:

$$k=\frac{k_{MTBE}\phi_{H_2O_2}N_0(1-10^{-AL})\varepsilon_{H_2O_2}[H_2O_2]}{VA\left(k_{H_2O_2}[H_2O_2]+\sum k_s[S]\right)}\text{, s-1} \quad (5.10)$$

and the "water factor" describing the average intensity is:

$$I_{avg.}=\frac{N_0\left(1-10^{-AL}\right)}{Area*A*L*Ln(10)} \quad (5.11)$$

therefore:

$$\frac{k_{MTBE}}{I_{avg.}}=k_{MTBE}'=\frac{Ln(10)\phi_{H_2O_2}\varepsilon_{H_2O_2\lambda}k_{MTBE}[H_2O_2]}{U_\lambda\left(k_{H_2O_2}[H_2O_2]+\sum k_s[S]\right)}\text{, cm2/mJ} \quad (5.12)$$

where,

U_λ is the energy (J) of one Einstein of photons of wavelength λ. (5.13)

The derivation of this fluence (or UV dose) based rate constant parallels the derivation presented by Cater, et al. (2000) for direct photolysis applications. This rate constant in units of inverse dose can be determined from Collimated Beam experiments by directly measuring $k(s^{-1})$ and dividing by the average irradiance (mW/cm^2) in the irradiated sample to arrive at k' (cm^2/mJ). This rate constant can be related to the electrical energy per order (E_{EO}) parameter for a given reactor by the following (Stefan 2000):

$$EE/O=\frac{P(kW)1000}{Q(gpm)60Log\left(C_0/C\right)} \quad (5.14)$$

where,

P is the power inputted (kW) from the wall to drive the UV-lamps
Q is the water flow rate (in gallons per minute) through the reactor and
C_o and C are the initial and the final concentrations of the contaminant (e.g., MTBE) under consideration.

and from the first order kinetics in the UV reactor:

$$Ln\left(C_0/C\right) = k'D \tag{5.15}$$

where, D is the UV dose and

$$Log\left(C_0/C\right) = \frac{Ln\left(C_0/C\right)}{Ln(10)} = \frac{k'D}{Ln(10)} \tag{5.16}$$

therefore:

$$EE/O = \frac{P(kW)1000Ln(10)}{Q(gpm)60k'D} = \frac{P(kW)38.38}{Q(gpm)k'D} \tag{5.17}$$

and since

$$D = I_{avg}\theta \text{ and } \theta = V_R/Q \tag{5.18}$$

where,

θ is the exposure time and
V_R is the reactor volume

then:

$$EE/O = \frac{0.6396P(kW)}{V_R(gal)k'\left(cm^2/mJ\right)I_{avg.}\left(mW/cm^2\right)} \tag{5.19}$$

The rate constant k' was calculated from the pilot data (i.e., E_{EO}) and the known reactor characteristics (i.e., $I_{avg.}$, V_R, and P). Having calculated the rate constant, the only unknown parameter in the equation for the rate constant is the scavenging term $\Sigma k_S[S]$. The other parameters (e.g., k_{OH}, $\varepsilon_{H2O2,\ \lambda}$, ϕ_{H2O2}, A_λ) are either easily measured or are available in the literature. Having thus determined the scavenging capacity of the water, the dependence of the rate constant, k', can be calculated for a range of hydrogen peroxide concentrations. The E_{EO} can then be calculated for any reactor with these rate constants provided that the average fluence rate and reactor volumes are known. It has to be mentioned that in the case of MP-UV applications the rate constant k' and average fluence rate I_{avg} are wavelength-dependent, through both the fraction of light absorbed by hydrogen peroxide and the molar absorption coefficient ε_{H2O2}.

Results and Discussion

This section presents a summary of the results and discussions from MP-UV/peroxide and LP-UV/peroxide testing.

Effect of Feed Water Quality on E_{EO}'s

Experiments were conducted to assess the impact of feed water quality on MP and LP-UV/peroxide processes using the following water types:

- Lake Huron and Santa Monica waters spiked with MTBE alone at 1.2 μM and this was defined as the "baseline condition"
- Lake Huron and Santa Monica waters spiked with MTBE (at 1.2 μM), TBA (at 4.0 μM), nitrate (at 36 mg/L as N) and bromide (at 0.5 mg/L) to simulate "poor water quality condition"

The E_{EO} (kWh/m^3-order) values measured for MTBE removal from Santa Monica groundwater and Lake Huron surface water are shown in Figure 5.48. The E_{EO}'s for MTBE removal in Santa Monica water spiked with 1.2 μM (or 100 μg/L) of MTBE and 200 μM (or 7 mg/L) hydrogen peroxide by MP-UV/peroxide and LP-UV/peroxidc treatment were respectively 2.20 kWh/m^3-order and 1.44 kWh/m^3-order. In tests conducted with Santa Monica water spiked with MTBE (~1.2 μM), H_2O_2 (~200 μM), TBA (~4.0 μM), nitrate (~36 mg/L as N) and bromide (~0.5 mg/L) to simulate poor quality groundwater, the E_{EO} values for MP-UV/peroxide and LP-UV/peroxide treatment were 2.58 kWh/m^3-order and 0.96 kWh/m^3-order, respectively. As indicated by the data, the LP-UV/peroxide treatment had lower E_{EO} values than the MP-UV/peroxide treatment for both the baseline and poor quality water conditions.

As shown in Figure 5.48, the E_{EO} values measured for the Lake Huron surface water were lower than those measured for Santa Monica groundwater. Some algae growth was observed in the Santa Monica groundwater that was transported from California and stored at Grand Bend, Ontario (test site). This algae growth may have resulted in an increase in the TOC content and the associated hydroxyl radical scavenging capacity of the water. The lower E_{EO} values for Lake Huron water is attributed primarily to the lower scavenging capacity of the surface water (no algae) as well as to the lower UV absorption coefficients. The E_{EO} values measured in this study for MP-UV/peroxide treatment were within the range of E_{EO}'s reported by Cater et al., 2000. The earlier studies were performed using batch reactors on Toronto tap water (Cater et al., 2000).

Radical Scavenging Capacities

The radical scavenging capacities were calculated using the approach described in the theory section. The scavenging capacity for Santa Monica water spiked with ~200 μM H_2O_2 and 1.2 μM MTBE was calculated to be $1.4 \times 10^5\ s^{-1}$, which is about 3 times the value of $4.72 \times 10^4\ s^{-1}$ that was calculated for ambient Santa Monica water. The effect of peroxide concentration on E_{EO}'s for LP-UV and MP-UV treatment at varying scavenging capacities are shown respectively

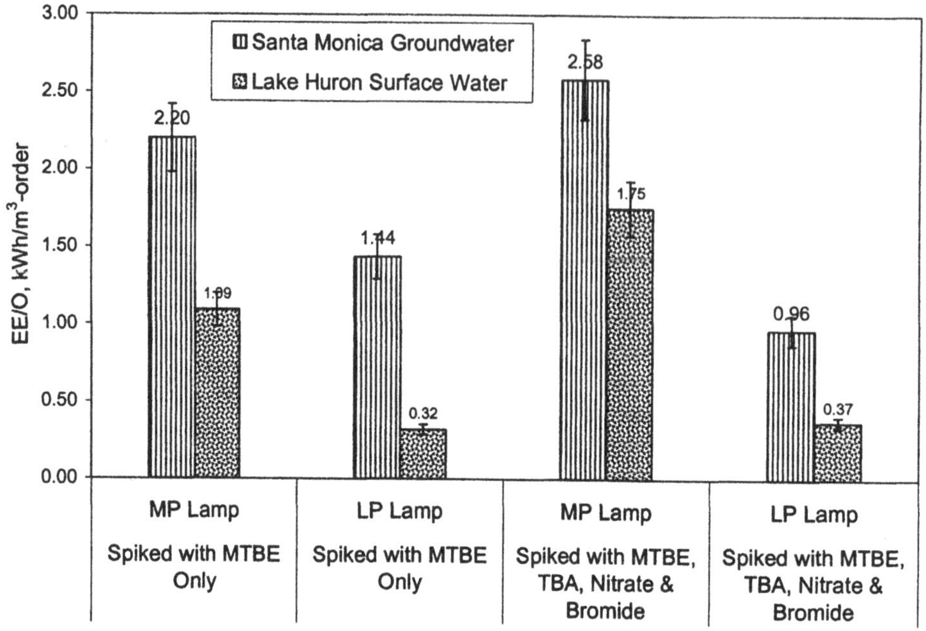

Figure 5.48 E_{EO}'s for MP-UV/peroxide and LP-UV/peroxide treatment of Santa Monica and Lake Huron waters

in Figures 5.49 and 5.50. The scavenging capacity for the Lake Huron surface water was calculated as 3.64×10^4 s^{-1} based on a comprehensive water analysis. This value is equal to 77% of that calculated for Santa Monica groundwater (no algae) and, therefore, the LP-UV or MP-UV results for algae-free Santa Monica water would be similar to those for Lake Huron surface water.

It is worthy to note the significant impact on the E_{EO} of the hydroxyl radical scavenging capacity of the water. The high alkalinity of the Santa Monica water results in a relatively high scavenging capacity of the water. Other natural waters with lower levels of scavenging species will have E_{EO}s that are correspondingly lower.

An important parameter that determines the efficiency of a photoreactor is its optical pathlength. Essentially, this is the average distance through which a UV photon may travel before it impacts an absorbing surface such as a reactor wall or other absorbing internal surface. For low absorbing waters, the longer the optical pathlength, the greater the opportunity for the photons to be absorbed by the target compound in the water (e.g., H_2O_2). When the pathlength is short much of the electrical energy powering the photoreactor is lost when the photons are absorbed by a surface. For low absorbing waters, the EEO varies inversely with the optical pathlength of the photoreactor. For highly absorbing waters, the radiation can be absorbed within a distance much shorter than either the physical or average path length. In such cases, the plot of EEO versus pathlength reaches a plateau at longer pathlengths. For this reason, large, full-scale photoreactors with longer pathlengths have lower EEO values than smaller, pilot photoreactors. This is especially true for LP-UV lamps whose monochromatic 254 nm emission travels further through the water (i.e., higher UV transmittance) than the shorter polychromatic wavelengths emitted by MP-UV lamps.

Although, the pilot reactors used in this study are commercial scale, they are much smaller than those that would be recommended for flows of 1 million gallons/day or more. Therefore, the

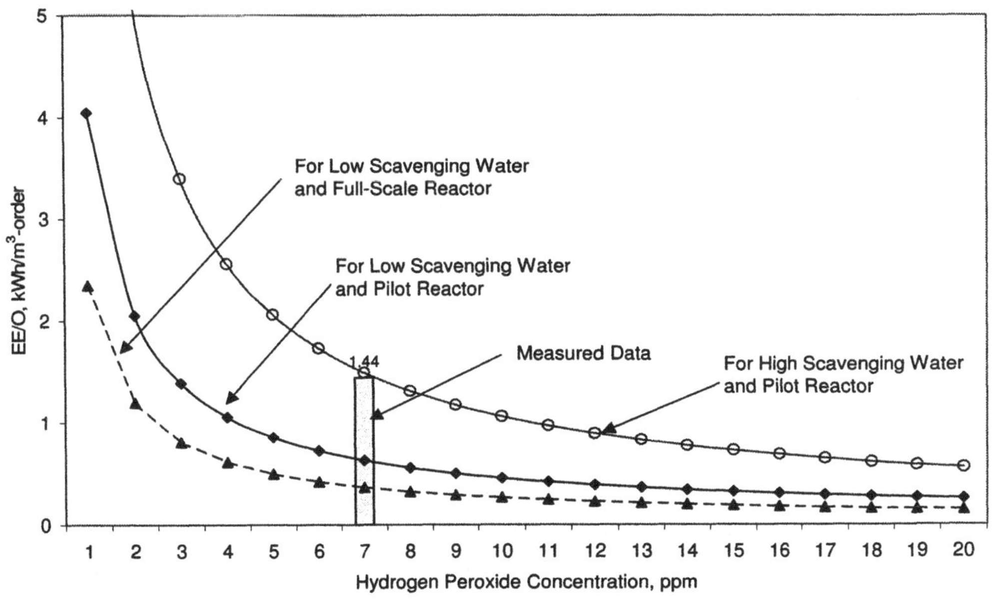

Figure 5.49 E_{EO}'s for MP-UV/peroxide treatment of MTBE plotted as a function of initial hydrogen peroxide concentration for low/high scavenging waters and pilot/full-scale reactors

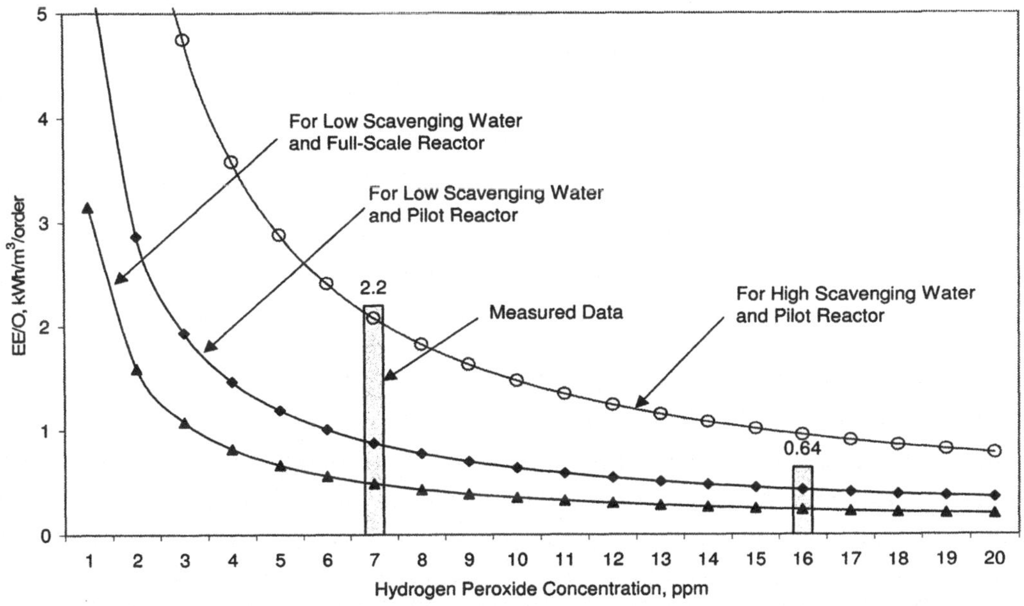

Figure 5.50 E_{EO}'s for MP-UV/peroxide treatment of MTBE plotted as a function of initial hydrogen peroxide concentration for low/high scavenging waters and pilot/full-scale reactors

E_{EO} values for full-scale photoreactors were projected to be lower than the E_{EO}'s for pilot reactors (Figures 5.49 and 5.50). These projected E_{EO} values for full-scale reactors were used as the basis for determining the treatment costs. Nevertheless, it should be noted that alternate full-scale reactor designs could result in still lower E_{EO}'s. The E_{EO} values for pilot scale reactors versus full-scale reactors can be up to 8-10 greater due to inefficiencies seen at smaller scales. This means that if a system were to be designed using the E_{EO} values from the pilot test data without adjusting for computational fluid dynamics, optical, photolysis and oxidation models, it would be oversized by a factor of 8 to 10.

UV Transmittances

Shown in Figure 5.51 are the transmittance scans for Santa Monica feed waters. As expected, higher transmittances were observed for baseline conditions (shown by dark lines) compared to the poor quality conditions water (shown by light lines). The transmittance scans on samples taken after ~2 min of UV exposure time are shown in Figure 5.52. No significant difference was observed in the transmittance data between the MP-UV and LP-UV treated waters. The baseline waters which initially had fewer light absorbing contaminants also had higher transmittances after LP or MP-UV/peroxide treatment compared to the poor quality waters (Figure 5.52).

MTBE and TBA Removal As a Function of Electrical Energy Dose

Shown in Figures 5.53 and 5.54 are the MTBE concentrations as a function of applied electrical energy dose (kWh/m^3) for MP-UV/peroxide and LP-UV/peroxide treatment, respectively. The MTBE removal was somewhat hindered in poor quality water compared to the baseline water for both MP-UV/peroxide and LP-UV/peroxide treatment processes. Approximately 2 min of exposure time resulted in applied electrical energy dosages of 1.25 kWh/m^3 and 0.35 kWh/m^3 respectively for the MP-UV and LP-UV, as indicated by the X-axis coordinates of the last data points (Figures 5.53 and 5.54). This difference in the electrical energy inputted was due to the difference in the LP-UV and MP-UV reactor set-ups. The LP-UV reactor had 3 lamps, each of 0.16 kW, operated at 100% power. In contrast, the MP-UV reactor had 2 lamps, each of 2.8 kW, operated at 30% power. The measured input power for the LP-UV and MP-UV reactors were 0.47 kW and 1.68 kW, respectively.

The TBA concentrations (in µM) as a function of applied electrical energy dose for MP-UV and LP-UV systems are shown in Figures 5.55 and 5.56, respectively. In MP-UV/peroxide testing with baseline water, the concentrations of TBA gradually increased with increasing electrical energy dose and this was possibly due to conversion of MTBE to TBA. In MP-UV/peroxide testing with poor quality water (data plotted on secondary axis), formation of TBA resulted in increased levels initially, followed by gradual removal with increasing exposure times (Figure 5.55). Formation of TBA was also observed in LP-UV/peroxide treatment (Figure 5.56). The possible accumulation of TBA in the MP-UV and LP-UV systems indicates the need for longer UV exposure times to enable complete oxidation of TBA. The TBA concentrations increased with increasing LP-UV dose for baseline water. The TBA levels also increased in MP-UV treatment of baseline water at applied electrical energy doses equivalent to the LP-UV process, as shown by the data in Figure 5.56.

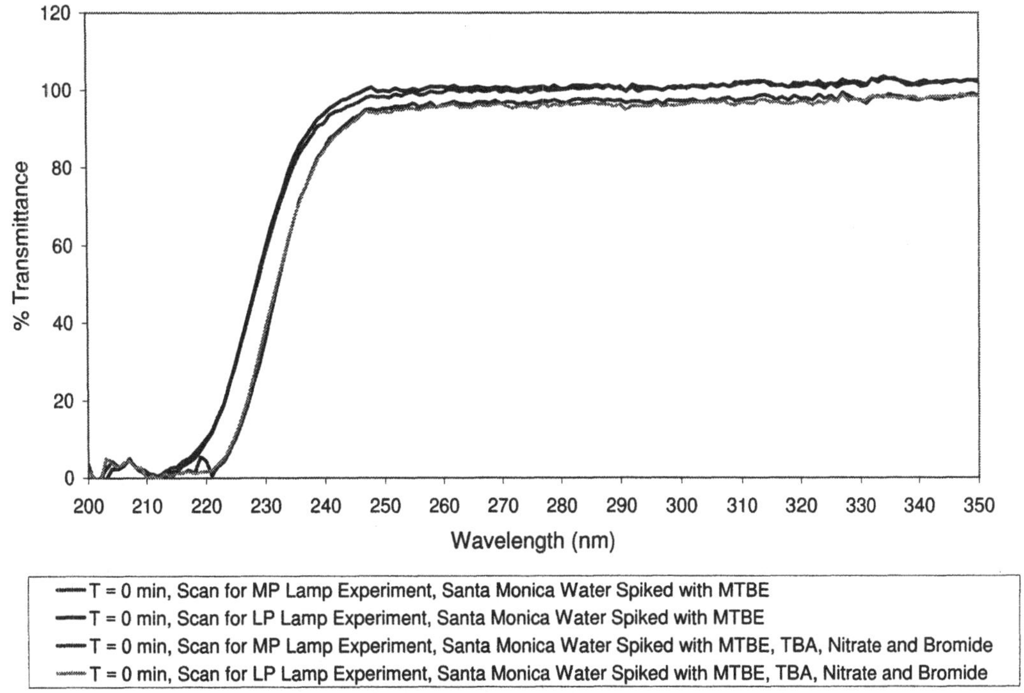

Figure 5.51 Percent UV transmittance of Santa Monica baseline and poor quality feed waters (after spiking), prior to LP-UV/peroxide and MP-UV/peroxide treatment

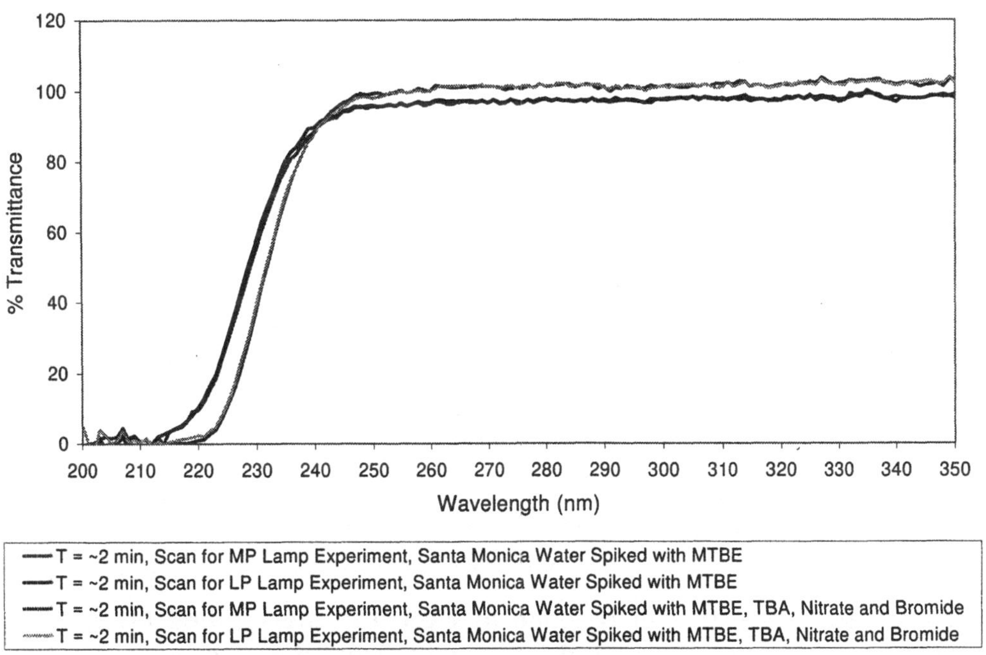

Figure 5.52 Percent UV transmittance for Santa Monica baseline and poor quality treated waters after ~ 2 minutes of LP-UV and MP-UV exposure

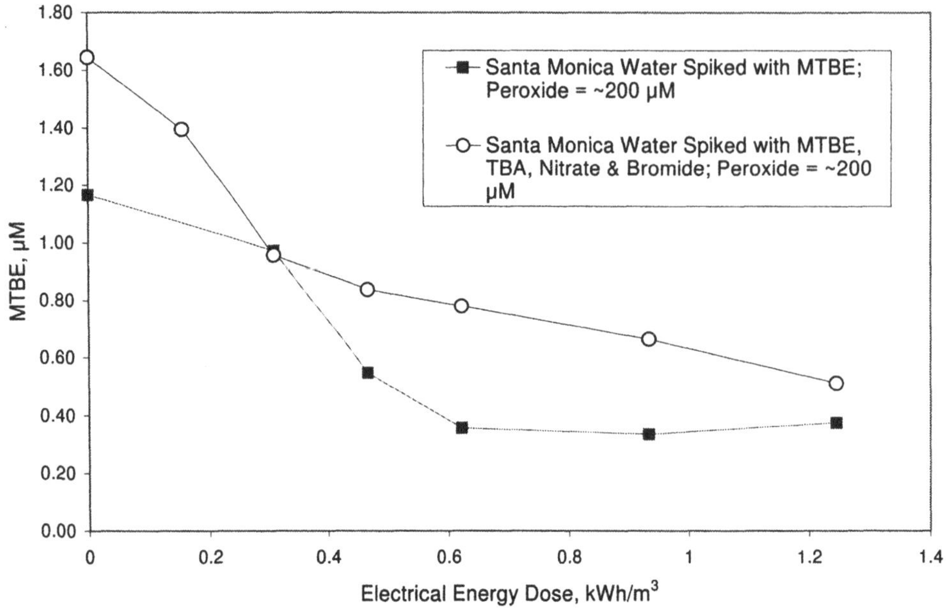

Figure 5.53 MTBE removal as a function of electrical energy dose for MP-UV/peroxide treatment of Santa Monica baseline and poor quality conditions water

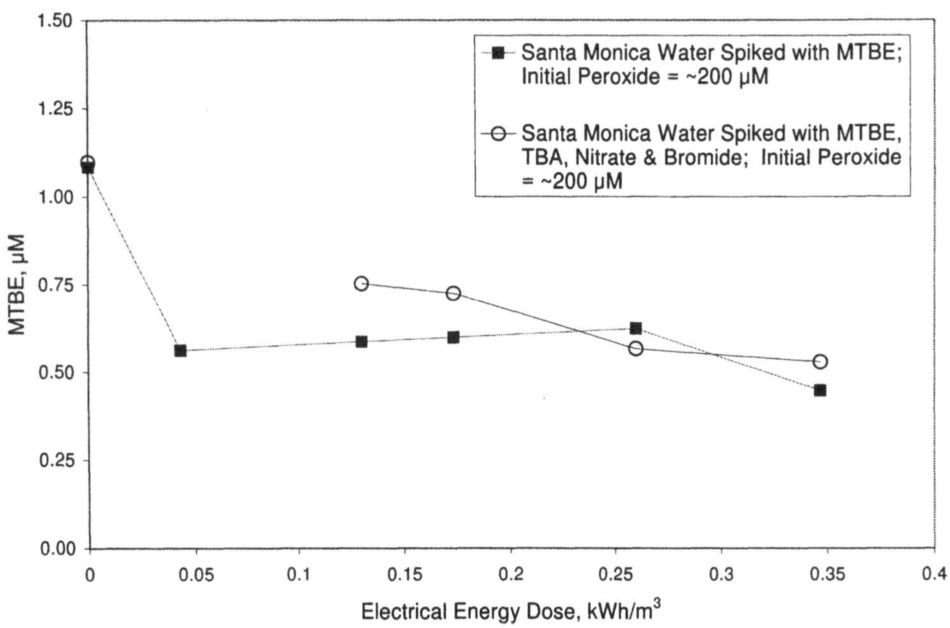

Figure 5.54 MTBE removal as a function of electrical energy dose for LP-UV/peroxide treatment of Santa Monica baseline and poor quality conditions water

Effect of UV Lamp on By-Products Formation

Shown in Figure 5.57 are the end products after exposure to LP-UV or MP-UV for 2 minutes. These tests were conducted on Santa Monica water spiked with 1.2 μM of MTBE. Higher concentrations of acetone and formaldehyde in waters treated with MP-UV/peroxide indicate possible oxidation of intermediates to these by-products. The data also indicates that for the amount of electrical energy applied in the MP-UV/peroxide process (~1.25 kW), which is 3.5 times that of the energy applied in LP-UV/peroxide treatment (~0.35 kW), the oxidation reactions may have progressed farther down the reaction pathway. Polishing treatment with biological activated carbon may be required to remove excess peroxide and end products such as acetone and formaldehyde.

Carbon Balances on Oxidation By-Products

In all the experiments, by-products were measured as a function of the applied electrical energy. Some of the most significant by-products measured for LP-UV/peroxide treatment of Santa Monica water spiked with 1.2 μM of MTBE are shown in Figure 5.58. In addition to the by-products that are shown in this figure, trace amounts (<10 μg/L) of glyoxals, M-glyoxals, butanal and propanal were also observed in some experiments. TBF was not detected in any of the pilot-scale experiments and this may be due to the shortcomings in the analytical methods employed. MTBE and its by-products (other than aldehydes) analysis was performed using direct aqueous injection method and purge and trap methods. Progress of oxidation resulted in the formation of formaldehyde and acetone (Figure 5.58). The carbon concentrations in MTBE and its by-products for the values shown in Figure 5.58 were calculated and plotted in Figure 5.59. The "total MTBE derived carbon" is obtained by adding the individual carbon concentrations in MTBE and its by-products. It should be noted that this is not a measured value and does not isolate possible conversion of natural organic matter in the water to by-products like aldehydes or acetone. No significant decline was observed in the total MTBE derived carbon concentration indicating partial oxidation.

MTBE and the by-products concentrations as a function of electrical energy dose for MP-UV/peroxide treatment of Santa Monica water spiked with 1.2 μM of MTBE is shown in Figure 5.60. More acetone formation was observed in MP-UV/peroxide treatment than LP-UV/ peroxide treatment (comparing acetone data in Figures 5.58 and 5.60). But, the amount of electrical energy expensed in the MP-UV process is more (about 3.5 times) than that inputted in LP-UV process. Similar to the LP-UV/peroxide treatment, no significant decline in MTBE derived carbon was observed in MP-UV/peroxide treatment (Figure 5.61). To completely oxidize MTBE and its by-products to carbon-di-oxide and water, more UV exposure or electrical energy dose and/or higher dose of hydrogen peroxide is required.

Nitrate Conversion to Nitrite

Presence of nitrate in high concentrations can interfere with UV/peroxide oxidation. As indicated earlier, nitrate absorbs UV light strongly in the wavelengths between 200-230 nm (UV-C region). Shown in Figure 5.62 are the nitrate concentrations before and after oxidation. The ambient Santa Monica water had nitrate levels of 21.5 mg-NO_3/L. In the poor quality water

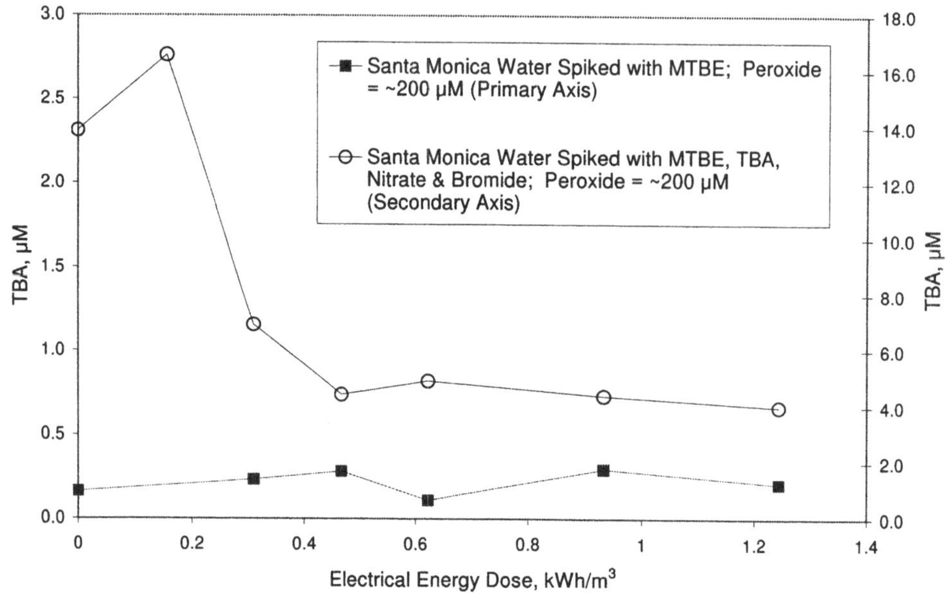

Figure 5.55 TBA profile as a function of electrical energy dose for MP-UV/peroxide treatment of Santa Monica baseline and poor quality conditions water

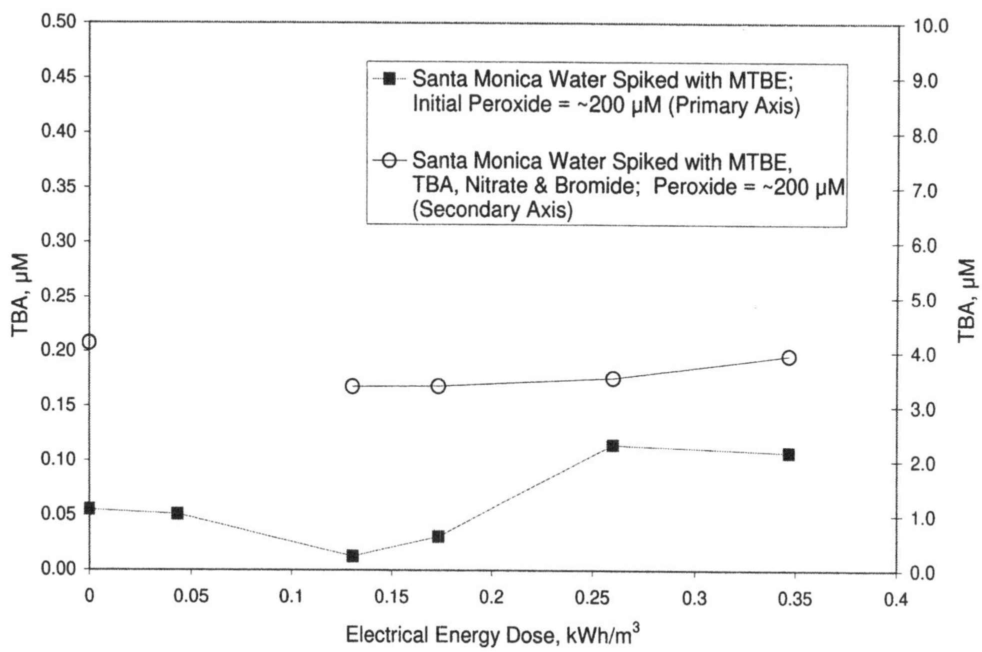

Figure 5.56 TBA profile as a function of electrical energy dose for LP-UV/peroxide treatment of Santa Monica baseline and poor quality conditions water

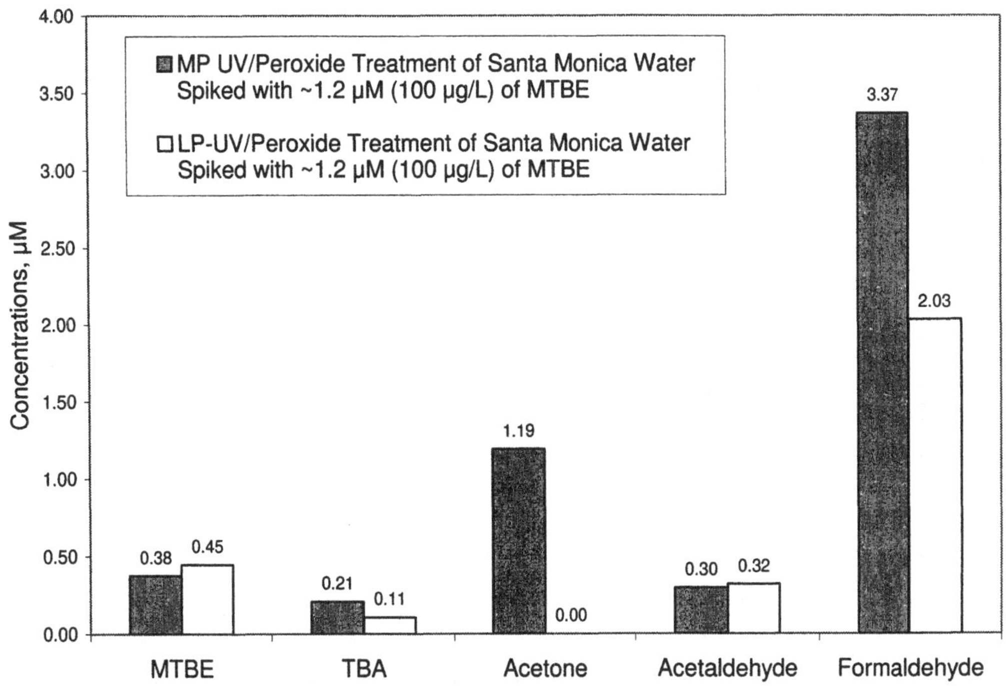

Figure 5.57 End products after ~2 minutes of LP-UV and MP-UV/peroxide treatment

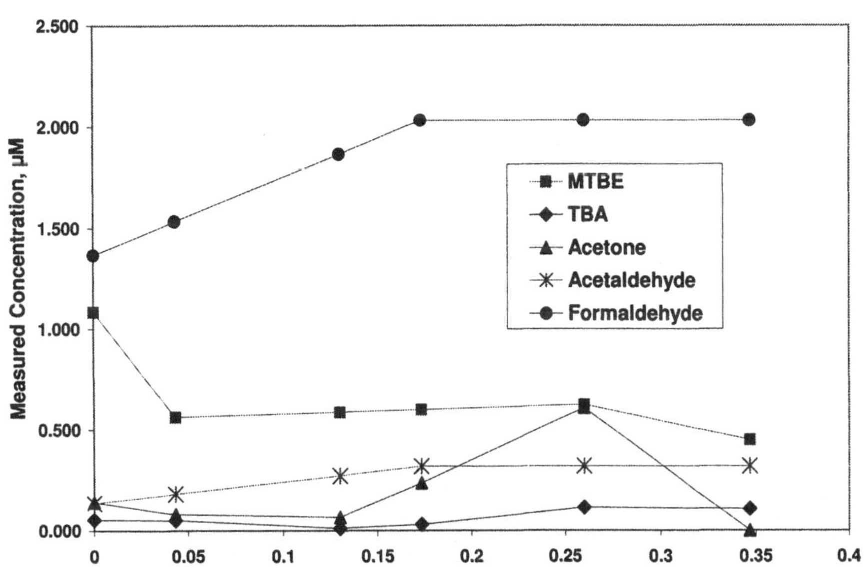

Figure 5.58 Energy-series of profiles of MTBE and its by-products for LP-UV/peroxide treatment of Santa Monica water spiked with 1.2 µM of initial MTBE (baseline conditions)

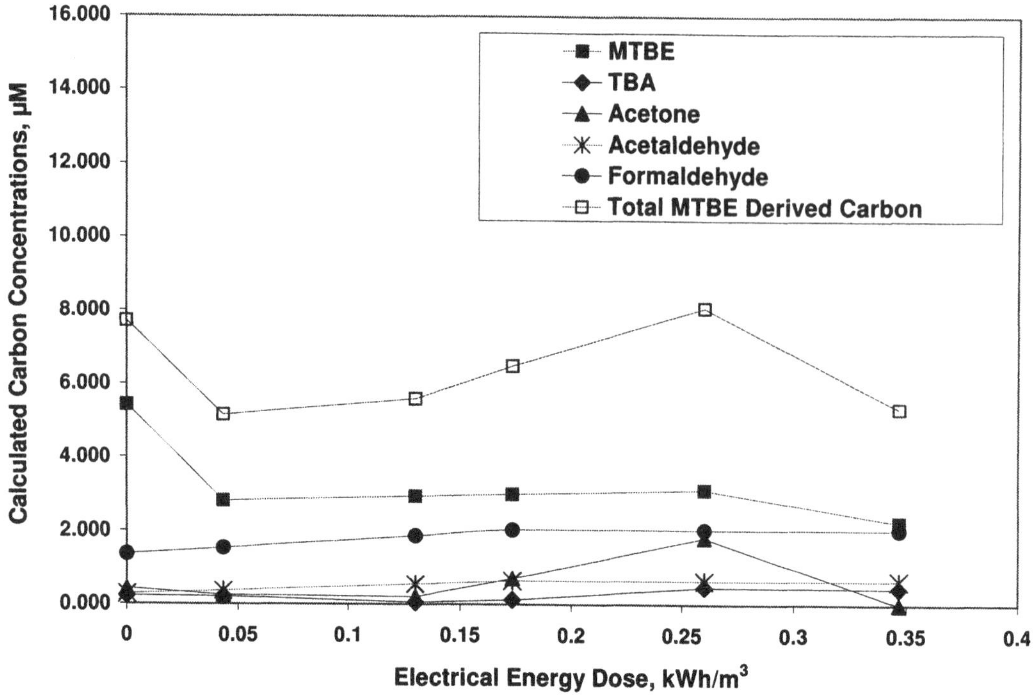

Figure 5.59 Calculated (organic) carbon profiles of MTBE and its by-products during the LP-UV/peroxide treatment of Santa Monica water spiked with 1.2 µM of initial MTBE (baseline conditions)

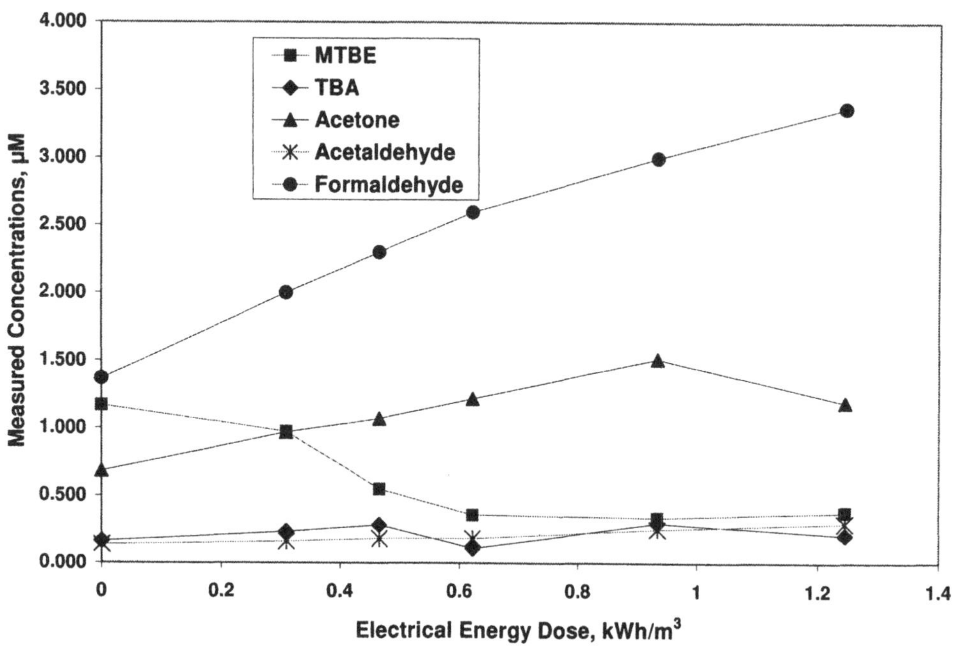

Figure 5.60 Energy-series profiles of MTBE and its by-products for MP-UV/peroxide treatment of Santa Monica water spiked with 1.2 µM of initial MTBE (baseline conditions)

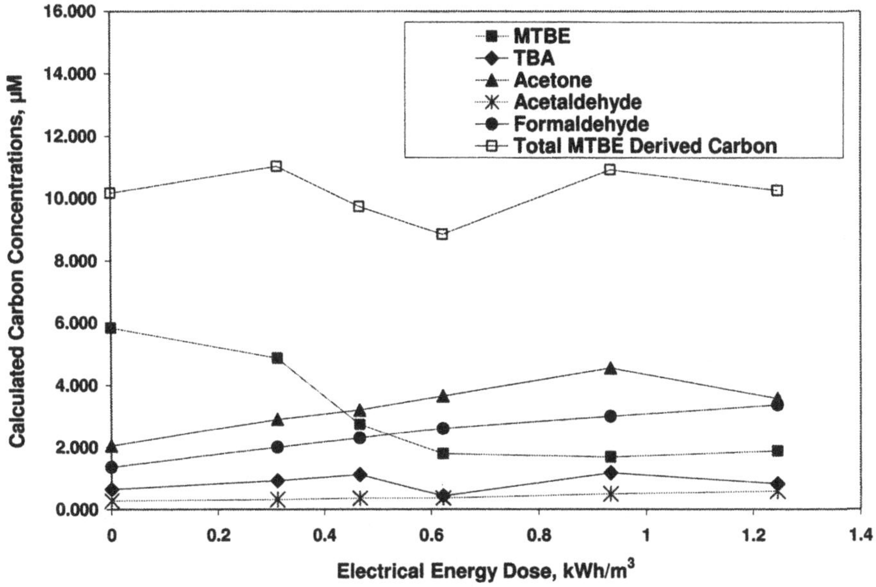

Figure 5.61 Calculated (organic) carbon concentration profiles of MTBE and its by-products during MP-UV/peroxide treatment of Santa Monica water spiked with 1.2 µM of initial MTBE (baseline conditions)

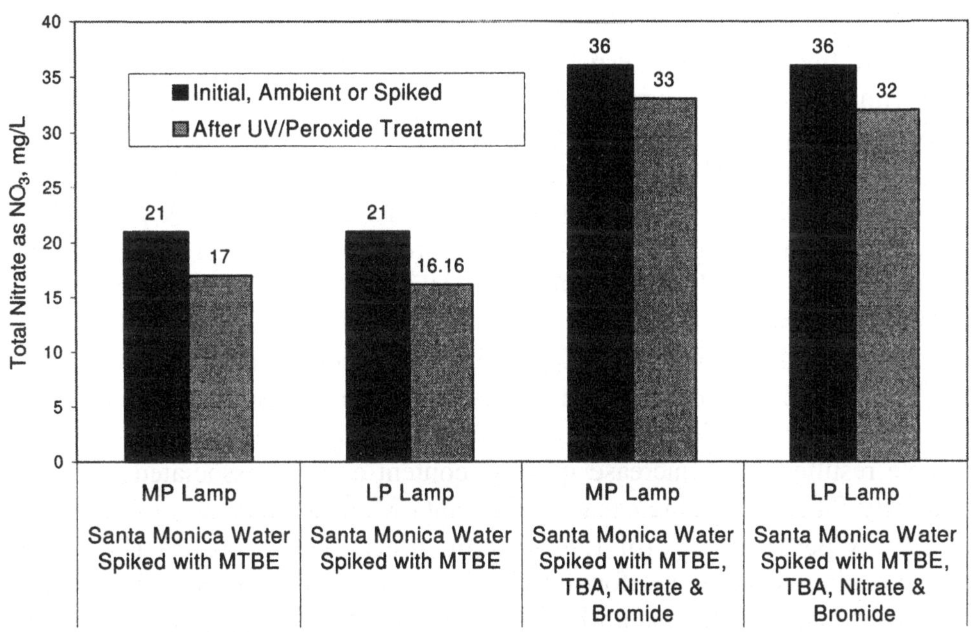

Figure 5.62 Total nitrate concentrations before and after (at ~2 minutes of UV exposure time) MP-UV/peroxide and LP-UV/peroxide treatment

experiments, nitrate was spiked into the feed water to an effective concentration of 36 mg/L as NO_3. Partial oxidation of nitrate results in nitrite and complete oxidation results in the formation of nitrogen dioxide. The MP/peroxide and LP/peroxide oxidation of the ambient Santa Monica water lowered the nitrate levels by 20% and 23%, respectively. In tests that were conducted with Santa Monica waters spiked with nitrate to 36 mg/L, approximately 8-10% nitrate removal was observed. The quantum yield of nitrite formation from nitrate photolysis is pH, wavelength and concentration dependent. Presence of nitrate had little impact on the end-products (data not shown here).

Bromate Formation

To assess the possibility of oxidation of bromide to bromate by continuous UV/peroxide processes, tests were conducted with Santa Monica water containing ambient (100 μg/L) and spiked (500 μg/L) levels of initial bromide. As shown by the data in Figure 5.63, no conversion of bromide to bromate was observed by continuous UV/peroxide treatment. This suggests that for waters with elevated levels of bromide for which the ozone-based oxidation processes (e.g., ozone/peroxide or ozone/UV) may not be appropriate, continuous UV/peroxide treatment would be a promising alternative to consider.

Summary of Findings

Continuous UV/peroxide pilot tests were conducted using LP and MP lamps on Santa Monica groundwater and Lake Huron surface waters. The E_{EO} (kWh/m^3-order) values for MTBE removal in Santa Monica water spiked with 1.2 μM (or 100 μg/L) of MTBE by MP-UV/peroxide and LP-UV/peroxide treatment were respectively 2.2 kWh/m^3-order and 0.7 kWh/m^3-order. In tests conducted with Santa Monica water spiked with MTBE (~1.2 μM), TBA (~4.0 μM), nitrate (~36 mg/L as N) and bromide (~0.5 mg/L) to simulate poor quality groundwater, the E_{EO} values for MP-UV/peroxide and LP-UV/peroxide treatment were 2.58 kWh/m^3-order and 0.96 kWh/m^3-order, respectively. As indicated by the aforementioned results, the LP-UV/peroxide treatment had slightly lower E_{EO} values than MP-UV/peroxide treatment. In a test conducted with Santa Monica water spiked with 33 μM (or 2900 μg/L) MTBE and 480 μM (or 16 ppm) H_2O_2, an E_{EO} of 0.64 kWh/m^3-order was measured for MP-UV/peroxide treatment of MTBE. This result illustrates the inverse relationship between E_{EO} and peroxide concentration.

Some algae growth was observed in the Santa Monica water that was stored in a large (3,000 gallon) storage tank at the Lake Huron Water Treatment Plant testing site. This algae growth may have resulted in an increase in TOC content and the associated hydroxyl radical scavenging capacity. The scavenging capacity for Santa Monica water was calculated based on the measured LP-UV/peroxide E_{EO} (of 1.44 kWh/m^3-order) with ~200 μM H_2O_2 and 1.2 μM MTBE. This scavenging capacity was found to be 1.4×10^5 s^{-1} or 2.5 times the value of 5.57×10^4 s^{-1} that was calculated for the Santa Monica water quality. The E_{EO} values measured for the Lake Huron surface water were lower than those for Santa Monica groundwater. This is attributed primarily to the lower scavenging capacity (no algae) as well as to the lower UV absorption coefficients for the Lake Huron water. The scavenging capacity of the Lake Huron surface water was calculated to be 4.7×10^4 s^{-1} based on a comprehensive water analysis.

It should be noted that the E_{EO} varies inversely with the optical pathlength of the photoreactor. Therefore, large, full-scale photoreactors with long pathlengths may result in lower

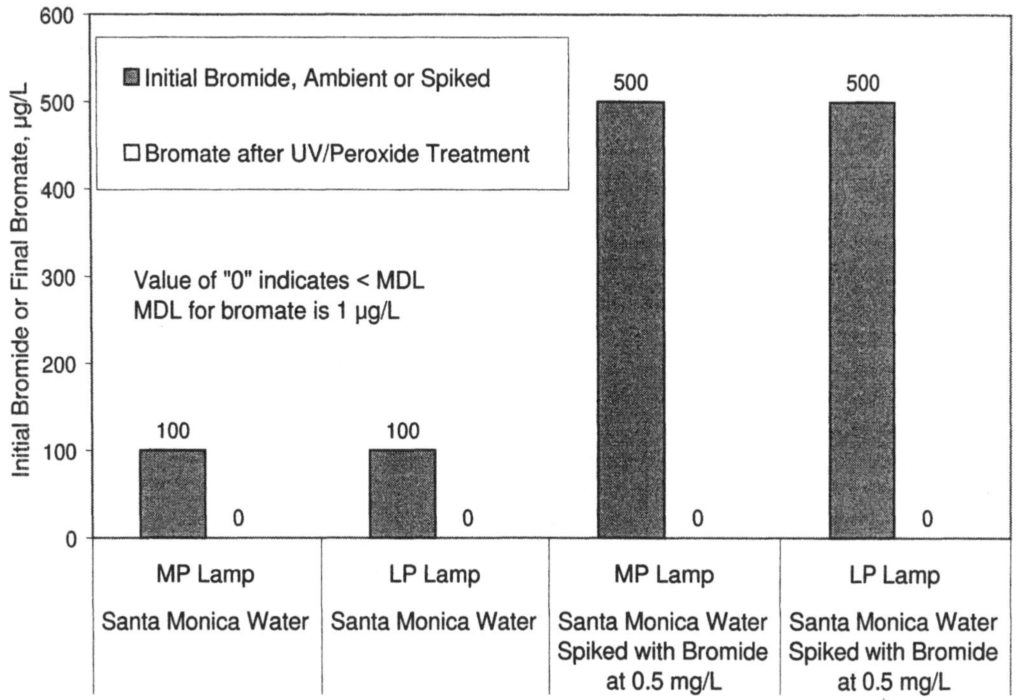

Figure 5.63 Bromate measurements, before and after MP-UV/peroxide and LP-UV/peroxide treatment

E_{EO} values than those measured in this study. The E_{EO} values observed in this study were slightly lower than those reported in the literature for MP-UV/H_2O_2 treatment (Cater et al., 2000). The E_{EO} values reported in the literature were developed using batch reactors and they varied between 0.2 kWh/m^3-order to 7.5 kWh/m^3-order (Cater et al., 2000).

Consistent with the bench scale test results, the MP-UV/H_2O_2 treatment resulted in higher acetone and formaldehyde generation compared to LP-UV/H_2O_2 treatment. This indicates that the MP-UV/peroxide treatment progressed farther in the oxidation pathway than the LP-UV/peroxide treatment.

IMPLICATIONS FOR FULL-SCALE USE

Process Design Criteria

The following design criteria were used in developing the UV/peroxide treatment costs for waters with quality comparable to Santa Monica groundwater (with scavenging capacity of ~5×10^4 s^{-1}):

- An E_{EO} of 0.35 kWh/m^3-order for MTBE removal using MP-UV/peroxide process, with an initial peroxide dose of 15 mg/L
- An E_{EO} of 0.13 kWh/m^3-order for MTBE removal using LP-UV/peroxide process, with an initial peroxide dose of 15 mg/L

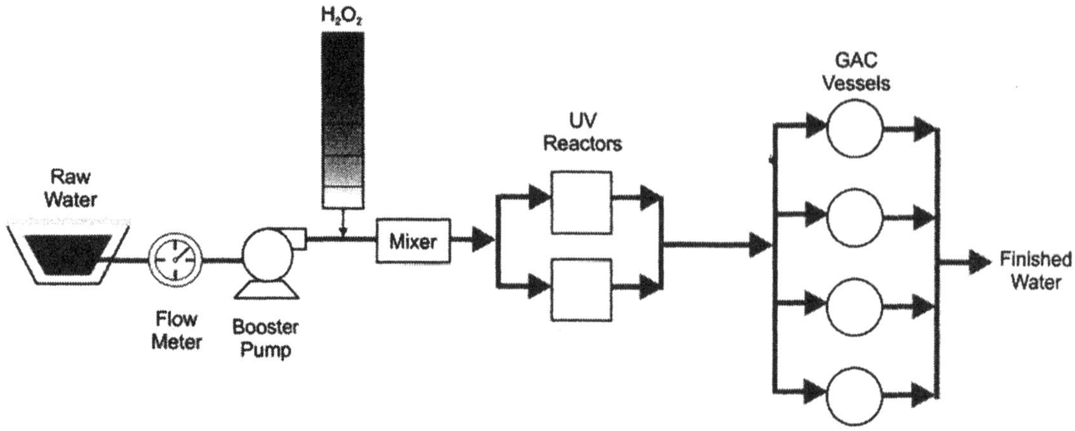

Figure 5.64 MP-UV/peroxide schematic

Similar to other AOP technologies, it was assumed that MTBE and its by-products will be further removed by biological GAC treatment. Test data indicated a design loading rate of 2 gpm/sf and a contact time of 20 minutes is necessary for biological GAC to remove MTBE and its by-products to very low (e.g., non-detect) concentrations. Costs for polishing, biological GAC are presented in Chapter 4 and they should be added to the UV/peroxide costs to obtain the total cost for sequential UV/peroxide-biological GAC treatment.

Costs were developed for both MP-UV/peroxide and LP-UV/peroxide treatment since the performance data indicated some differences for LP-UV compared to the MP-UV performance. Unlike for other AOPs, costs for UV/peroxide treatment were not developed for plant capacities of 0.5 mgd, since there was insufficient information available for systems of this size.

Process Schematics and Layout

MP-UV/Hydrogen Peroxide System

A schematic of the MP-UV/hydrogen peroxide system is shown in Figure 5.64. Raw water supply is metered prior to pumping. Booster pumps provide the hydraulic head required for flow through the entire system. Following the pumps, hydrogen peroxide is added to the water at a concentration of 15 mg/L. Hydrogen peroxide can either be fed through a venturi-type injector device which facilitates mixing or metered into the process stream and mixed downstream of the injection point via a static or dynamic mixing device. The MTBE contaminated water is then directed through the MP-UV reactors.

The final unit process consists of biologically enhanced GAC polishing to destroy any remaining MTBE and the by-products. Allowing biological growth within the carbon beds enhances the ability of the process to destroy the remaining chemicals and extends the life of the carbon. Pilot testing of GAC resulted in a recommended hydraulic loading of 2 gpm/sf and a contact time of 20 minutes. The design parameters require multiple pre-manufactured GAC vessels, depending on the flow rate through the system.

Based on the process schematic, potential site layouts were prepared for 1 mgd (Figure 5.65) and 10 mgd (Figure 5.66) treatment systems. Major equipment is shown on the site layout. MTBE contaminated water is pumped from the well pump base and boosted through the

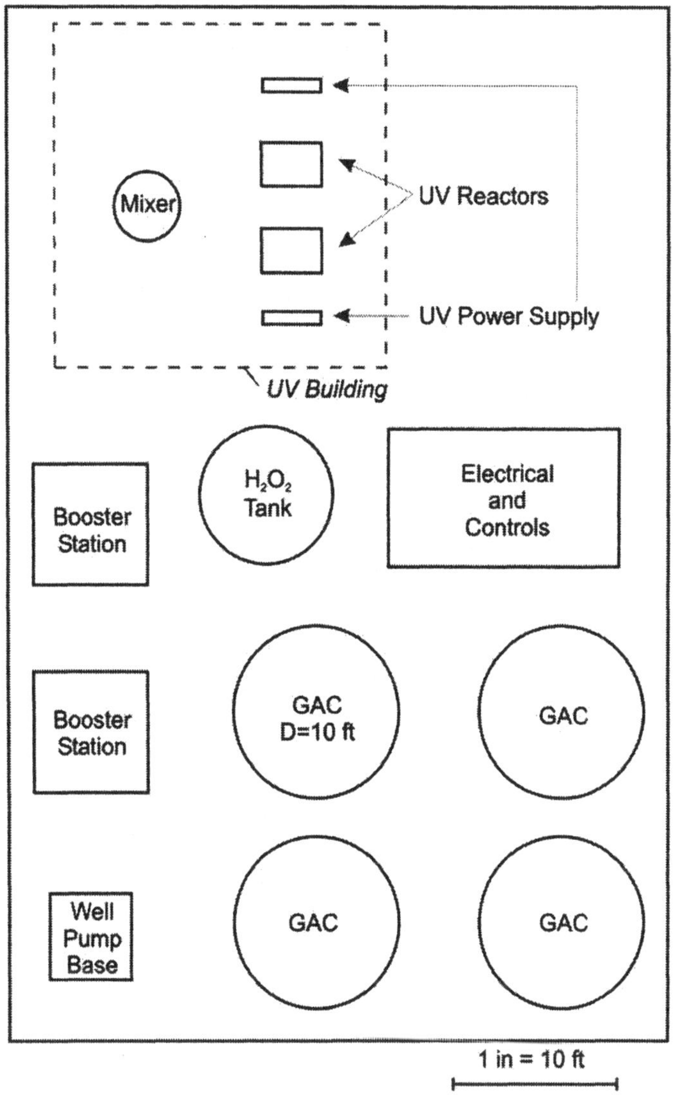

Figure 5.65 MP-UV/peroxide layout for 1 mgd system

booster pump stations. From the booster pumps, the water flows through a mixing device where the hydrogen peroxide is added and mixed. For a 1 mgd system, an 8 feet diameter hydrogen peroxide tank will be required and a 10 feet diameter hydrogen peroxide tank will be required for the 10 mgd system. The process water then flows through a header system and into the UV reactors. The volumes and the number of lamps present in each MP-UV reactor varied with the reactors rated capacity, the assumed E_{EO} and UV dose. One standby MP-UV reactor train was assumed for all system sizes (0.5 mgd to 10 mgd). The number of MP-UV reactors, including standby, for 0.5 mgd, 1 mgd, 3 mgd and 10 mgd system sizes are 3, 2, 2 and 4 with rated capacities of 0.3 mgd, 1.2 mgd, 3.5 mgd and 4.5 mgd, respectively. Approximate footprint sizes for 0.3 mgd and 4.5 mgd capacity UV reactors and associated power supply are shown in Figures 5.65 and 5.66, respectively.

The process water is recombined after the UV reactors. Process water would then flow through a distribution header to the GAC vessels. Large pre-manufactured vessels are sold in

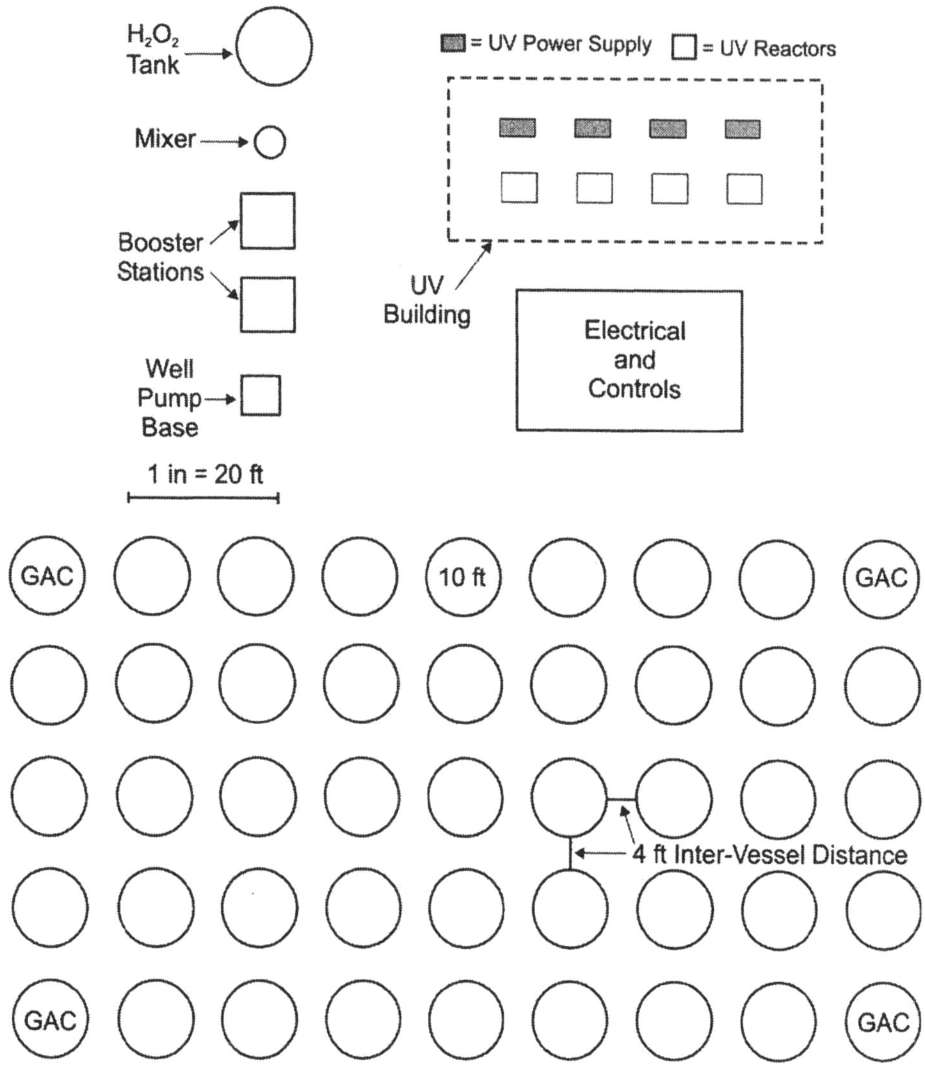

Figure 5.66 MP-UV/peroxide layout for 10 mgd system

pairs of 20,000 lb carbon capacity. The size of each vessel is 10 feet in diameter. To meet the hydraulic loading requirement, 4 GAC vessels are required for the 1 mgd system and 40 GAC vessels are required for the 10 mgd system. After determining the number of GAC vessels required to meet the hydraulic loading requirement, the volume of carbon required for each vessel was calculated to meet the required 20 minute contact time. An electrical control room was included to house electrical components and process control equipment.

Technology Costs

Shown in Tables 5.12 and 5.13 are the costs for MP-UV/peroxide and LP-UV/peroxide treatment, respectively. Every row of the attached cost tables are numbered and the assumptions are described below by row number.

Table 5.12
Capital, O&M, and total costs for MP-UV/peroxide treatment of MTBE

Design Flow (mgd)		1	3	10
Design Flow (gpm)		695	2084	6945
Sizing Calculations				
Number of UV Reactor Trains	(1)	2	2	3
Number of UV Reactor Trains in Duty	(2)	1	1	2
Number of UV Reactor Trains in Standby	(3)	1	1	1
Number of UVT Monitors	(4)	1	1	1
Number of UV Intensity Sensors/Reactor	(5)	2	2	2
Number of Reference UV Intensity Sensors/Reactor	(6)	1	1	1
Hydrogen Peroxide Required (gal/d)	(7)	38	114	379
UV Building Size (ft × ft)	(8)	20 × 20	20 × 20	40 × 20
Capital Cost Summary				
Process Capital Cost				
UV Reactor Cost	(9)	$400,000	$600,000	$900,000
UV Lamp Cost	(10)	$4,400	$8,800	$17,600
Sleeves Cost	(11)	$4,800	$9,600	$14,400
Ballast Cost	(12)	$7,200	$14,400	$28,800
UVT Monitor Cost	(13)	$15,000	$15,000	$15,000
UV Intensity Sensor Cost	(14)	$500	$500	$750
Reference UV Intensity Sensor Cost	(15)	$250	$250	$375
Pipes, Valves, Tees, Couplings, Adapters	(16)	$40,000	$50,000	$55,000
Uninterrupted Power Supply	(17)	$30,000	$40,000	$60,000
Peroxide Storage Tank	(18)	$36,000	$36,000	$40,000
Peroxide Injection System	(19)	$9,000	$9,000	$10,000
Subtotal Process Cost	(20)	$530,000	$750,000	$1,080,000
Direct Capital Cost	(21)	$1,325,000	$1,875,000	$2,700,000
Indirect Capital Cost Summary				
Piloting	(22)	$65,000	$65,000	$65,000
Permitting	(23)	$15,900	$22,500	$32,400
Land	(24)	$15,900	$22,500	$32,400
Operator Training	(25)	$1,800	$1,800	$1,800
Housing	(26)	$24,800	$24,800	$49,600
Indirect Capital Cost	(27)	$123,400	$136,600	$181,200
Total Capital Cost	(28)	$1,448,400	$2,011,600	$2,881,200
Annual O&M Summary				
Chemicals (Peroxide)	(29)	$76,048	$228,143	$760,478
Materials (Lamps, Sensors, Sleeves, Ballast Replacement)	(30)	$12,193	$24,235	$60,292
Electricity	(31)	$33,360	$100,079	$333,597
Labor	(32)	$36,500	$54,750	$73,000
Total Annual O&M Cost	(33)	$158,100	$407,208	$1,227,367
For a project period of 20 years with 7% interest				
Amortized Total Annual Cost	(34)	$294,250	$596,298	$1,498,200
Cost/1,000 gallons Treated	(35)	$0.81	$0.54	$0.41

Table 5.13
Capital, O&M, and total costs for LP-UV/peroxide treatment of MTBE

Design Flow (mgd)		1	3	10
Design Flow (gpm)		695	2084	6945
Sizing Calculations				
Number of UV Reactor Trains	(1)	2	5	13
Number of UV Reactor Trains in Duty	(2)	1	4	12
Number of UV Reactor Trains in Standby	(3)	1	1	1
Number of UVT Monitors	(4)	1	1	1
Number of UV Intensity Sensors/Reactor	(5)	1	1	1
Number of Reference UV Intensity Sensors/Reactor	(6)	1	1	1
Hydrogen Peroxide Required (gal/d)	(7)	38	114	379
UV Building Size (ft × ft)	(8)	20 × 20	20 × 20	40 × 20
Capital Cost Summary				
Process Capital Cost				
UV Reactor Cost	(9)	$260,000	$650,000	$1,690,000
UV Lamp Cost	(10)	$10,800	$43,200	$129,600
Sleeves Cost	(11)	$7,200	$18,000	$46,800
Ballast Cost	(12)	$10,800	$43,200	$129,600
UVT Monitor Cost	(13)	$15,000	$15,000	$15,000
UV Intensity Sensor Cost	(14)	$250	$625	$1,625
Reference UV Intensity Sensor Cost	(15)	$250	$625	$1,625
Pipes, Valves, Tees, Couplings, Adapters	(16)	$40,000	$50,000	$55,000
Uninterrupted Power Supply	(17)	$30,000	$40,000	$60,000
Peroxide Storage Tank	(18)	$36,000	$36,000	$40,000
Peroxide Injection System	(19)	$9,000	$9,000	$10,000
Subtotal Process Cost	(20)	$390,000	$800,000	$1,870,000
Direct Capital Cost	(21)	$975,000	$2,000,000	$4,675,000
Indirect Capital Cost Summary				
Piloting	(22)	$65,000	$65,000	$65,000
Permitting	(23)	$11,700	$24,000	$56,100
Land	(24)	$11,700	$24,000	$56,100
Operator Training	(25)	$1,800	$1,800	$1,800
Housing	(26)	$24,800	$24,800	$49,600
Indirect Capital Cost	(27)	$115,000	$139,600	$228,600
Total Capital Cost	(28)	$1,090,000	$2,139,600	$4,903,600
Annual O&M Summary				
Chemicals (Peroxide)	(29)	$76,048	$228,143	$760,478
Materials (Lamps, Sensors, Sleeves, Ballast Replacement)	(30)	$10,869	$42,248	$125,923
Electricity	(31)	$17,253	$51,760	$172,532
Labor	(32)	$36,500	$54,750	$73,000
Total Annual O&M Cost	(33)	$96,885	$245,545	$694,082
For a project period of 20 years with 7% interest				
Amortized Total Annual Cost	(34)	$199,345	$446,668	$1,155,020
Cost/1,000 gallons Treated	(35)	$0.55	$0.41	$0.32

Sizing Calculations

(1) Row 1 shows the total number of UV reactor trains, including one redundant train, assumed for each system size.
(2) Number of UV reactor trains in duty at a given time is shown in this row.
(3) One standby UV reactor was assumed for each system.
(4) One UV transmittance monitor was assumed for each system size.
(5) Two UV intensity sensors were assumed per reactor.
(6) Two reference UV intensity sensors were assumed per reactor.
(7) Hydrogen peroxide requirements per day were calculated assuming a dose of 15 mg/L and concentration strength of 35% v/v.
(8) UV building sizes were estimated based on reactor sizes and accessory equipment sizes shown in the layout drawings.

Capital Cost Assumptions

(9) The total cost per UV reactors was obtained by multiplying the number of reactors required (row 1) with the unit cost per reactor. The costs per reactor were obtained from reputed UV manufacturers.
(10) UV manufacturers provided the costs for lamps based on the assumed E_{EO}'s.
(11) The cost for quartz sleeves was also provided the UV manufacturers. Assumed each UV lamp to be provided with a quartz sleeve.
(12) Ballast cost was obtained from manufacturers. Assumed one ballast per two lamps.
(13) One UV transmittance monitor was assumed at unit cost of $15,000. This is also based on vendor quotes.
(14) UV intensity sensor costs were obtained from vendors.
(15) Reference UV intensity sensor costs were also obtained from manufacturers.
(16) Costs for pipes, valves, tees and adapters were estimated based on the amount of piping and valving necessary to install the UV reactors.
(17) Uninterrupted power supply costs were also provided by the manufacturers.
(18) Peroxide storage tank costs were provided by the tank vendors.
(19) Peroxide injection system costs were provided by vendors.
(20) Process cost subtotal consists of the sum of all the above-discussed process costs (rows 9 - 19).
(21) Process cost multiplier of 2.5 for flow rates equal to or below 1 mgd and 2.0 for flow rates above 1 mgd was used to obtain direct capital cost from process cost. Direct capital cost was obtained by multiplying the process cost (row 20) with the indicated multipliers. The multipliers account for: site work, contractor's overhead and profit, contingencies, engineering and design, mobilization and bonding, legal and administrative, interest during construction, and installation.
(22) Piloting costs were assumed to be $10,000 for flow rates below 1 mgd and $65,000 for flow rates equal to and above 1 mgd. For large systems, the piloting costs were calculated at $65,000 and for small systems it was assumed that limited piloting would be provided for by $10,000 (EPA Stage-2/LT-2 Technology and Cost Document, 2002).

(23) Permitting costs were assumed to be similar to those calculated for the EPA Arsenic Technology and Cost document (2001). Permit costs equal 3% of the process cost subtotal (row 20), with a minimum of $2500.

(24) Land costs were assumed to be 3% of the process cost subtotal (row 20).

(25) Operator training costs were based on 40 hours of supervisory training at a rate of $45/hour.

(26) The building size for housing was estimated from the footprints (row 8). Costs were developed from the RS Means Building Construction Cost (2002) data for warehouse/office space and equal to $62/sq.ft.

(27) Indirect capital cost subtotal is a sum of all the indirect process costs described in rows 22 through 26.

(28) Total capital cost equals the sum of indirect (row 27) and direct (row 21) capital costs.

O&M Cost Assumptions

(29) Chemical costs for hydrogen peroxide was assumed to be $5.50/gallon.

(30) The material cost was developed assuming once a year replacement for UV lamps and once in five years replacement of sleeves, ballasts, intensity sensors and reference sensors.

(31) Electricity costs were computed based on E_{EO}'s and unit cost of power of $0.08/kWh. The building maintenance power, assumed at 20% of the UV process power, is also included.

(32) Labor costs were estimated using the following assumptions:

for 0.5 mgd system need 2 hours/day of operator/maintenance time,
for 1.0 mgd system need 4 hours/day of operator/maintenance time,
for 3.0 mgd system need 6 hours/day of operator/maintenance time and
for 10 mgd system need 8 hours/day of operator/maintenance time.

The labor rate was assumed at a typical technician rate of $25/hour.

(33) Annual O&M subtotal consists of the sum of all the O&M costs (rows 29 - 32).

Total Cost Assumptions

(34) The total capital costs (row 28) were amortized over 20 years at a 7% interest rate and then annual O&M (row 33) was added to derive the amortized total annual cost.

(35) Cost per thousand gallons treated was calculated by dividing the amortized total annual cost (row 34) by the flow rate in thousands of gallons and the days in a year.

CHAPTER 6
PULSED UV/PEROXIDE TREATMENT

BENCH SCALE TESTING

Introduction

Pulsed-UV Process

Unlike a low- or a medium-pressure UV lamp, which provides continuous UV light, a pulsed-UV lamp produces high-intensity, pulsed radiation by accumulating electrical energy in a capacitor and releasing the stored energy via a high-speed electric switch in thousandths or millionths of a second. This process of "pulsing" provides the extremely intense flashes (or pulses) of polychromatic radiation, which produces hydroxyl radicals to remove MTBE in water.

MTBE does not absorb UV light above the wavelength of 200 nm (Figure 6.1). Thus, the MTBE degradation by a monochromatic, low-pressure UV lamp, which emits light at 254 nm, is primarily done via oxidation of MTBE by hydroxyl radicals generated during the UV process. At the wavelength below 200 nm, MTBE weakly absorbs UV light, and thus with a polychromatic UV lamp, such as a pulsed UV lamp, which emits UV light below 200 nm may directly photolyzed MTBE in the absence of hydroxyl radicals. At the wavelength below 200 nm, however, water can also absorb UV light to produce hydroxyl radicals by photolysis of water. Although MTBE weakly absorbs UV light below 200 nm, the amount of MTBE is very small compared with the amount of water. Thus the reduction of MTBE in pure water when the MTBE concentration is low is likely to be the result of the oxidation by hydroxyl radicals (from photolysis of water) rather than by the direct photolysis of MTBE. Both of these mechanisms may play roles in reduction of MTBE at the low wavelength depending on the characteristics of water and the experimental conditions. However, it may be difficult to differentiate whether the

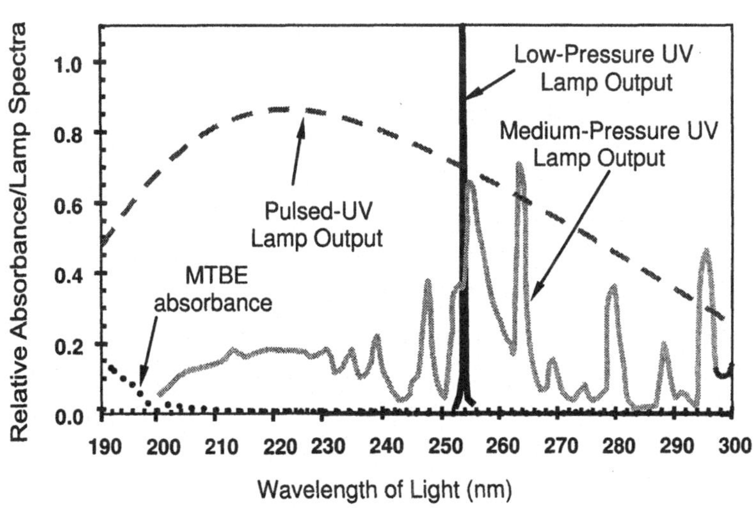

Figure 6.1 MTBE absorption spectrum and characteristics of various UV lamps

reduction of MTBE is actually initiated by the direct photolysis of MTBE by UV light or by hydroxyl radicals generated from dissociation of water below the wavelength of 200 nm. More controlled experiments may be necessary to identify the different MTBE removal pathways in pure water where no initial H_2O_2 or radical generating precursors are available.

During the advanced oxidation treatment of MTBE, certain by-products are formed. These include *tert*-butyl formate (TBF), *tert*-butyl alcohol (TBA), isopropyl alcohol, acetone, formaldehyde, and methyl acetate. Currently, these contaminants are not regulated for the drinking water supplies, but some of them pose health concerns to the public. In fact, the CDHS has proposed an action level for TBA for its cancerous effect in 1999. The currently proposed CDHS level for TBA is 12 μg/L. Consequently, the process design was focused not only on achieving the best MTBE reduction, but also on reducing the by-product levels during the MTBE treatment process.

The waters tested include groundwater from the City of Santa Monica, California; the Suffolk County Water Authority, New York; the City of Long Beach, California; and effluent from the Miami Dade Wastewater Treatment Plant in Florida. These four waters were chosen based on the level of total organic carbon (TOC) and alkalinity to investigate the effects of these constituents on removal of MTBE during the pulsed-UV process. Some waters also contained other variables such as NO_3^-, which affect the pulsed-UV/H_2O_2 process by filtering out some of the available UV light.

Measurement of Hydroxyl Radical Production From Pulsed UV

As a part of this project, a method was developed to consistently estimate the UV dose by measuring the hydroxyl radical production by the pulsed UV system under the same experimental conditions as the MTBE experiment (same frequency, same water matrix, same hydrogen peroxide dose, same reaction time, etc.). Due to the differences in 1) lamp efficiency, 2) reactor design, 3) water matrix, 4) hydraulic retention time, and 5) mixing intensity, comparison of MTBE reduction efficiency from different UV platforms can be inconsistent or unfeasible. Depending on the reactor design, the same lamp can also perform differently with the similar amount of energy applied. This process of estimating hydroxyl radicals may allow the data to be compared from different types of UV system, so that more accurate estimation can be made for calculating the cost of treatment. This approach can complement the electrical energy per order of magnitude reduction ($E_{E/O}$) values conventionally used for comparing the effectiveness of the UV, by providing information on the actual production of hydroxyl radical for each set of tests.

The approach for estimating hydroxyl radical production was based on the radiolysis using alcohol such as ethanol and hydrogen peroxide to estimate the production of hydroxyl radicals by measuring the oxidation of alcohol to aldehyde, which is done by hydroxyl radicals formed during the pulsed UV process. Ethanol was chosen since UV absorption by ethanol is minimum in the wavelength range from 190 to 400 nm.

The hydroxyl radical is known to react with aliphatic alcohols via abstraction of a hydrogen atom from a C-H bond (Eq (6.1)) [Asmus et al., 1973]. The following reactions describe the process of oxidation of alcohol to aldehyde, which then can be quantified to correlate the production of hydroxyl radical to the production of aldehyde.

$$\cdot OH + C_2H_5OH \rightarrow H_2O + \cdot C_2H_4OH \qquad (6.1)$$

$$\cdot C_2H_4OH + O_2 \rightarrow \cdot O_2C_2H_4OH \quad (6.2)$$

$$\cdot O_2C_2H_4OH \rightarrow CH_3CHO + HO_2 \quad (6.3)$$

More detail is provided by Asmus et al., (1973) who have investigated the reaction of hydroxyl radicals with alcohols in pulse-irradiated aqueous solutions. At a higher level of hydroxyl radicals, however, the aldehydes can also react with hydroxyl radicals, thus underestimating the hydroxyl radical production. The detailed discussion on this topic is presented in the result and discussion section.

Experimental Plan and Objectives

The main objective of this project was to investigate the applicability of pulsed UV on removing MTBE from contaminated water. The effects of the background water matrix on the performance of the pulsed-UV process on the MTBE treatment and the by-product formation were also investigated using different types of water. The waters tested include the groundwater from the City of Santa Monica, California; the Suffolk County Water Authority, New York; the City of Long Beach, California; and effluent from the Miami Dade Wastewater Treatment Plant in Florida. Also the removal of TBA using the pulsed-UV/H_2O_2 process was investigated. The approach was to optimize the process to meet the primary and possibly the secondary MTBE standards of 13 and 5 μg/L set by the State of California Department of Health Services (CDHS).

Specific objectives of the study are listed below.

- Determine the required pulsed-UV dose and H_2O_2 ratio for MTBE oxidation
- Decide the optimum retention time for MTBE destruction
- Develop MTBE decay rate information
- Evaluate the effects of water quality on the performance of the pulsed-UV process for the removal of MTBE using different types of water containing various levels of TOC and alkalinity
- Investigate the evolution of MTBE by-product formation for different types of water with varying levels of TOC and alkalinity, which can scavenge hydroxyl radicals to affect the treatment efficiency
- Assess the performance of the pulsed-UV/H_2O_2 process in removing water spiked with TBA only
- Develop a method for estimating •OH radical production

Materials and Methods

Bench-Scale Pulsed-UV Reactor Description

The pulsed-UV system used (Pulsar UV Technologies, Roseville, Calif.) utilized a stainless steel batch reactor (0.5 gal capacity), which included a 1.6 kW power source and a 3.6 in. xenon and inert gas-filled, tungsten-electrode flashlamp inside a quartz jacket (a conduit for deionized cooling water). The blackbody flashlamp design and the high-voltage electronics can reliably generate up to thousands of times more UV light than the conventional lamps. The high electrical conversion efficiency of the new flashlamp allowed more efficient treatment of

MTBE than conventional UV technologies. With power applied to the lamp, a standby "simmer mode," consisting of a steady-state partial ionization of the xenon gas, is maintained with a low current arc between the electrodes. Pulses are generated by an electrical discharge that quickly heats the xenon and inert gas and creates a plasma of ionized gas, which reaches a temperature (approximately 9,800 Kelvin [K]) high enough to emit blackbody radiation. As the plasma inside the flashlamp expands during a pulse, the spectrum of light emission ranges across UV, visible, and infrared wavelengths. The electric switch allows the bulb to operate from a minimum pulse setting of 0.1 to 3 Hz, in the increments of 0.1 Hz.

Pulsed-UV Bench-Scale Test Setup

Water spiked with MTBE (99.5 percent purity; Burdick and Johnson, Muskegon, Mich.) and H_2O_2 (3 percent by weight, USP grade) was allowed to mix in the system for several minutes prior to each test. Since a pulsed-UV lamp emits the low-intensity light in a "simmer" mode even when there is no pulsing of the lamp, the initial samples (t_0) were taken before the pulsed-UV power was turned on to preclude any premature formation of hydroxyl radicals.

The test setup consisted of a 0.5 gal pulsed-UV reactor and a 2-gal container for a semi-batch setup (Figure 6.2). For the 2 gal reservoir, a tube was connected to the influent line and extended to below the minimum water level (indicated by a dotted line in the figure) to minimize perturbation in the reservoir during the tests.

Volatilization Losses

A series of control tests were done prior to bench and pilot study to account for losses due to volatilization during pulsed-UV tests. In order to mimic actual run condition, temperature rise was measured with pulsed lights being on. For the bench scale pulsed UV control test, water was heated to the same temperature range to mimic volatilization during re-circulation within a small reactor. Preliminary data indicated no significant MTBE loss due to temperature rise with tightly sealed bench scale reactor.

Sampling

The actual pulsed-UV exposure time for the bench-scale, semi-batch setup was about one-fifth of the sampling time used for the semi batch setup. For each experiment, six samples were collected at sampling times of 0, 1, 2.5, 5, 30, and 60 min, which corresponded to the irradiation time of up to 12 min.

For each sample collected, the following analyses were done: MTBE and the by-products (including TBA, TBF, and acetone), aldehydes, bromate, and residual H_2O_2. Both nitrate (NO_3^-) and nitrite (NO_2^-) were also analyzed where NO_3^- was initially presented in the water. MTBE and the by-products samples were frozen before they were shipped to Oregon Graduate Institute (OGI) for analyses. The freezing of the samples was necessary to slow down the hydrolysis rate of *tertiary* butyl formate (TBF) to *tertiary* butyl alcohol (TBA), thus preserving the concentration of the intermediate by-products (TBF and TBA) at any given sampling time. Upon arrival at OGI, the samples were kept frozen until they were analyzed.

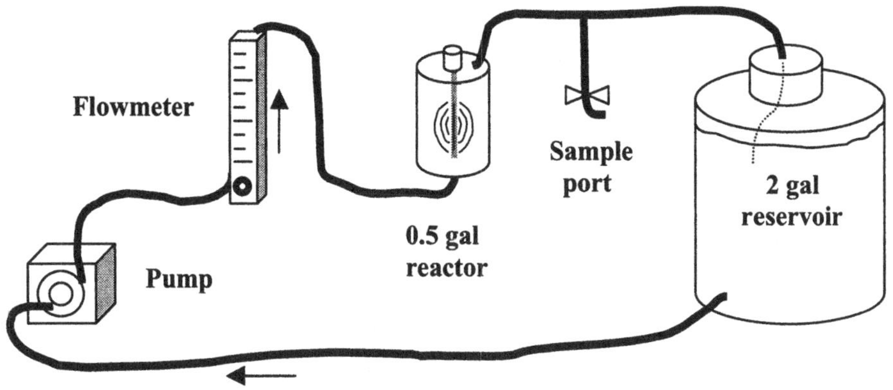

Figure 6.2 Schematic diagram of the semi-batch pulsed-UV setup

Analytical Methods

All analyses were conducted in accordance with the procedures described in *Standard Methods* (APHA, AWWA, and WEF, 1995). More detailed descriptions of the analytical method for residual H_2O_2, bromide, bromate, MTBE, TBA, TBF, acetone, aldehydes, and H_2O_2 are presented elsewhere [Liang et al., 1999b]. Aldehydes, NO_3^-, NO_2^-, H_2O_2, and bromate were analyzed by the Metropolitan Water District of Southern California's (Metropolitan's) Water Quality Laboratory.

Results and Discussion

Raw Water Quality

Each water matrix was analyzed for turbidity, pH, alkalinity, conductivity, bromide, iron, manganese, NO_3^-, UV light absorbance at 254 nm (UV_{254}), and TOC. The selection of the four specific water sources was based on the TOC and alkalinity levels. The approach was to test the effects of hydroxyl racial scavenging species by selecting water sources with varying levels of TOC (either low or high) and alkalinity (either low or high). Thus, four types of water were tested to study the effects of these parameters. However, some waters contained other variables such as NO_3^-, which affects the pulsed-UV/H_2O_2 process by filtering out some the UV light. Other variables such as, iron (Fe) (depending on the oxidation state of Fe), pH (controls the speciation of carbonates), turbidity, etc., could have played roles in the performance of the pulsed-UV process for a specific type of water. The typical characteristics of the water matrixes used for the tests are listed in Table 6.1.

In general, high concentration of alkalinity serves as a hydroxyl radical scavenger by reaction of bicarbonate (HCO_3^-) and carbonate (CO_3^{2-}) with hydroxyl radicals. Background natural organic matter (NOM) could also serve as an effective hydroxyl radical scavenger, but depending on the characteristics of NOM, a photochemical process, such as a pulsed-UV process, could generate small amount of hydroxyl radicals [Schwarzenbach et al., 1993] and organic peroxy radicals [Hoigné, 1989]. The background organic matter, however, is be more important sink than a source of hydroxyl radicals in general.

Table 6.1
Typical characteristics of water used for testing

Parameters	Units	Santa Monica Groundwater	Long Beach Groundwater	Suffolk County Groundwater	Miami Dade Plant Effluent
TOC	(mg/L)	0.5	3.5	0.56	3.86
Total alkalinity	(mg/L)	327	150	13	48
NO_3^-	(mg/L)	20.5	ND	9.61	1.89
Bromide	(mg/L)	0.23	0.14	0.14	0.07
Turbidity	(NTU)	0.15	0.33	4.7	0.08
Fe	(μg/L)	ND	20	2,320	167
Manganese	(μg/L)	10	3	128	1
UV_{254}	(abs/cm)	0.02	0.23	0.017	0.095
pH		7.13	8.62	6.62	9.33
Conductivity	(μmhos/cm)	1,328	357	307	267
Sulfate	(mg/L)	210	ND	17.4	NA
Total hardness	(mg/L)	538	27	45	65

ND—Not detected; NA—Not analyzed

Earlier researchers demonstrated the effect of alkalinity and dissolved organic carbon (DOC) concentration on the efficiency of hydroxyl radical attack on a target compound. As the DOC and alkalinity concentrations increase, hydroxyl radicals produced during the advanced oxidation processes are scavenged by DOC and carbonate, and consequently the relative efficiency of hydroxyl radicals decreases significantly compared to the efficiency at the reference point (no DOC and no alkalinity).

The other parameter that plays a significant role in the UV treatment process is NO_3^-. Water with NO_3^- can produce hydroxyl radicals by a photochemical reaction of NO_3^- without the addition of H_2O_2. The irradiation of NO_3^- results in the primary photochemical processes as described below [Zepp et al., 1988].

$$NO_3^- + UV \rightarrow NO_3^{-*} \mapsto NO_2^- + O(^3P) \tag{6.4}$$

$$\rightarrow NO_2^- + \cdot O^- \tag{6.5}$$

$$\cdot O^- + H_2O \rightarrow \cdot OH + OH^- \tag{6.6}$$

Overall, the steady-state concentration of hydroxyl radicals, which are responsible for MTBE degradation during an advanced oxidation process, can be correlated by the simplified general equation shown below [Schwarzenbach et al., 1993].

$$[\cdot OH]_{ss} \propto [NO_3^-] / \{[HCO_3^-] + [DOC]\} \tag{6.7}$$

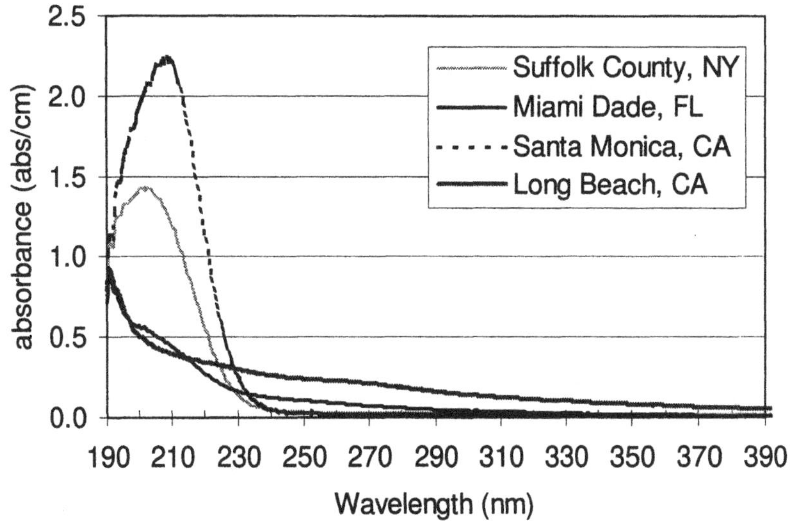

Figure 6.3 UV scans of four types of water used for pulsed-UV experiments

Also, photolysis of Fe(III) is the other possible source of hydroxyl radical from natural water during an UV process [Stumm and Morgan, 1996]. Depending on the concentration of these hydroxyl radical generating species in a specific water matrix, the addition of H_2O_2 may not contribute significantly in improving the MTBE degradation since significant amount of hydroxyl radicals may be generated from the background constituents.

Figure 6.3 shows the UV scans of four types of waters used in the pulsed UV experiments. Groundwater from the City of Santa Monica exhibited the highest absorbance below 230 nm followed by groundwater from the Suffolk County Water Authority. These two water sources have peaks around 210 nm, possibly as a result of the presence of high NO_3^- levels in these water matrixes. Water from the Miami Dade Wastewater Treatment Plant and the City of Long Beach showed increased absorbance as the wavelength decreased from 230 to 190 nm. The levels of absorption, however, were much lower than that for water with the high level of background NO_3^-. Unlike other water sources, City of Long Beach water with the highly colored characteristics (high TOC) showed highest absorbance for most of the UV region from 230 to 400 nm, as seen in Figure 6.3.

The parameters mentioned above are only some of those which may affect the efficiency of the pulsed-UV treatment of MTBE in different types of water. Thus, although attempts were made to attribute certain MTBE decay characteristics to the specific parameters, full understanding of the MTBE removal mechanisms for each type of water would require further investigations including full characterization of a specific water matrix related to the UV process.

Removal of MTBE by Pulsed-UV/H_2O_2

Pulsed-UV only. The hypothesis of the photolysis dominant removal of MTBE by the pulsed-UV process in the absence of any radical generating initiators in the water is based on the possibility of either a) directly photolyzing MTBE or b) photolyzing water to produce hydroxyl radicals which will then react with MTBE. In natural waters where high absorption occurs in the low-wavelength region, however, it may be difficult to directly photolyze MTBE or water

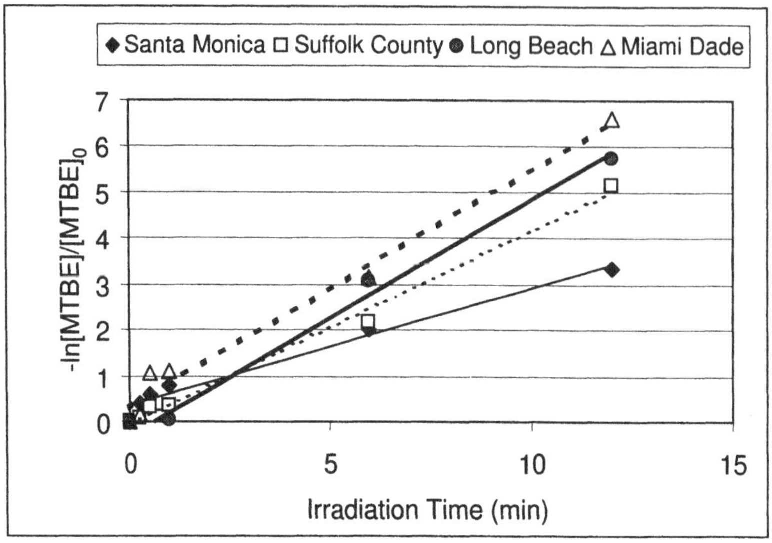

Figure 6.4 MTBE decay for different types of water, by pulsed-UV only, at 3 Hz

because of the higher absorption or filtering of UV by the background water matrix. The background constituents, which absorbs UV light may then produce hydroxyl radicals or organic peroxy radicals when subjected to the UV light, and these radicals can oxidize MTBE in the absence of H_2O_2 addition.

Groundwater from the City of Santa Monica, California, was used as the benchmark water for investigating the effects of various experimental parameters. To compare the rates of MTBE decay for different types of water, the results on the MTBE removal at the pulsed-UV frequency of 3 Hz without H_2O_2 were plotted in Figure 6.4. The data without H_2O_2 were used to compare the role of background constituents in MTBE degradation. Figure 6.4 is plotted in a natural logarithmic scale as a function of the irradiation time to obtain the pseudo-first order rate constants for different water matrixes. The plot shows that the water with high level of background NO_3^- (i.e., Santa Monica and Suffolk County water) resulted in the slower rates of MTBE degradation. Water with lower level of NO_3^- (i.e., Long Beach and Miami Dade waters) exhibited the faster MTBE degradation rates. The calculated rate constants and the measured background constituents that may contribute in controlling the efficiency of the pulsed-UV process are presented in Table 6.2.

In general, the comparison of the rate constants shows that the Santa Monica water with the highest NO_3^- concentration showed the slowest rate of MTBE reduction. If the photochemically mediated MTBE degradation process in the absence of pre-spiked H_2O_2 is primarily governed by the direct photolysis of MTBE or by photolysis of water to produce hydroxyl radicals, then high concentration of background NO_3^- present in the Santa Monica water could filter out the UV light in the low wavelength by strong absorption around 190 to 220 nm where MTBE or water absorption occurs. Since the UV absorption of NO_3^- is much stronger than that of MTBE or water, the high concentrations of NO_3^- will strongly diminish the efficiency of MTBE decay when H_2O_2 is not present. Although the absorption of NO_3^- produces hydroxyl radicals, the net overall effect of high NO_3^- concentration may be slower MTBE decay if photolysis (of either MTBE or water) is the controlling mechanism. Bicarbonate also absorbs the UV light, but UV scan of water with high alkalinity (~150 mg/L, Long Beach water) showed

Table 6.2
Rate constants for MTBE degradation by pulsed-UV with different water sources

Conditions	Kinetics Rate	Santa Monica Groundwater*	Suffolk County Groundwater	Long Beach Groundwater	Miami Dade Plant Effluent
3 Hz UV frequency	k (s^{-1})	4.3×10^{-3}	7.0×10^{-3}	8.6×10^{-3}	8.6×10^{-3}
0 mg/L H_2O_2	r^2	0.9759	0.9922	0.9919	0.9822
3 Hz UV frequency	k (s^{-1})	7.8×10^{-3}*	7.6×10^{-3}	8.3×10^{-3}	6.3×10^{-3}
7 or 20 mg/L H_2O_2	r^2	0.9544	0.9857	0.9327	0.9156
Hydroxyl Radical Controlling Factors		**Santa Monica Groundwater**	**Suffolk County Groundwater**	**Long Beach Groundwater**	**Miami Dade Plant Effluent**
TOC (mg/L)		0.5	0.56	3.5	3.86
Total alkalinity (mg/L)		327.0	13.00	150.0	48.00
NO_3^- (mg/L)		20.5	9.61	ND	1.89

* With 20 mg/L H_2O_2, ND – Not Detected

much less absorption near 200 nm than the water with high concentrations of NO_3^- (10 – 20 mg/L, Suffolk County or Santa Monica water) (Figure 6.3). Thus, NO_3^- may play much more important role in hindering removal of MTBE than alkalinity in the absence of H_2O_2.

The control tests done (data not shown) with deionized (DI) water (Super Q©, Millipore Corp, Bedford, Mass.) without the H_2O_2 spike under the same experimental setup showed effective removal of MTBE ($k = 3.7 \times 10^{-2}$). However, with DI water, it is less likely to produce hydroxyl radicals from the water matrix since the background constituent levels are low. Thus, the tests done with DI water without H_2O_2 may also suggest that the photolysis of MTBE or water by the pulsed-UV process may have played roles in removing MTBE. Consequently, it may be difficult to conjecture what the dominant mechanism is for different types of water without developing a consistent methodology to determine the nature of water matrix.

Both the Long Beach groundwater and the Miami Dade Wastewater Treatment Plant effluent contain high TOC, and the Long Beach groundwater also contains high alkalinity. According to Eq. 6.7, Long Beach groundwater and Miami Dade water, with low NO_3^- and high alkalinity or high TOC, would yield less available hydroxyl radicals than the other water by scavenging available hydroxyl radicals, and thus show the slower MTBE degradation rates. However, the results indicate the opposite trend, which may suggest that MTBE decay is dominated by the photolysis and assisted by the hydroxyl radical oxidation in the tested water matrix by the pulsed-UV. TOC and alkalinity can scavenge hydroxyl radicals which are produced during the pulsed-UV process and slow down the hydroxyl radical mediated decay of MTBE, but these two water sources showed the faster kinetics rates than the other waters with low TOC values in the absence of H_2O_2 (Figure 6.4). In fact the kinetics rate plot shows that the Suffolk County water with the lowest levels of TOC and alkalinity (but higher NO_3^- level) has a slower kinetics rate than that for the Long Beach or Miami Dade water. The rates of MTBE decay seem to correlate with the amount of NO_3^- in the water. That is, Santa Monica groundwater with the highest amount of NO_3^- showed the slowest rate of MTBE decay followed by the Suffolk County groundwater, which contains the second highest amount of NO_3^- in the water. The other waters

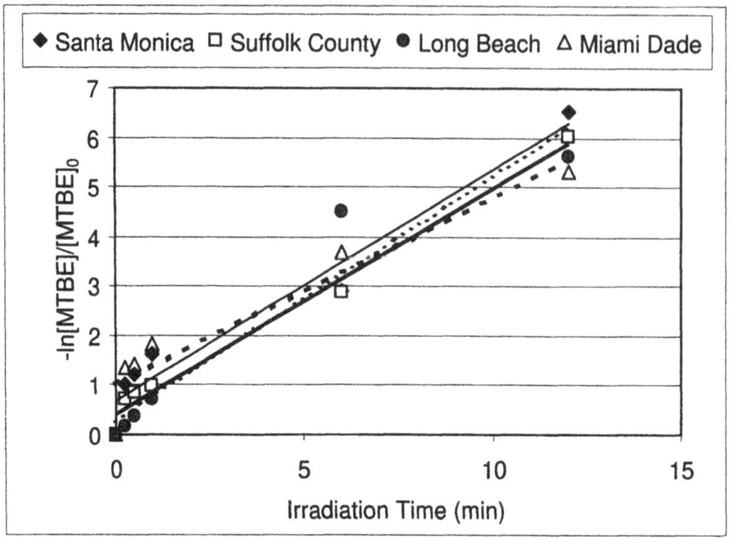

Figure 6.5 MTBE decay for different types of water by pulsed-UV at 3 Hz with 7 mg/L H_2O_2

with a lower or non-detectable level of NO_3^-, but with higher level of TOC, showed a faster rate of MTBE reduction. Thus this set of data seems to suggest that the presence of high concentration of NO_3^- does not promote the reduction of MTBE by producing hydroxyl radicals according to the equation 3, but rather inhibit the MTBE decay by filtering out the low wavelength UV light.

Pulsed-UV/H_2O_2. When H_2O_2 was added (Figure 6.5), some changes in the rate constants were observed. This may have been caused by the shift from generation of radicals by photolysis of the background constituents in the absence of H_2O_2 (or photolysis-mediated mode) to generation of hydroxyl radicals from H_2O_2. For Santa Monica groundwater where a high NO_3^- level (20 mg/L) is found, the addition of H_2O_2 would have no impact on filtering the UV light since the absorption of UV light by H_2O_2 is much less than that of NO_3^-. The addition of H_2O_2 on the other hand would have produced more hydroxyl radicals and yielded a faster MTBE degradation rate for Santa Monica water. In fact, the rate of MTBE reduction increased by 80 percent when 20 mg/L of H_2O_2 was added while all the other conditions were kept constant.

Suffolk County groundwater with approximately 10 mg/L of NO_3^- showed only small increase in the rate of MTBE reduction when 7 mg/L of H_2O_2 was added. Also, Long Beach groundwater with 150 mg/L of alkalinity showed small decrease in the calculated rate constant, but the difference was only about 3 percent (Table 6.2). For both of these water sources, there were no significant net changes in the MTBE degradation by addition of H_2O_2.

For Miami Dade Wastewater Treatment Plant effluent, where the absorption of UV light was the lowest in the wavelength range between 190 nm and 230 nm, the addition of H_2O_2 decreased the rate of MTBE reduction by approximately 27 percent. The presence of H_2O_2 may have filtered out some UV light where photolysis might have played more dominant role in the reduction of MTBE in this particular water matrix, which is quite different from all other raw waters tested.

Regardless of the specific characteristics of water, however, effective removal of MTBE was achieved with the addition of H_2O_2 for all water matrixes tested. With irradiation time of 12 min at the UV pulse rate of 3 flashes per second (Hz), less than 5 - 6 μg/L of MTBE was measured in the effluent (greater than 99 percent removal of MTBE) for all waters spiked with initial MTBE concentrations of 1,000 to 2,000 μg/L and H_2O_2 concentration of 7 to 20 mg/L. The typical MTBE removal efficiency with 6-min irradiation time at 3 Hz was greater than 94 percent, which is indicative of the efficiency of the process for removing MTBE.

The addition of H_2O_2 ensured consistent reduction of MTBE by producing high steady-state concentrations of hydroxyl radicals from the H_2O_2 spiked to the water rather than the background constituents, which varied depending on the source water. In the absence of H_2O_2, the characteristics of water played more important role in the effectiveness of MTBE reduction by the pulsed UV process.

Fate of MTBE Degradation By-Products

The effects of pulse frequency (rate of energy input) and the H_2O_2 initial spike have been investigated for each source water for the production of the MTBE degradation by-products. In order to provide the consistent and more representative energy input value, the applied UV dose in kWh/1,000 gal was used throughout this paper. The energy input to the system was calculated using the typical energy output from the lamp, which is 367 W per pulse at 1 Hz. Thus at 3 Hz, the applied energy from the lamp was approximately 1.1 kW or three times the per pulse value. Actual energy input to the system is calculated by multiplying the irradiation time by the average power input at an appropriate frequency as described below.

$$\text{UV dose} = \text{average power input (kW)} \times \frac{\text{irradiation time (h)}}{\text{reactor volume (gal)}} \times \frac{1{,}000\ \text{gal}}{1{,}000\ \text{gal}} \tag{6.8}$$

MTBE can be oxidized by hydroxyl radicals to form transient intermediate by-product species, namely TBF and TBA. TBF, however, undergoes a rapid hydrolysis reaction to TBA, so as TBF concentration is decreased, a typical increase in TBA is observed given enough reaction time. As the reaction is proceeded further, TBA is eventually converted to acetone, giving a rise to the acetone concentration towards the end of the test run. Formaldehyde is also formed either from oxidation of acetone or TBF.

The general trend based on the hydroxyl radical removal pathway of MTBE observed for the ozone/H_2O_2 process was also observed for the Santa Monica water spiked with H_2O_2 (Figure 6.6). The effect of presence of H_2O_2 was much more pronounced for the higher energy setting of 3 Hz than for 1 Hz. TBF was not formed in the absence of H_2O_2, but with 20 mg/L of H_2O_2, as much as 150 μg/L of TBF was formed with the initial MTBE concentration of approximately 2,000 μg/L. Unlike the ozone/H_2O_2 process where hydroxyl radical mediated pathway is dominant mechanism for the MTBE removal, the UV process adds another pathway, at least for TBF, which strongly absorbs UV light at the wavelength below 230 nm (peak absorbance at 195 nm). Thus, in the absence of H_2O_2, TBF may be efficiently photolyzed to yield low level of TBF in the UV treated effluent. When H_2O_2 is added to the water, H_2O_2 may compete for UV light, so TBF may not be directly photolyzed, but rather oxidized by hydroxyl radicals, similar to the case for ozone/H_2O_2 process. In fact, this type of behavior is shown in all three natural water sources tested (Figures 6.6–6.8).

Without the initial H_2O_2 spike, the TBA level peaked after the applied UV dose of more than 200 kWh/1,000 gal. On the contrary, with the spike of 20 mg/L H_2O_2, the TBA level peaked much more earlier with the applied UV dose of about 20 kWh/1,000 gal. Since the CDHS recommends 12 μg/L for TBA in drinking water because of its cancerous effect, it is important to

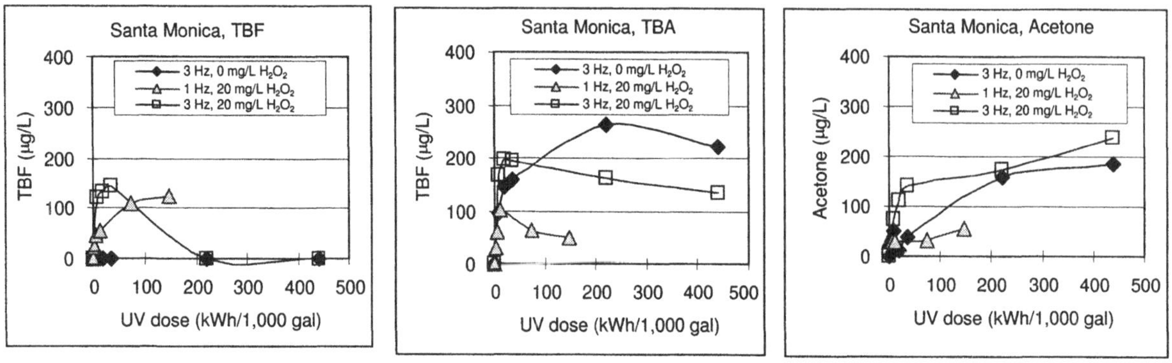

[a] TBF [b] TBA [c] Acetone

Figure 6.6 Formation of MTBE by-products for Santa Monica water

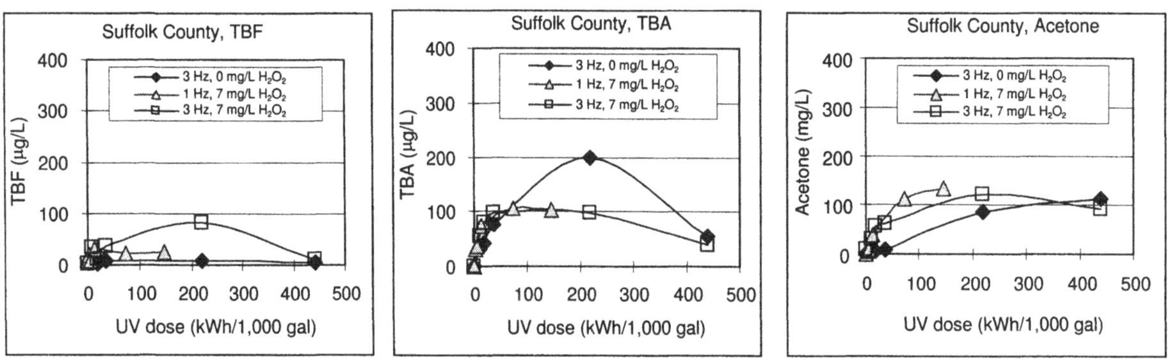

[a] TBF [b] TBA [c] Acetone

Figure 6.7 Formation of MTBE by-products for Suffolk County water

monitor the level of TBA in the treated water. At the slower pulse rate of 1 Hz with 2,000 µg/L of initial MTBE, the TBF levels showed steady increase within the applied UV dose of 150 kWh/ 1,000 gal. Similar to the 3 Hz case, the TBA level peaked with only small amount of the applied UV dosage, but the level peaked at a much lower level. The lower concentrations of acetone were also observed at 1 Hz compared with 3 Hz with or without the addition of H_2O_2. Formaldehyde also showed increase in its level when applied UV dosage increased (data not shown), but the effect of the operating parameters on the formation for formaldehyde was much less than that for the other degradation by-products.

Figure 6.7 shows the evolution of TBF, TBA and acetone for the Suffolk County water. After the applied UV dose of approximately 200 kWh/1,000 gal, the level of TBF started to decline with the pulse rate of 3 Hz and 7 mg/L of H_2O_2. Without the addition of H_2O_2, the highest level (~200 µg/L) of TBA was measured for the Suffolk County water. Thus, H_2O_2 addition is also recommended to reduce the level of TBA in the effluent. This may be indicative of a slow TBA conversion to acetone, and in fact, the least amount of acetone was observed for this case suggesting a lag phase of oxidation of the by-products for this case. When there is no added H_2O_2 in the water, the effective photolysis of TBF may also occur, to decrease the residual amount of

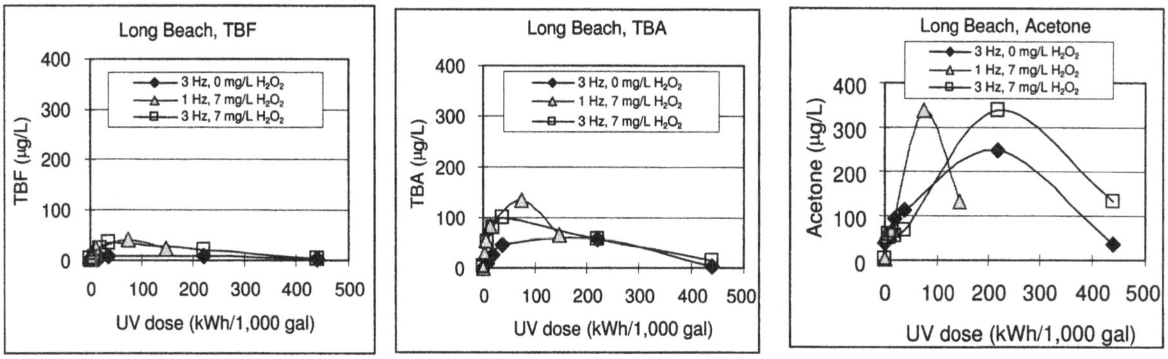

[a] TBF [b] TBA [c] Acetone

Figure 6.8 Formation of MTBE by-products for Long Beach groundwater

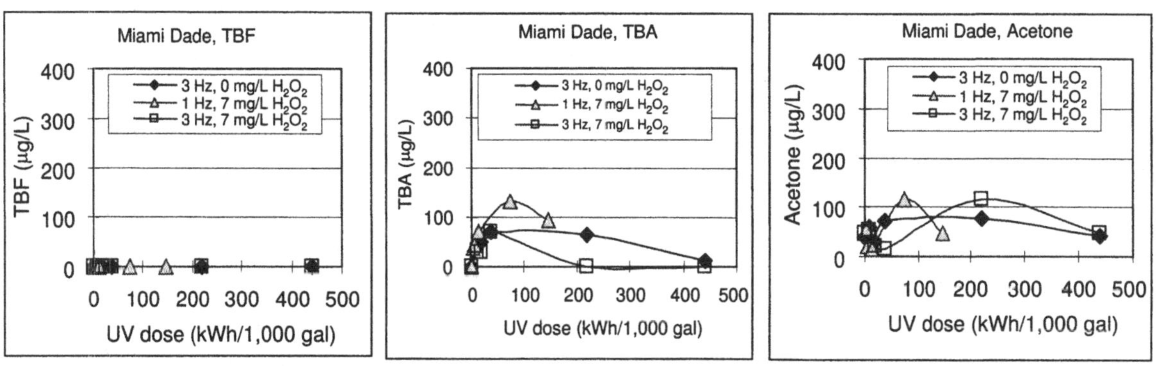

[a] TBF [b] TBA [c] Acetone

Figure 6.9 Formation of MTBE by-products for Miami Dade water

TBF (Figure 6.7 [a]). Overall, the gradual increase in the amount of acetone was observed for all the conditions tested.

The Long Beach water contains high level of background TOC (3.5 mg/L) which may contribute as a sink of photochemically induced hydroxyl radicals for the case when H_2O_2 is added. Similar to the Suffolk County water, the addition of H_2O_2 increased the rate of production of TBF (Figure 6.8). The lower level of TBA and the higher level of acetone compared with the results for the Suffolk County water indicate that the oxidation of the by-products was faster for the Long Beach water. This result is in agreement with the MTBE removal result when the faster MTBE removal rate was observed in Long Beach water than Suffolk County and Santa Monica water.

One distinct difference for the Miami Dade effluent was the non-detectable amount of TBF observed under all conditions tested (Figure 6.9 [a]). The Miami Dade effluent may contain different types of the background constituents, depending on the treatment process used at the wastewater treatment plant. Certain background constituents in the treated water may contain precursors to produce high amount of hydroxyl radicals. Thus TBF may have effectively converted to TBA at a much faster rate than that for other water. In fact, the level of hydroxyl radicals estimated for the Miami Dade water was approximately 10 times higher than the other water matrixes, indicating the highly effective production of hydroxyl radicals for the Miami

Dade effluent during the pulsed-UV process. This may be a reason for non-detectable level of residual TBF in the effluent regardless of H_2O_2 spike, since TBF is oxidized completely by the presence of the large amount of hydroxyl radicals generated during the UV process for this particular water. The inherent differences in the characteristics of water may have been much more important than the apparent TOC and alkalinity values measured for the Miami Dade effluent and other natural water matrixes when making the comparisons. Consequently, more detailed analyses of the background water characteristics may be needed to understand the different behavior of the by-product formation in different types of water.

One interesting observation for the by-product results is the characteristics of intermediate by-product evolution, especially when there was no addition of H_2O_2. With high level of the background NO_3^-, contribution of the photolysis of MTBE or water may be small since the UV light is absorbed strongly by NO_3^- as discussed previously. In turn NO_3^--induced hydroxyl radicals may be generated, and oxidation of MTBE may occur by hydroxyl radicals. For this case, the evolution of the intermediate by-products is similar to the case where H_2O_2 is spiked. For the hydroxyl radical-controlled case, the sequential oxidation of MTBE and MTBE degradation by-products would give rise to each species followed by some lag time for the next by-products to develop peak concentration (Figures 6.10 [a] and [b]) similar to the case for ozone/H_2O_2 process. For this hydroxyl radical-mediated MTBE degradation, as the TBA concentration decrease, acetone concentrations increase in an agreement with the sequential oxidation of the parent compound and the intermediate by-products.

The photolysis-mediated MTBE degradation may not follow the model for ozone/H_2O_2 where degradation of MTBE is dominated by the oxidation by hydroxyl radicals. Figures 6.10 [c]–[e] show such case where NO_3^- levels are low. In the absence of NO_3^-, alkalinity may play a role in filtering UV since bicarbonate absorbs UV near 200 nm. Thus, the Long Beach water with low level of NO_3^- but with high level of alkalinity (150 mg/L) may exhibit a different evolution of by-products as shown in Figure 6.10 [c]. Miami Dade effluent and DI water contain low levels of NO_3^- and alkalinity, and these water matrixes demonstrate the different characteristics of the intermediate by-product evolution compared with that for water with high level of NO_3^- and/or high level of alkalinity. In fact, the evolutions of TBA and acetone are almost identical as shown in Figures 6.10 [d] and [e].

One other interesting observation for these results for no H_2O_2 addition is the absence of TBF for all water matrixes tested except for DI water. As addressed previously, TBF can be photolyzed directly in the absence of H_2O_2. When DI is subject to the UV process, the photolysis of MTBE or water itself is likely to be the mechanism since there is no other background radical initiators, and thus yield different degradation pathway. In fact, the evolutions of the by-products in DI water are quite different from those of other water sources tested. All the by-products measured have the same peaks, suggesting that MTBE is broken into these by-products at the same time, possibly because of the photolysis-mediated pathway for MTBE in the DI water spiked with MTBE. However, it is not clear why the TBF level is high for DI water, since TBF strongly absorbs UV light, and it was expected to be degraded fast. It is not clear whether the difference in the pH of DI and the natural water may have affected the hydrolysis rate of TBF to TBA. Further studies are recommended to investigate the controlling mechanism(s) for TBF degradation for these cases.

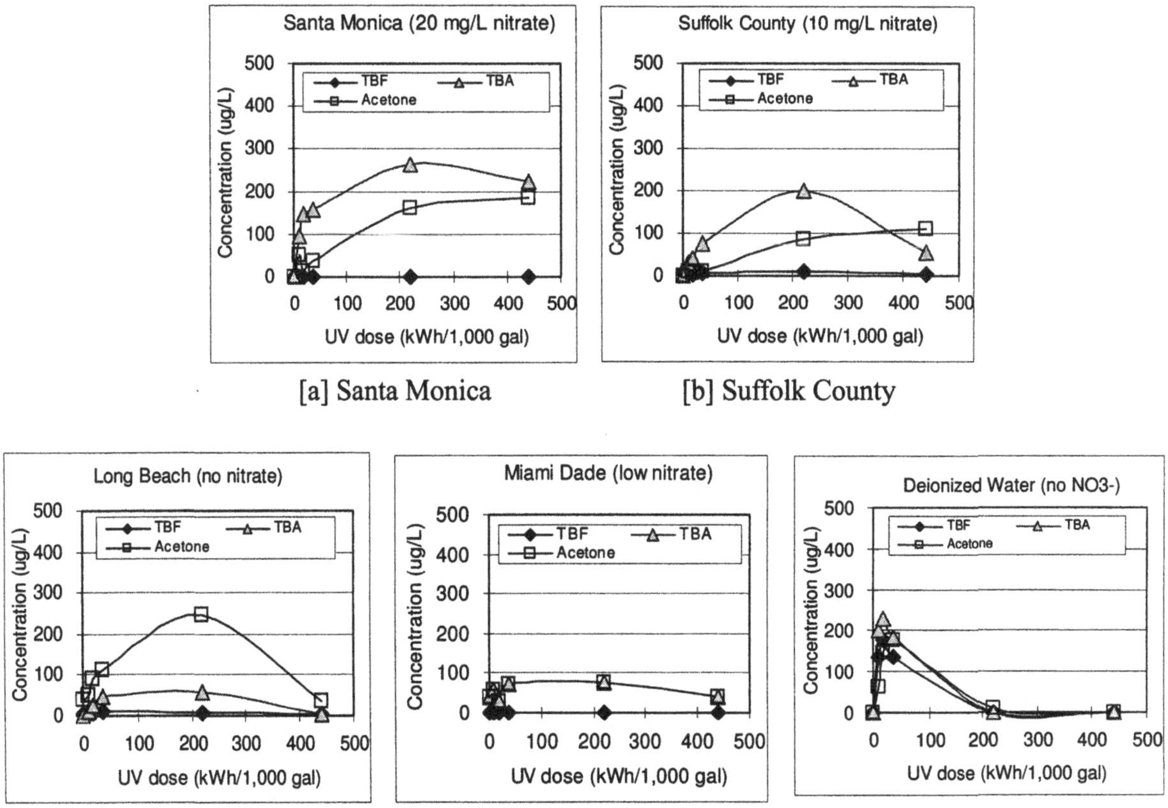

[a] Santa Monica [b] Suffolk County

[c] Long Beach [d] Miami Dade [e] DI water

Figure 6.10 Effect of water matrixes on evolution of intermediate by-products by pulsed-UV

Residual H_2O_2

The tests done with the Santa Monica groundwater with 20 mg/L of NO_3^- produced the excess amount of H_2O_2 during the pulsed-UV test. This was especially true and very consistent when the initial H_2O_2 level was low (less than 10 mg/L) (Figure 6.11 [a]). Five- to ten-fold increase from the initial H_2O_2 was typically measured with low concentration of H_2O_2 spike.

The iodide titration method used for all hydrogen peroxide samples can distinguish between H_2O_2 and organic peroxides since the latter react more slowly with the molybdenum-iodide system [Kormann et al., 1988]. The short mixing of chemicals (5 minutes) was used for the iodometric titration, so the measured peroxide levels for all the cases were mostly H_2O_2, and an insignificant amount of it was likely to be attributed to organic peroxide, which reacts over a period of hours.

The Suffolk County water with 10 mg/L NO_3^- also showed increase in the H_2O_2 levels as shown in Figure 6.11 [b], but to a much less extent than that observed for the Santa Monica water. On the other hand, the decrease in the H_2O_2 levels were observed for the City of Long Beach water and the Miami Dade effluent (except for two data points) as shown in Figure 6.11 [c] and [d].

In natural water, there are certain constituents that may produce H_2O_2 during the advanced oxidation process. Also, hydroxyl radicals produced from the background constituents may be recombined to produce H_2O_2. The consistency in the excess amount of residual peroxide

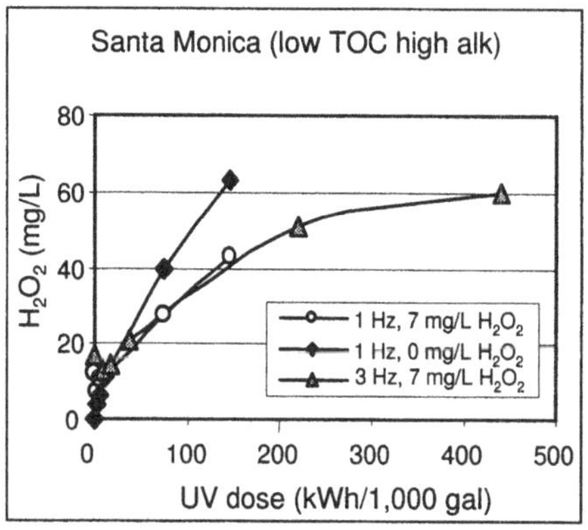

[a] Santa Monica, Calif.

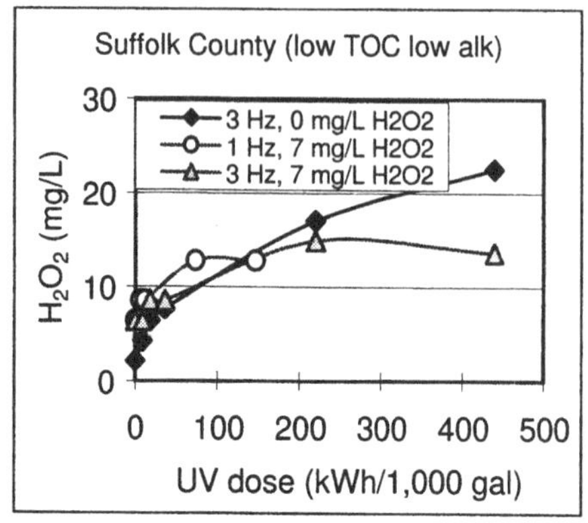

[b] Suffolk County, N.Y.

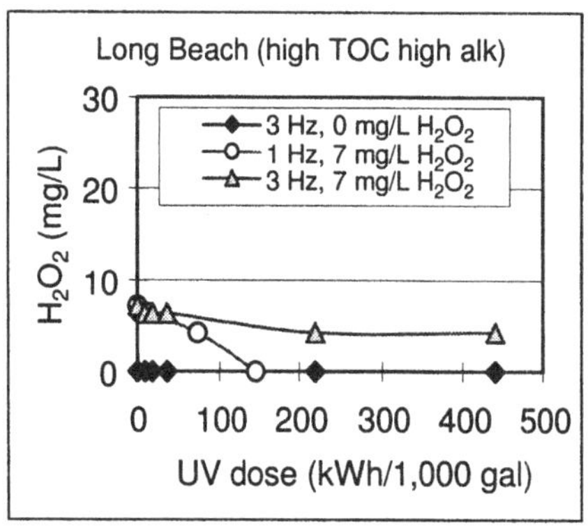

[c] Long Beach, Calif.

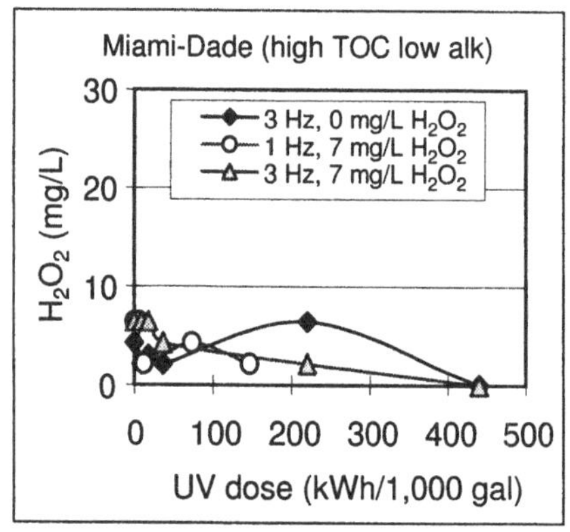

[d] Miami Dade, Fla.

Figure 6.11 Residual H_2O_2 levels for the pulsed-UV experiments

produced with Santa Monica and Suffolk County groundwater might be the result of the presence of high levels of NO_3^- in water. The groundwater from the City of Long Beach contained no detectable level of NO_3^-, and for this case, no increase in the residual H_2O_2 was measured when H_2O_2 was not spiked. When 7 mg/L of H_2O_2 was spiked to the Long Beach water, consistent decrease in the residual peroxide was observed unlike the waters containing high NO_3^-. Miami Dade Wastewater Treatment Plant effluent has low levels of NO_3^- (2 mg/L) and showed a generally decreasing trend of residual peroxide.

The rate of hydroxyl radical recombination has been reported to be on the order of 3.6 to 5.5×10^9 $M^{-1}s^{-1}$ [Buxton et al., 1988]. Thus, it is possible that NO_3^--induced hydroxyl radicals contribute to the increases in the production of H_2O_2 for the Santa Monica and Suffolk County water. Water matrix with low level of NO_3^- did not show any significant production of the residual H_2O_2.

Fate and Effect of NO_3^- in the Water

A consistent trend of NO_3^- to NO_2^- conversion was observed for the tests done with Suffolk County water with approximately 10 mg/L of NO_3^-. A good mass balance was obtained as indicated in Figure 6.12, where total nitrogen was calculated by adding NO_3^--N and NO_2^--N

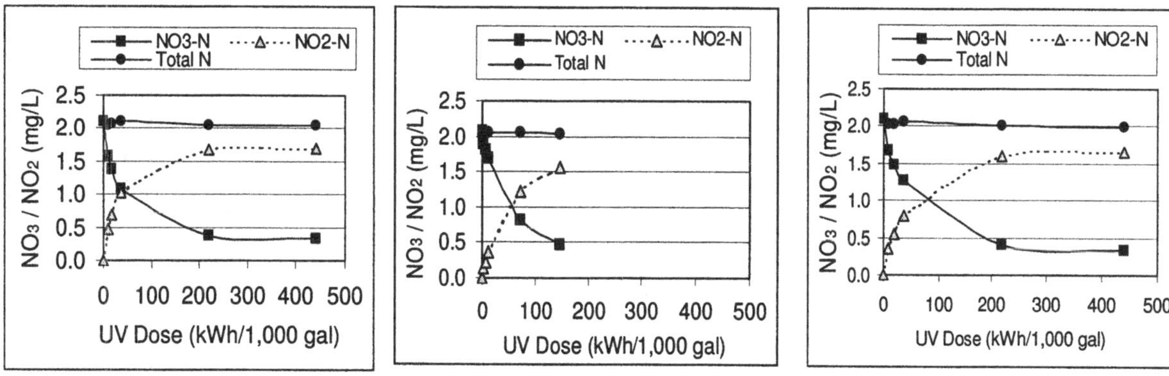

[a] 3 Hz, 0 mg/LH_2O_2 [b] 1 Hz, 7 mg/LH_2O_2 [c] 3 Hz, 7 mg/LH_2O_2

Figure 6.12 NO_3^- and NO_2^- levels in Suffolk County groundwater

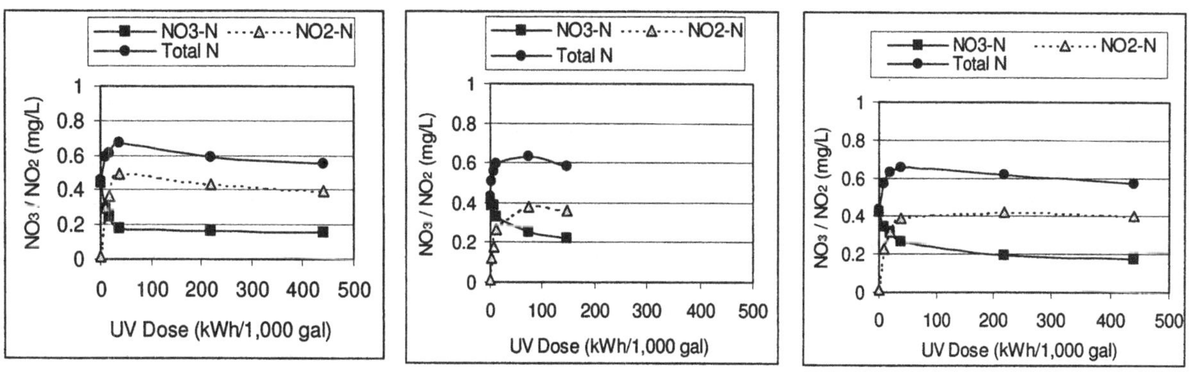

[a] 3 Hz, 0 mg/LH_2O_2 [b] 1 Hz, 7 mg/LH_2O_2 [c] 3 Hz, 7 mg/LH_2O_2

Figure 6.13 NO_3^- and NO_2^- levels in Miami Dade water

concentrations. These results are similar to the ones observed for Santa Monica water with 20 mg/L of NO_3^- (data not shown). The applied UV dose of about 50 kWh/1,000 gal was needed to convert approximately 50 percent of NO_3^- to NO_2^-. The results also suggest that the high concentration of NO_3^- may produce hydroxyl radicals during this conversion process.

Although fast conversion of NO_3^- to NO_2^- was also observed for the tests with Miami Dade Wastewater Treatment Plant effluent, with 2 mg/L of initial NO_3^-, the total nitrogen level was not consistent during the course of experiments as indicated in Figure 6.13.

The level of total nitrogen increased as much as 50 percent at the beginning of the tests and then gradually decreased, as shown in Figure 6.13. It is not clear how the total nitrogen level changed for Miami Dade effluent with low amount of initial NO_3^-, but it is likely that this wastewater treatment effluent contain some precursors that can produce NO_2^- species during the pulsed-UV process, thus contributing to the higher yield of NO_2^- than the value expected from NO_3^- alone as a source.

Effect of Pulsed-UV Process for Removing TBA

TBA is not only one of the major MTBE degradation by-products but also one of the constituents in the reformulated gasoline. Although the amount of TBA is much lower than that of MTBE in the reformulated gasoline, it is likely that the groundwater contaminated with MTBE may also contain low level of TBA in the contaminated water. Since TBA is a suspected carcinogen, tests have been done to study the effect of the pulsed-UV process on treatment of TBA without MTBE spikes.

Effect of H_2O_2 on the removal of low level of TBA in the Santa Monica water is shown in Figure 6.14. The addition of 4 mg/L H_2O_2 increased the rate of TBA removal slightly. Since MTBE was not spiked to the water for these tests, no formation of TBF is shown, but slow increase in the level of acetone is observed as TBA is oxidized by hydroxyl radicals.

The effect of H_2O_2 is also shown for the Santa Monica water with higher level of TBA spike at approximately 1,500 μg/L (Figure 6.15). There are some scattered data points for TBA, but the general decreasing trend is shown for both cases. The addition of 4 mg/L H_2O_2 also increased the efficiency of TBA reduction with higher level of TBA, but the overall percentage removal was much less than that observed for MTBE with similar applied UV dosage. TBA is known to react with hydroxyl radicals at the rate constants in the order of 4.8×10^8 to 7.6×10^8 $M^{-1}s^{-1}$, so the slower rate of reduction for TBA compared with MTBE was expected. However, increase in the pulse frequency (to 3 Hz) and the level of H_2O_2 spike may effectively reduce the TBA level from typical contamination level of 200 μg/L to below the action level of 12 μg/L. Further optimization of the test parameter is recommended to investigate the effective treatment alternative for TBA with the pulsed-UV/H_2O_2 process.

Measurement of Hydroxyl Radicals Production

The average rate of reaction of hydroxyl radicals with ethanol reported in recent literatures is approximately 2.0×10^9 $M^{-1}s^{-1}$ [Buxton et al., 1988]. This rate is within same order of magnitude, so the same sampling time used as in the MTBE treatment study would have produced sufficient amount of acetaldehyde, which is the oxidation product of primary probe compound. Acetaldehyde was monitored to correlate the steady state hydroxyl radical concentrations.

There were two issues related to adopting this process for measuring the production of steady state hydroxyl radicals concentrations: 1) aldehyde interference with UV and 2) aldehyde interference with hydroxyl radicals.

Both acetaldehyde and formaldehyde samples were prepared in Super Q® laboratory water and the samples were scanned using Shimadzu UV/Vis 2401 PC. Four different concentration samples were scanned from 0 up to 20 mg/L spike range, which is far above the expected level of production during the tests. This was done to make sure that there is no absorption of UV by the aldehydes in this range. Absorption of UV light by acetaldehyde and formaldehyde was negligible.

The level of acetaldehyde can increase as the oxidation of ethanol occurs by the hydroxyl radicals, which are produced during pulsed-UV process. As the acetaldehyde concentration increases above certain critical concentration, the level of acetaldehyde can be affected by the reaction of acetaldehyde with hydroxyl radicals. It has been reported that the rate of reaction of acetaldehyde with hydroxyl radicals is 3.6×10^9 $M^{-1}s^{-1}$. The scope of the proposed method was to investigate if the proposed method has a potential to be refined as a method for estimating

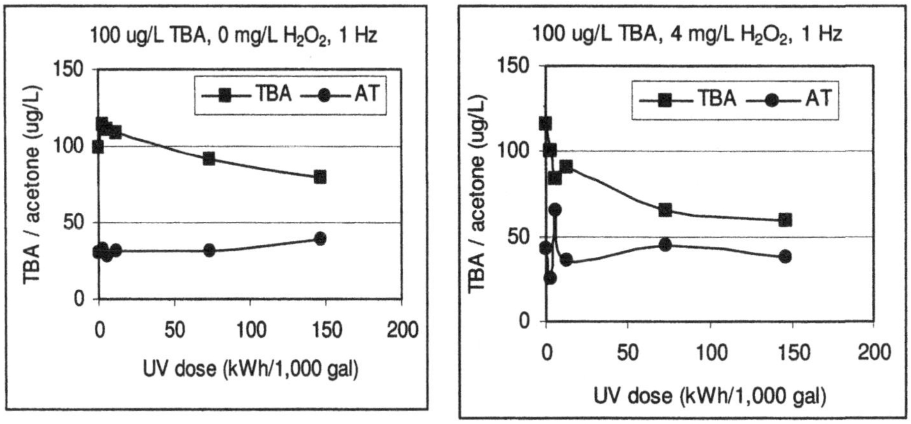

[a] 1 Hz, 0 mg/LH_2O_2 [b] 1 Hz, 4 mg/LH_2O_2

Figure 6.14 Effect of H_2O_2 on removal of low concentration of TBA in Santa Monica water (Note: AT = acetone)

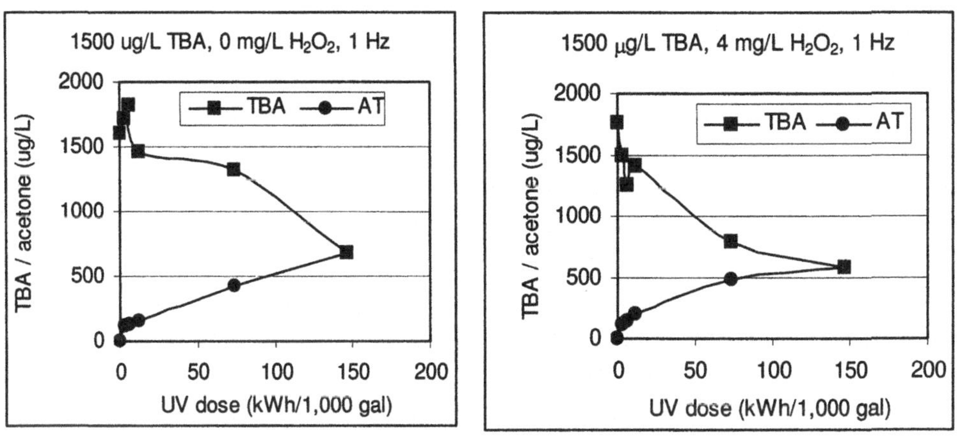

[a] 1 Hz, 0 mg/LH_2O_2 [b] 1 Hz, 4 mg/LH_2O_2

Figure 6.15 Effect of H_2O_2 on removal of high concentration of TBA in Santa Monica water

production of hydroxyl radicals, and thus compare different type of UV lamps and the reactor design. More accurate estimation could have been made by directly measuring the decay of ethanol instead of measuring the production of acetaldehyde, but no reliable analytical method was available at the time of the testing. However, the preliminary results indicate that the reliable production of acetaldehyde may be obtained to quantitatively compare the effect of various parameters and the effect of background constituents affecting the production of hydroxyl radicals.

The steady state hydroxyl radicals concentration ($[OH]_{ss}$) can be estimated using the following equation,

$$[OH]_{ss} = [acetaldehyde]/t \times 1/(k \times [EtOH]) \times (1/Y) \quad (6.9)$$

where

k	=	1.90×10^9 L/mol-sec
$[OH]_{SS}$	=	steady state hydroxyl radicals concentration
[acetaldehyde]	=	measured acetaldehyde concentration
t	=	time in sec
[EtOH]	=	initial ethanol concentration (14 mg/L or 0.30 mM)
Y	=	ethanol to acetaldehyde yield (0.85 to 0.95)

The assumptions made for this approximation is that 1) ethanol concentration is constant with respect to the acetaldehyde production during the tests, 2) the production of acetaldehyde will be high enough to be monitored but low enough so that acetaldehyde does not scavenge hydroxyl radicals, and 3) that acetaldehyde is first order below this critical acetaldehyde concentration.

One more factor that can interfere with this approach, especially with high alkalinity water, is the ethanol oxidation by carbonate radicals, which are generated during hydroxyl radicals scavenging by carbonates. Production of carbonate radicals from high alkalinity water may interfere with ethanol oxidation by hydroxyl radicals.

Some of the hydroxyl radicals tests have been cut down since there were problems with using high concentration of ethanol (140 mg/L) and hydrogen peroxide (20 to 70 mg/L). The level of aldehyde production was extremely high even for the low concentration of ethanol (14 mg/L) and hydrogen peroxide (2 to 7 mg/L) used, so it was not possible to obtain a reliable measurements for tests with high concentration of ethanol due to high dilution factor needed even for the low level ethanol tests. Also vaporization of aldehyde after 5 minutes of sampling time posed a problem for 1) the safety of operators from exposure to aldehyde vapor and 2) the reliability of accurately measuring aqueous concentration of aldehydes because the loss of aldehydes in gas phase could not be accounted for. Consequently, except for a few initial tests, the hydroxyl radicals estimation tests were carried out only up to five minutes of sampling time. The initial four samples taken during the first five minutes of sampling time, however, would have provided enough information as to the validity of the proposed method.

Figure 6.16 shows the production of acetaldehyde during the pulsed-UV experiments with 14 mg/L of ethanol and varying amount of initial hydrogen peroxide spikes for Santa Monica groundwater. The level of acetaldehyde reached plateau, which may be indicative of acetaldehyde starting to react with hydroxyl radicals and thus limiting the further production of acetaldehyde. For Santa Monica water, no clear trend of the effect of pulsed UV input or the initial hydrogen peroxide was observed on acetaldehyde production. Smaller amount of formaldehyde, which is not directly related to ethanol oxidation by hydroxyl radicals, was also measured during the tests as shown in Figure 6.16 (b). The trend was much more linear than acetaldehyde, but the production of formaldehyde is not associated with direct oxidation of ethanol. The background constituents may have contributed to the production of small amount of formaldehyde during the pulsed-UV process. Increase in the level of residual hydrogen peroxide was observed for Santa Monica water at 3 Hz. With 1 Hz pulse frequency, a slight decrease in the residual peroxide level was observed as shown below.

For Suffolk County water, much more linear and well correlated acetaldehyde production results were obtained as shown below (Figure 6.17). Under the similar test conditions, acetaldehyde level consistently increased with the applied UV dose of up to 40 kWh/1,000-gallons for all test conditions used. The rate of acetaldehyde production showed good correlation

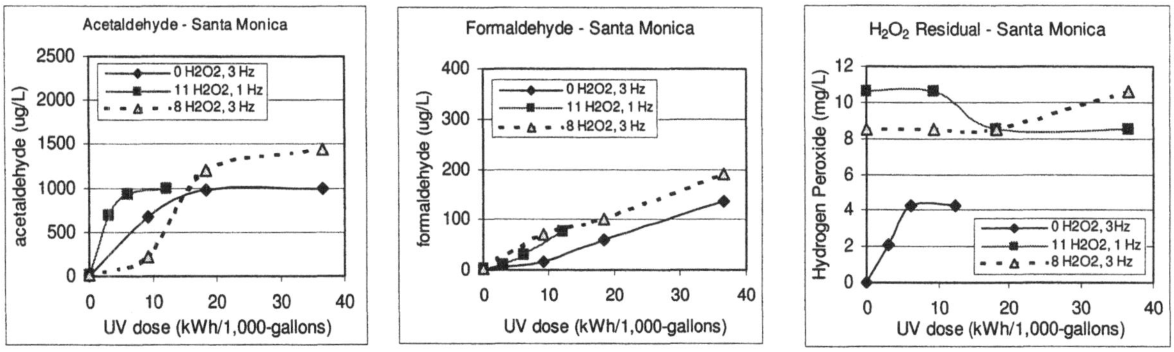

[a] acetaldehyde, Santa Monica [b] formaldehyde, Santa Monica [c] H_2O_2 residual, Santa Minica

Figure 6.16 Acetaldehyde, formaldehyde, and H_2O_2 residual for hydroxyl measurement test for Santa Monica groundwater using pulsed-UV

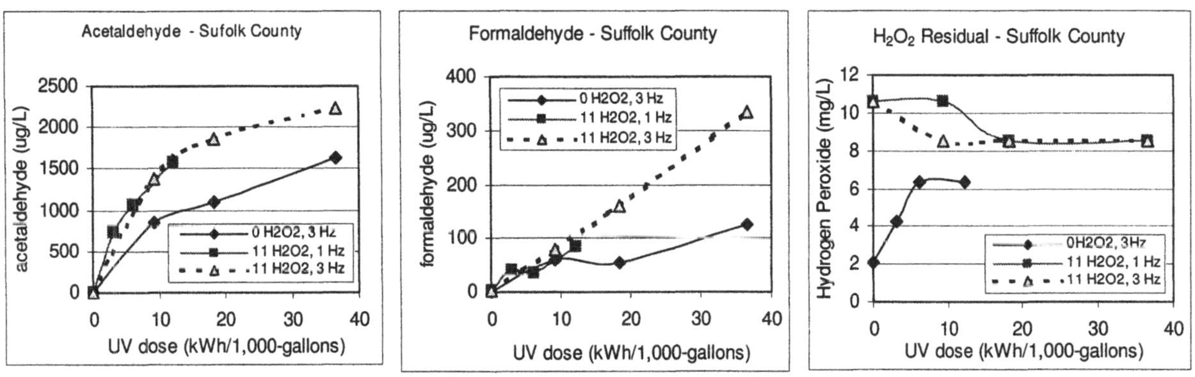

[a] acetaldehyde, Suffolk Co. [b] formaldehyde, Suffolk Co. [c] H_2O_2 residual, Suffolk Co.

Figure 6.17 Hydroxyl estimation for Suffolk County groundwater using pulsed-UV

for the cases with H_2O_2 spikes regardless of pulse frequency. The pulse frequency is normalized to the applied UV dose, so with the same irradiation time, the applied UV dose at 3 Hz is three times the UV dose at 1 Hz as indicated in the Figure. Similar trends of production of formaldehyde and hydrogen peroxide residual were observed for Suffolk County water compared with Santa Monica water.

Similar to the Suffolk County water, a well-established trend of acetaldehyde and formaldehyde were also observed for Long Beach water (Figure 6.18). Again clear effect of the presence of hydrogen peroxide was observed by the enhanced production of aldehydes. It was noted in a previous section on hydrogen peroxide, that the level of H_2O_2 consistently decreased during the MTBE treatment tests with Long Beach water. However, the results here indicate that the level of hydrogen peroxide increased during the hydroxyl radicals estimation experiments. The different experimental setup with ethanol as a probe compound might have caused the differences in terms of residual hydrogen peroxide level.

In the absence of hydrogen peroxide, the level of acetaldehyde production for Long Beach water with low nitrate, possible source of hydroxyl radicals in the absence of hydrogen peroxide

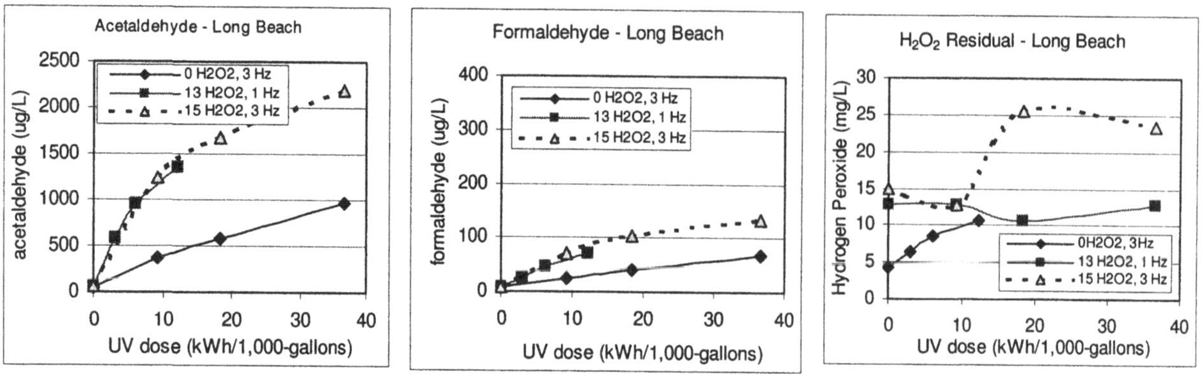

[a] acetaldehyde, Long Beach   [b] formaldehyde, Long Beach   [c] H_2O_2 residual, Long Beach

Figure 6.18 Hydroxyl estimation test for Long Beach groundwater using pulsed-UV

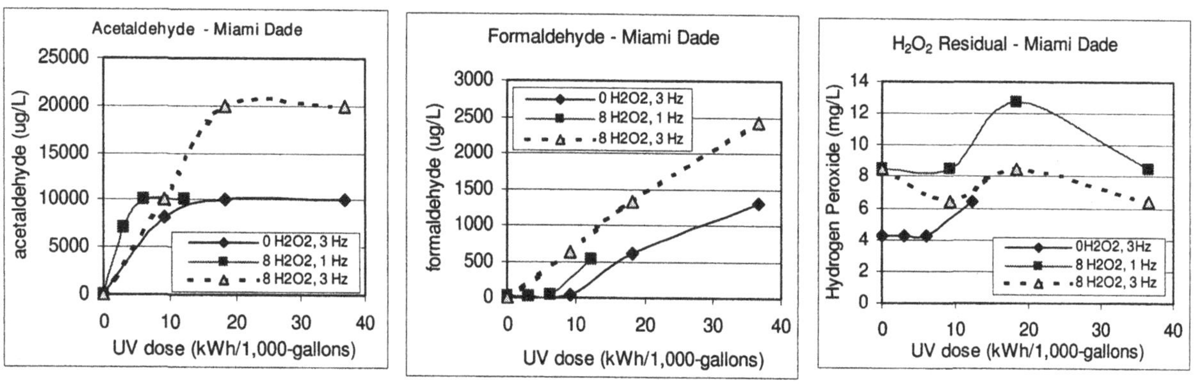

[a] acetaldehyde, Miami Dade   [b] formaldehyde, Miami Dade   [c] H_2O_2 residual, Miami Dade

Figure 6.19 Hydroxyl estimation test for Miami Dade wastewater effluent using pulsed-UV

spike, was lower than that of Santa Monica and Suffolk County waters. In fact, the comparison of initial slopes of acetaldehyde production with Long Beach water suggests lower production of hydroxyl radicals compared with other waters with high nitrate.

One distinct aspect of the results obtained on Miami Dade Wastewater Treatment Plant effluent is the level of aldehyde production during the hydroxyl estimation test (Figure 6.19). Approximately 10 times higher level of acetaldehyde and formaldehyde were produced. The data has been reconfirmed by the Organic Research Chemistry Team of the Water Quality Section at Metropolitan, so the results are not likely due to the analytical error, although high dilution rate was necessary to measure the aldehydes with the analytical linear range of approximately 100 μg/L. The level of acetaldehyde reached a plateau as shown in Figure 6.19 [a], but the strong linear production of formaldehyde was shown (Figure 6.19 [b]). As discussed earlier, there may be differences in inherent characteristics of wastewater treatment effluent compared with natural waters that may be attributed for this observation.

Although it has been shown that the acetaldehyde production is high and may correlate under certain conditions with ethanol oxidation, the model needs to be modified to incorporate the scavenging effect of acetaldehyde, oxidation of ethanol by carbonate radicals during the hydroxyl radicals estimation tests.

More accurate results may be obtained by also measuring decrease in ethanol concentration along with production of acetaldehyde for this approach to be more reliable since both of these constituents are involved with the experimental conditions used for hydroxyl radicals estimation tests. Although it may be difficult to estimate the steady state hydroxyl radical concentrations for field waters due to number of parameters affecting the process, this approach may be used to compare different types of UV lamps by using laboratory grade Super Q® water, where background interference is small compared with filed waters.

Summary of Findings

Although the batch tests results confirm that the background water quality plays an important role in the mechanism(s) involved in the reduction of MTBE, effective removal of MTBE was observed for all water matrixes tested even without the addition of H_2O_2. Generally, the water with high background NO_3^- inhibited the decay of MTBE much more than the water with high levels of TOC and alkalinity, which are known to scavenge hydroxyl radicals. Santa Monica water and Suffolk County water with high levels of NO_3^- exhibited slower MTBE removal rates when H_2O_2 was not added to the system. The addition of H_2O_2 increased the rates of MTBE decay, especially for the Santa Monica water. Thus in designing the flow through pilot scale testing, it may be advantageous to add H_2O_2 to decrease the retention time needed for higher throughput from the system. Regardless of the water characteristics, it was possible to remove MTBE below 5 – 6 µg/L from 1,000 – 2,000 µg/L initial MTBE levels (99 percent reduction) when H_2O_2 was added with the irradiation time of 12 minutes. The effects of water quality characteristics were much more evident for the formation and fate of MTBE degradation by-products. The full characterization of the water to be tested would provide valuable information in predicting the effectiveness of the pulsed-UV system since some background constituents in a specific water matrix have important ramifications with respect to the effectiveness of MTBE treatment.

PILOT SCALE TESTING

Introduction

Similar to the ozone pilot scale tests, the pulsed UV with hydrogen peroxide experiments were setup to utilize pulsed UV as a primary treatment method for MTBE and biological GAC as a post treatment for MTBE degradation by-products such as TBA, TBF, and aldehydes. Although the pilot scale UV used the same type of lamp used in the bench scale pulsed UV unit, different reactor configuration and the power supply used required new optimization of the operating conditions to meet the treatment criteria. However, the bench scale results provided valuable information as the basis for correlating the amount of energy and hydrogen peroxide applied to the removal efficiency of MTBE and the by-products. Thus, a series of flow through pilot scale tests were performed to validate the MTBE and by-product treatment results from the bench scale experiments using the pilot scale (10 gallon) pulsed UV reactor and the biological GAC (BAC) columns, located at the Metropolitan's La Verne Pilot Plant (La Verne, CA). The same BAC used for the ozone pilot scale tests were also used for the pilot scale pulsed UV tests, so for the pilot scale pulse UV tests, which were performed after the ozone tests, BAC probably had less chemical adsorption capacity for some of the constituents. The experimental conditions were initially chosen to treat MTBE to below the secondary standard of 5 μg/L, and although some tests conditions did not meet this criteria, it provided information as to how much more energy or retention time would be required to meet the goal in an actual application. The operational conditions for pulsed UV, such as frequency (applied energy) and flow rate, can easily modified to accommodate a desired level of treatment.

Experimental Plan and Objectives

The pilot scale pulsed UV tests were designed to investigate the effect of water matrix (surface water vs. groundwater), applied UV dose by varying the frequency (the frequency of UV application), hydrogen peroxide dose, and the initial MTBE concentration. Similar to the ozone pilot tests, effect of the presence of TBA was also studied in removing MTBE during the advanced oxidation process for the groundwater used. Although the removal of TBA only, (also a primary contaminant in the gasoline contaminated water) without MTBE by the pulsed UV was investigated in the bench scale studies, it was not addressed for the pilot scale tests. However, some of the tests using both MTBE and TBA can provide information as to how much TBA may be removed during the pilot scale pulsed UV treatment.

The main objective of the tests done during the pilot scale phase was to validate the performance of the bench scale advanced oxidation processes, specifically pulsed UV with hydrogen peroxide in pilot scale, flow through tests. Specific objectives for this portion of the project are listed below.

1. Investigate and validate the performance of pulsed UV/H_2O_2 process for the removal of MTBE with flow through pilot scale setup
2. Investigate the performance of biological GAC column in removing MTBE degradation by-products in combination with pulsed UV/H_2O_2 process
3. Evaluate the effects of water source (groundwater and surface water) on pulsed UV/ H_2O_2 process in pilot scale

Materials & Methods

Pilot Scale Pulsed UV System

The pilot scale pulsed-UV system (Pulsar UV Technologies, Roseville, CA) used was capable of providing up to 8 Hz of pulsed UV (see Figure 6.20). The pulsed-UV pilot test setup also involved influent equalization tank to be used for pumping MTBE spiked water. For a limited number of control tests where no UV is applied, the effluent was collected and treated before discharging as in the case of ozone control tests. The reactor was approximately 10 gallons, and the flow rate was determined based on the preliminary pilot tests based on the results on the bench scale unit, which had smaller power supply and the energy delivery. For all tests, 1 gpm flow rate was used to provide 10 minute retention time in the pulsed UV reactor. The adequate flow rate was used to allow sufficient mixing of the water in the UV reactor. When appropriate hydrogen peroxide was injected prior to the pulsed UV reactor. The tests were done at two energy levels to compare the effect of applied UV in removing MTBE as well as varying level of hydrogen

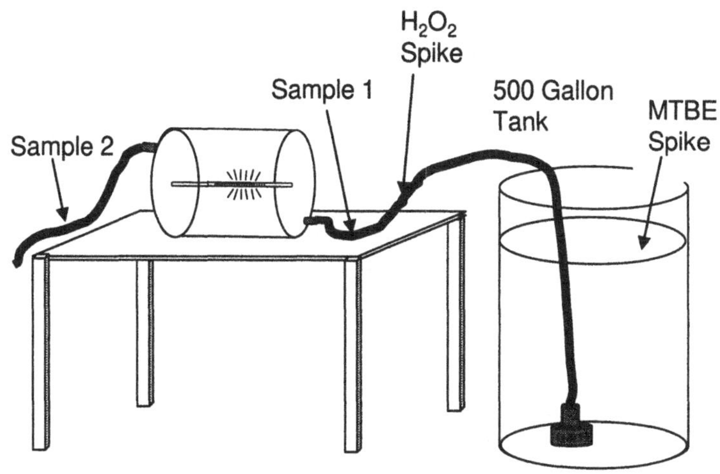

Figure 6.20 Schematic of pilot scale pulsed-UV/H_2O_2 setup

peroxide. The pulsed UV frequency of 4 Hz and 7 Hz (0.5 kWh and 0.58 kWh respectively based on the manufacturer's data) were used for all the pilot scale pulsed UV tests.

As mentioned before, pulsed UV was primarily used in treating MTBE, and the retrofitted BAC was used to complement the pulsed UV by removing the MTBE degradation by-products, such as TBA, TBF, acetone, and aldehydes. The carbon used to pack the biological columns was virgin Calgon Carbon (Pittsburgh, PA) Filtrasorb 400 with the apparent density of 0.48 g/cm^3. Filtrasorb 400 has some MTBE adsorption capacity as shown in the following isotherm (Figure 6.21), but during three months of operation with ozone pilot scale system, some of the adsorption capacity for MTBE as well as for some other by-products (with lower adsorption capacity than MTBE) may have been decreased.

The schematic of biological GAC equipped with dechlorinated water tank for backwashing, is shown in Figure 6.22. The flow rate to the biological filters was adjusted to achieve 20-minute retention time in the filters. As in the case for the ozone pilot scale setup, 100% raw SPW was pumped through the biological columns continuously during non-test period and the filters were typically backwashed once every two days.

Sampling

For each pilot scale flow through test, influent, pulsed UV effluent, and BAC effluent samples were taken for analyses of MTBE, MTBE degradation by-products, and hydrogen peroxide residual. Since elevated levels of aldehydes are typically found only in the effluent as a result of the breakdown of MTBE or natural organic matter in the water matrix during the treatment process, aldehydes samples were only analyzed for pulsed UV effluent and BAC effluent. All samples were quenched for residual hydrogen peroxide (except for the hydrogen peroxide samples) as soon as the samples were taken. For the cases, where initial MTBE concentration is in the order of few hundred µg/L, TBF levels are typically low, and the analyses were not performed for TBF. Also, nitrate, and nitrite were analyzed for the tests with Santa Monica groundwater where the initial nitrate level is greater than 4.5 mg/L nitrate as nitrogen.

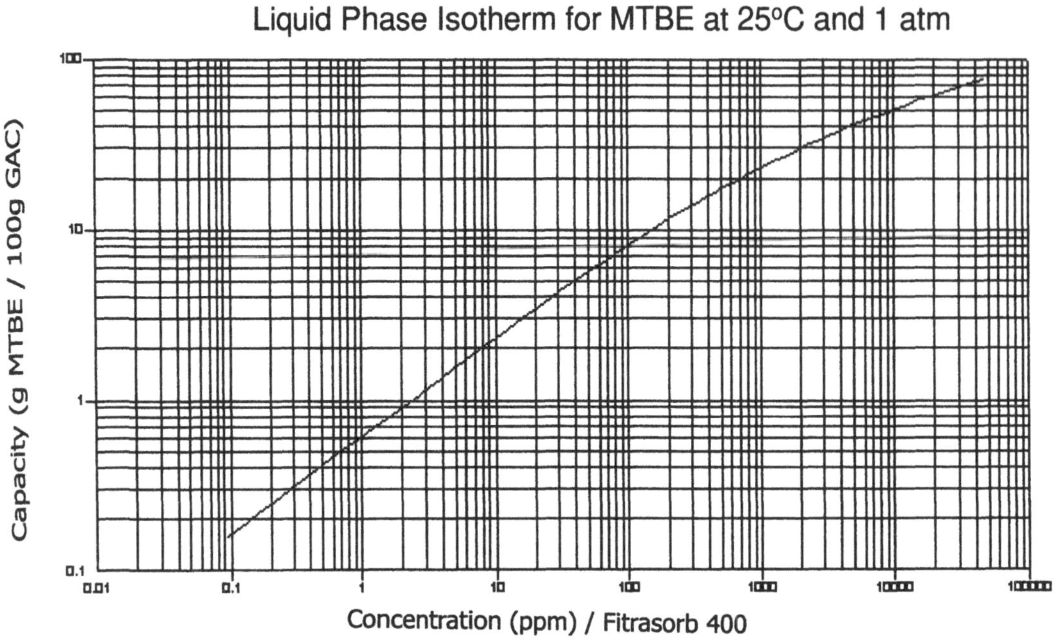

Figure 6.21 Calgon Carbon GAC (Filtrasorb 400) isotherm for MTBE used for biological columns

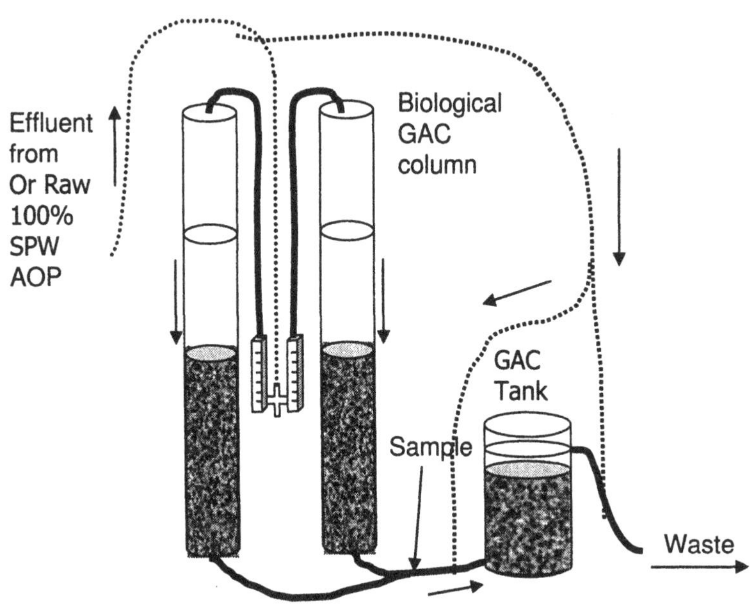

Figure 6.22 Schematic of the post treatment biological GAC setup

Table 6.3
Typical background water quality characteristics

Parameters	Unit	State Project Water Surface Water	Santa Monica Groundwater
TOC	(mg/L)	3.33	0.52
EC	(μmhos)	500	1319
Nitrate as NO_3	(mg/L)	1.98	21.5
Alkalinity	(mg/L)	82	326
Turbidity	(NTU)	1.7	0.14
Total Hardness	(mg/L)	114	538
pH		8.05	7.21
UV_{254}	(abs/cm)	0.087	0.01
Bromide	(mg/L)	0.2	0.18

Volatilization Losses

A series of control tests were done prior to pulsed UV pilot-scale tests to measure the MTBE losses from volatilization. During these tests, no significant temperature rise was observed in control tests that were done without pulsed UV input. No significant MTBE loss was observed during pilot-scale flow through setup.

Results & Discussion

Raw Water Quality

As it was mentioned briefly in the previous section, one groundwater and one surface water were used as the source water. Table 6.3 provides the background water quality characteristics of the groundwater from the City of Santa Monica, CA and 100% State Project Water used for the pilot scale tests. Typical contamination level for the surface water is up to a few hundred μg/L range, while the level of groundwater contamination is in the order of a few thousand μg/L. Thus, the raw water and the groundwater were spiked up to 200 and 2,000 μg/L of MTBE respectively for the tests.

MTBE and By-Products Removal With Pulsed UV/H_2O_2 in Surface Water

Pulsed UV only without the addition of hydrogen peroxide showed some degree of MTBE removal at the bench scale. The water quality matrix played important role in terms of the MTBE removal efficiency since some water matrix may contain higher level of initiators to generate hydroxyl radicals, which oxidize MTBE. At the higher applied UV frequency of 7 Hz without any hydrogen peroxide, the MTBE removal efficiency of 40 ~ 57% were observed (Tables 6.4 and 6.5) with retention time of 10 minutes in the flow through setup. During the ozone pilot scale tests, it was demonstrated that the biological GAC exhibited removal of MTBE as well as by-products, and similar trends were observed with the combination of pulsed UV and the BAC.

Table 6.4
Removal of 200 μg/L MTBE and by-products with pulsed UV only at 7 Hz

Sample Location	MTBE (μg/L)	TBA (μg/L)	TBF (μg/L)	Acetone (μg/L)	Acetaldehyde (μg/L)	Formaldehyde (μg/L)	Glyoxal (μg/L)	M-Glyoxal (μg/L)	H_2O_2 (mg/L)
Detection Limit	0.1	0.1	5	10	1	5	1-2	1-2	0.5
Influent	223	8	NA	11	NA	NA	NA	NA	0
Effluent	97	28	NA	52	22	84	16	15	2
BAC Eff.	ND	ND	NA	5	ND	9	ND	ND	1

ND – Not Detected, NA – Not Analyzed

Table 6.5
Removal of 50 μg/L MTBE and by-products with pulsed UV only at 7 Hz

Sample Location	MTBE (μg/L)	TBA (μg/L)	TBF (μg/L)	Acetone (μg/L)	Acetaldehyde (μg/L)	Formaldehyde (μg/L)	Glyoxal (μg/L)	M-Glyoxal (μg/L)	H_2O_2 (mg/L)
Influent	44	ND	NA	36	NA	NA	NA	NA	NA
Effluent	27.0	6.9	NA	65	21	53	15	14	NA
BAC Eff.	9.7	ND	NA	61	ND	8	ND	ND	NA

Although the MTBE remained in the pulsed UV effluent were completely removed by the BAC column for one case (Table 6.4), only 64% was removed for the case with the lower MTBE concentration of 44 μg/L (Table 6.5). When the complete removal of MTBE was observed in the BAC column, greater than 90% removal of acetone was also observed (Table 6.4). On the other hand, when the limited MTBE removal efficiency was observed, only 7% reduction of acetone was achieved in the BAC column, which is far less than the typical removal efficiency observed during the ozone pilot testing.

The primary degradation by-products of MTBE during an advanced oxidation process are TBA, TBF, acetone, and formaldehyde. Depending on the level of oxidation, typical break down of MTBE by oxidation causes the concentrations of primary MTBE degradation by-products (TBA and TBF) to increase [Liang et al., 1999a]. Further application of AOP then reduces the level of TBA and TBF to form the secondary by-products such as acetone and formaldehyde. TBF, however, also goes through hydrolysis reaction to TBA, which affect the level of both TBA and TBF in the effluent. The smaller concentrations of acetaldehyde, glyoxal, and M-glyoxal observed throughout this section may have been due to the degradation of other constituents, such as natural organic matter, in the surface water, but not directly from the MTBE itself.

For the control study done without applying any pulsed UV light, no loss of MTBE was observed in the UV reactor as expected, indicating negligible amount of MTBE adsorption in the system. All of MTBE found in the effluent was removed by the BAC column as seen in Table 6.6. Thus implementing biological GAC as a post treatment for AOP can provide additional treatment barrier for MTBE during the possible upset of the system.

As in the case of ozone, addition of hydrogen peroxide improved the MTBE removal efficiency by the pulsed UV system. With the applied hydrogen peroxide dose of 2 mg/L, 70 ~ 80%

Table 6.6
Removal of 200 µg/L MTBE and by-products with biological GAC (no UV)

Sample Location	MTBE (µg/L)	TBA (µg/L)	TBF (µg/L)	Acetone (µg/L)	Acetaldehyde (µg/L)	Formaldehyde (µg/L)	Glyoxal (µg/L)	M-Glyoxal (µg/L)	H_2O_2 (mg/L)
Influent	198	3	NA	18	NA	NA	NA	NA	6
Effluent	201	5	NA	50	NA	NA	NA	NA	5
BAC Eff.	1	ND	NA	22	NA	NA	NA	NA	0

Table 6.7
Removal of 200 µg/L MTBE with pulsed UV of 4 Hz and 2 mg/L H_2O_2

Sample Location	MTBE (µg/L)	TBA (µg/L)	TBF (µg/L)	Acetone (µg/L)	Acetaldehyde (µg/L)	Formaldehyde (µg/L)	Glyoxal (µg/L)	M-Glyoxal (µg/L)	H_2O_2 (mg/L)
Influent	199	6	NA	ND	NA	NA	NA	NA	2
Effluent	35	17	NA	65	21	106	21	21	1
BAC Eff.	3	ND	NA	3	ND	9	ND	ND	0

Table 6.8
Removal of 200 µg/L MTBE with pulsed UV of 7 Hz and 2 mg/L H_2O_2

Sample Location	MTBE (µg/L)	TBA (µg/L)	TBF (µg/L)	Acetone (µg/L)	Acetaldehyde (µg/L)	Formaldehyde (µg/L)	Glyoxal (µg/L)	M-Glyoxal (µg/L)	H_2O_2 (mg/L)
Influent	186	6	NA	ND	NA	NA	NA	NA	3
Effluent	45	18	NA	70	19	101	19	20	2
BAC Eff	9	ND	NA	12	ND	9	ND	ND	0

Table 6.9
Removal of 50 µg/L MTBE with pulsed UV of 4 Hz and 2 mg/L H_2O_2

Sample Location	MTBE (µg/L)	TBA (µg/L)	TBF (µg/L)	Acetone (µg/L)	Acetaldehyde (µg/L)	Formaldehyde (µg/L)	Glyoxal (µg/L)	M-Glyoxal (µg/L)	H_2O_2 (mg/L)
Influent	49	4	NA	55	NA	NA	NA	NA	NA
Effluent	14	9	NA	442	19	66	18	18	NA
BAC Eff	0.9	ND	NA	39	ND	8	ND	1	NA

of MTBE was removed during the pulsed UV process (Tables 6.7 and 6.8) compared with 40 ~ 60% removal observed without hydrogen peroxide. At the initial MTBE concentration of approximately 200 µg/L, the level of applied UV energy (frequency) did not significantly affect the MTBE removal efficiency (Tables 6.7 and 6.8). At a lower MTBE concentration of 50 µg/L similar MTBE and by-products removal efficiency was also observed (Tables 6.9 and 6.10). Again the applied pulsed UV energy (frequency, 4 Hz and 7 Hz) did not play any significant role

Table 6.10
Removal of 50 μg/L MTBE with pulsed UV of 7 Hz and 2 mg/L H_2O_2

Sample Location	MTBE (μg/L)	TBA (μg/L)	TBF (μg/L)	Acetone (μg/L)	Acetaldehyde (μg/L)	Formaldehyde (μg/L)	Glyoxal (μg/L)	M-Glyoxal (μg/L)	H_2O_2 (mg/L)
Influent	45	4	NA	36	NA	NA	NA	NA	NA
Effluent	14	6	NA	71	19	62	17	17	NA
BAC Eff	ND	ND	NA	16	ND	6	ND	1	NA

Table 6.11
Removal of 200 μg/L MTBE with pulsed UV of 4 Hz and 4 mg/L H_2O_2

Sample Location	MTBE (μg/L)	TBA (μg/L)	TBF (μg/L)	Acetone (μg/L)	Acetaldehyde (μg/L)	Formaldehyde (μg/L)	Glyoxal (μg/L)	M-Glyoxal (μg/L)	H_2O_2 (mg/L)
Influent	195	5	NA	ND	NA	NA	NA	NA	4
Effluent	20	10	NA	4	12	97	18	24	3
BAC Eff	3	ND	NA	ND	ND	8	ND	ND	1

Table 6.12
Removal of 200 μg/L MTBE with pulsed UV of 7 Hz and 5 mg/L H_2O_2

Sample Location	MTBE (μg/L)	TBA (μg/L)	TBF (μg/L)	Acetone (μg/L)	Acetaldehyde (μg/L)	Formaldehyde (μg/L)	Glyoxal (μg/L)	M-Glyoxal (μg/L)	H_2O_2 (mg/L)
Influent	186	5	NA	11	NA	NA	NA	NA	5
Effluent	8	7	NA	74	13	100	20	25	4
BAC Eff	4	ND	NA	12	ND	8	ND	ND	1

in the removal efficiency. With 2 mg/L of hydrogen peroxide, the level of MTBE after the BAC column was all below the secondary regulation of 5 μg/L except for one case (Table 6.8).

As the amount of hydrogen peroxide spiked increased above 4 mg/L, increased in the removal efficiencies were observed. Greater than 90% reduction of MTBE was consistently observed with at least 4 mg/L of hydrogen peroxide regardless of the applied UV energy (frequency) (Tables 6.11 and 6.12). Thus, the hydrogen peroxide level was far more sensitive than the UV dose in removing the MTBE and the by-products. The BAC also removed all the residual level of MTBE and the by-products to low levels.

A test done at the adjusted pH of 6.5 (Table 6.13) showed greater than 90% removal efficiency with hydrogen peroxide of 4 mg/L. This is consistent with the previous results at the typical pH range of 8 with the similar level of hydrogen peroxide. Thus, the application of the pulsed UV technology for the MTBE treatment for surface water requires sufficient amount of hydrogen peroxide to oxidize MTBE. The levels of by-products from the water matrix (acetaldehyde and glyoxals) were not decreased at the lower level of pH. Lower pH can be used to control the level of bromate during the ozone process, however, since no detectable level of bromate has been observed in the pulsed UV effluent, the lower pH didn't have any effect on the pulsed UV/H_2O_2 process.

Table 6.13
Removal of 200 µg/L MTBE with pulsed UV of 7 Hz and 4 mg/L H_2O_2 at pH 6.5

Sample Location	MTBE (µg/L)	TBA (µg/L)	TBF (µg/L)	Acetone (µg/L)	Acetaldehyde (µg/L)	Formaldehyde (µg/L)	Glyoxal (µg/L)	M-Glyoxal (µg/L)	H_2O_2 (mg/L)
Influent	207	2	NA	13	NA	NA	NA	NA	4
Effluent	8	5	NA	73	13	71	23	34	0
BAC Eff	ND	ND	NA	11	ND	6	ND	ND	0

Table 6.14
Removal of 200 µg/L MTBE in groundwater with pulsed UV of 7 Hz

Sample Location	MTBE (µg/L)	TBA (µg/L)	TBF (µg/L)	Acetone (µg/L)	Acetaldehyde (µg/L)	Formaldehyde (µg/L)	Glyoxal (µg/L)	M-Glyoxal (µg/L)	H_2O_2 (mg/L)
Influent	196	5	NA	20	NA	NA	NA	NA	1
Effluent	65	12	NA	22	6	28	3	3	7
BAC Eff	5	ND	NA	10	ND	6	ND	ND	1

Table 6.15
Removal of 2,000 µg/L MTBE in the presence of TBA in groundwater with pulsed UV of 7 Hz

Sample Location	MTBE (µg/L)	TBA (µg/L)	TBF (µg/L)	Acetone (µg/L)	Acetaldehyde (µg/L)	Formaldehyde (µg/L)	Glyoxal (µg/L)	M-Glyoxal (µg/L)	H_2O_2 (mg/L)
Influent	2727	555	NA	137	NA	NA	NA	NA	1
Effluent	1924	565	NA	371	7	140	3	5	7
BAC Eff	51	15	NA	142	1	7	ND	ND	1

MTBE and By-Products Removal With Pulsed UV/H_2O_2 in Groundwater

The groundwater used for the pulsed UV experiments contained high levels of nitrate in the range of 4.5 mg/L as nitrogen (NO_3^--N). Nitrate strongly absorbs UV light, and the bench scale experiments showed that the presence of high nitrate inhibited the MTBE reduction by filtering out some UV light that may have otherwise used to generate hydroxyl radicals from hydrogen peroxide. With initial MTBE concentration in the range of 200 µg/L, about 67% of MTBE was removed with pulsed UV only at 7 Hz (Table 6.14). With higher initial MTBE concentration of 2,700 µg/L, however, only 30% MTBE removal was observed (Table 6.15). For both cases, greater than 93% of residual MTBE found in the UV effluent was removed in the BAC columns. The aldehyde levels in the BAC effluent were close to detection limits regardless of the MTBE concentration. However, with the 2,700 µg/L of initial MTBE, TBA and MTBE in the effluent exceeded the action level and the secondary limit respectively even with the combination of pulsed UV and the BAC. Increase in the retention time in the pulsed UV process,

Table 6.16
Removal of 2,000 μg/L MTBE with BAC only (no pulsed UV)

Sample Location	MTBE (μg/L)	TBA (μg/L)	TBF (μg/L)	Acetone (μg/L)	Acetaldehyde (μg/L)	Formaldehyde (μg/L)	Glyoxal (μg/L)	M-Glyoxal (μg/L)	H_2O_2 (mg/L)
Influent	2261	ND	NA	176	NA	NA	NA	NA	0
Effluent	2456	ND	NA	205	NA	NA	NA	NA	1
BAC Eff	14	102	NA	179	NA	NA	NA	NA	1

Table 6.17
Removal of 200 μg/L MTBE with pulsed UV of 4 Hz and 1 mg/L H_2O_2

Sample Location	MTBE (μg/L)	TBA (μg/L)	TBF (μg/L)	Acetone (μg/L)	Acetaldehyde (μg/L)	Formaldehyde (μg/L)	Glyoxal (μg/L)	M-Glyoxal (μg/L)	H_2O_2 (mg/L)
Influent	185	5	NA	11	NA	NA	NA	NA	1
Effluent	98	15	NA	17	6	31	3	4	7
BAC Eff	ND	ND	NA	7	ND	6	ND	ND	1

may provide conditions for higher removal efficiency, especially when the initial MTBE concentration is in the range of 2,000 μg/L.

The levels of acetaldehyde, glyoxal, and M-glyoxal, were much lower for the groundwater than those for the surface water tested even with the higher levels of initial MTBE, which indicates that these by-products are the results of degradation of constituents in the water matrix. In fact, the typical level of TOC in the surface water (3.33 mg/L) was more than six times that of groundwater (0.52 mg/L).

The control test without the pulsed UV light showed insignificant change of the MTBE level in the pulsed UV reactor as expected, but close to 99% MTBE reduction in the BAC column (Table 6.16) was observed. Although non-detectable amount of TBA was shown in the effluent, elevated level was found in the BAC effluent. Some of the adsorbed TBA in the BAC may have leached out when the concentration in the effluent is lower than the adsorbed concentration. Since TBA poses a health concern, more understanding of the cause for the elevated level of TBA should be addressed by MTBE treatment by BAC alone.

With the addition of small amount of hydrogen peroxide (~1 mg/L), the benefit of added hydrogen peroxide was not realized for Santa Monica groundwater possibly because high nitrate strongly masks pulsed UV light, and about 50 ~ 60% MTBE removal was achieved when the initial MTBE concentration was in 200 μg/L range (Tables 6.17 and 6.18). As in the previous case, higher applied pulsed UV energy (frequency) did not make any significant improvement in the MTBE reduction. This level of reduction is less than that observed for SPW where the nitrate level is low. The level of all the by-products including TBA was low in the BAC effluent, and greater than 95% removal of residual MTBE was observed. Slightly higher MTBE removal was observed with higher applied UV frequency of 7 Hz (Table 6.19).

Similar to the previous case with ozone, the presence of TBA did not interfere significantly with MTBE reduction for the pilot scale pulsed UV experiments. Approximately

Table 6.18
Removal of 200 μg/L MTBE with pulsed UV of 7 Hz and 1 mg/L H_2O_2

Sample Location	MTBE (μg/L)	TBA (μg/L)	TBF (μg/L)	Acetone (μg/L)	Acetaldehyde (μg/L)	Formaldehyde (μg/L)	Glyoxal (μg/L)	M-Glyoxal (μg/L)	H_2O_2 (mg/L)
Influent	203	13	NA	13	NA	NA	NA	NA	1
Effluent	84	25	NA	44	6	29	3	3	4
BAC Eff	5	ND	NA	16	ND	6	ND	ND	1

Table 6.19
Removal of 200 μg/L MTBE in the presence of TBA with pulsed UV of 7 Hz and 1 mg/L H_2O_2

Sample Location	MTBE (μg/L)	TBA (μg/L)	TBF (μg/L)	Acetone (μg/L)	Acetaldehyde (μg/L)	Formaldehyde (μg/L)	Glyoxal (μg/L)	M-Glyoxal (μg/L)	H_2O_2 (mg/L)
Influent	270	324	NA	6.8	NA	NA	NA	NA	1
Effluent	110	299	NA	126.0	8	47	4	6	5
BAC Eff	5.1	65	NA	7.1	ND	6	ND	ND	2

Table 6.20
Removal of 200 μg/L MTBE with pulsed UV of 4 Hz and 4 mg/L H_2O_2

Sample Location	MTBE (μg/L)	TBA (μg/L)	TBF (μg/L)	Acetone (μg/L)	Acetaldehyde (μg/L)	Formaldehyde (μg/L)	Glyoxal (μg/L)	M-Glyoxal (μg/L)	H_2O_2 (mg/L)
Influent	288	18	NA	7	NA	NA	NA	NA	4
Effluent	68	31	NA	56	8	ND	6	6	5
BAC Eff	5	9	NA	8	ND	5	ND	ND	1

60% MTBE reduction was observed with the applied pulsed UV frequency of 7 Hz in the presence of TBA. However, TBA itself, which reacts much slower with hydroxyl radicals than MTBE, showed less than 8% removal during the pulsed UV process. Majority of TBA (~78%) was achieved in the BAC column, but the TBA level in the final BAC effluent still exceeded the action level of 12 μg/L.

When the amount of hydrogen peroxide added was increased to 3 ~ 5 mg/L during the pulsed UV system to generate more hydroxyl radicals, increased in the MTBE reduction of 75% was achieved (Tables 6.20 and 6.21). Again, increase in the pulsed UV frequency from 4 Hz to 7 Hz did not improve the removal efficiency as observed for the previous case. With increased in the hydrogen peroxide level, TBA removal efficiency increased slightly (37%) as shown in Table 6.22.

Since the increased in the applied UV energy could not provide enough treatment for MTBE, greater than 10 mg/L of hydrogen peroxide was used to accommodate higher than 2,000 μg/L level of MTBE contamination sometimes found in the groundwater. As shown in

Table 6.21
Removal of 200 μg/L MTBE with pulsed UV of 7 Hz and 5 mg/L H_2O_2

Sample Location	MTBE (μg/L)	TBA (μg/L)	TBF (μg/L)	Acetone (μg/L)	Acetaldehyde (μg/L)	Formaldehyde (μg/L)	Glyoxal (μg/L)	M-Glyoxal (μg/L)	H_2O_2 (mg/L)
Influent	259	20	NA	11	NA	NA	NA	NA	5
Effluent	64	30	NA	58	8	15	6	6	6
BAC Eff	4.6	11	NA	11	ND	5	ND	ND	1

Table 6.22
Removal of 200 μg/L MTBE in the presence of TBA with pulsed UV of 7 Hz and 4 mg/L H_2O_2

Sample Location	MTBE (μg/L)	TBA (μg/L)	TBF (μg/L)	Acetone (μg/L)	Acetaldehyde (μg/L)	Formaldehyde (μg/L)	Glyoxal (μg/L)	M-Glyoxal (μg/L)	H_2O_2 (mg/L)
Influent	256	349	NA	7.2	NA	NA	NA	NA	3
Effluent	57	221	NA	159.0	8	94	6	10	4
BAC Eff	5	101	NA	8.1	ND	5	ND	ND	1

Table 6.23
Removal of 2,000 μg/L MTBE with pulsed UV of 4 Hz and 11 mg/L H_2O_2

Sample Location	MTBE (μg/L)	TBA (μg/L)	TBF (μg/L)	Acetone (μg/L)	Acetaldehyde (μg/L)	Formaldehyde (μg/L)	Glyoxal (μg/L)	M-Glyoxal (μg/L)	H_2O_2 (mg/L)
Influent	2305	ND	NA	184	NA	NA	NA	NA	11
Effluent	308	237	NA	430	4	211	5	44	6
BAC Eff	88	51	NA	179	ND	6	ND	ND	2

Tables 6.23 and 6.24, approximately 90% MTBE reduction was achieved during the pulsed UV experiments with higher than 10 mg/L of hydrogen peroxide level. Even with higher removal efficiency, however, the final concentration of MTBE did not meet the treatment goal of 5 μg/L under the operational parameter used.

The hydrogen peroxide level beyond 10 mg/L provided marginal improvement in MTBE removal. With approximately 30 mg/L of hydrogen peroxide, about 95% MTBE removal was achieved (Tables 6.25 and 6.26). At this condition, the removal efficiency of BAC column was actually reduced as the hydrogen peroxide level increased. Thus, the final concentration after BAC column for MTBE, TBA, and acetone were high. As in the case for the other cases, the presence of TBA had no effect on the removal efficiencies. Based on the data provided in this section, optimization for each source water and the typical treatment goal need to be re-evaluated.

As it was discussed previously, when MTBE is oxidized, MTBE degradation by-products are formed as a result of breakdown of MTBE by hydroxyl radicals. When the higher oxidant dose is used (higher level of hydrogen peroxide, for example), the primary degradation by-products,

Table 6.24
Removal of 2,000 μg/L MTBE with pulsed UV of 7 Hz and 11 mg/L H_2O_2

Sample Location	MTBE (μg/L)	TBA (μg/L)	TBF (μg/L)	Acetone (μg/L)	Acetaldehyde (μg/L)	Formaldehyde (μg/L)	Glyoxal (μg/L)	M-Glyoxal (μg/L)	H_2O_2 (mg/L)
Influent	3303	ND	NA	339	NA	NA	NA	NA	11
Effluent	313	244	NA	450	4	221	5	ND	7
BAC Eff	102	NA	NA	183	ND	6	ND	ND	2

Table 6.25
Removal of 2,000 μg/L MTBE with pulsed UV of 4 Hz and 28 mg/L H_2O_2

Sample Location	MTBE (μg/L)	TBA (μg/L)	TBF (μg/L)	Acetone (μg/L)	Acetaldehyde (μg/L)	Formaldehyde (μg/L)	Glyoxal (μg/L)	M-Glyoxal (μg/L)	H_2O_2 (mg/L)
Influent	2714	ND	NA	224	NA	NA	NA	NA	28
Effluent	138	96	NA	680	4	181	4	ND	13
BAC Eff	74	90	NA	562	ND	5	ND	ND	2

Table 6.26
Removal of 2,000 μg/L MTBE with pulsed UV of 7 Hz and 29 mg/L H_2O_2

Sample Location	MTBE (μg/L)	TBA (μg/L)	TBF (μg/L)	Acetone (μg/L)	Acetaldehyde (μg/L)	Formaldehyde (μg/L)	Glyoxal (μg/L)	M-Glyoxal (μg/L)	H_2O_2 (mg/L)
Influent	1474	ND	NA	741	NA	NA	NA	NA	29
Effluent	40	86	NA	791	5	166	3	ND	14
BAC Eff	69	59	NA	50	ND	ND	ND	ND	2

such as TBA and TBF levels, are further oxidized to increase the level of secondary by-products such as acetone and formaldehyde. In fact, the elevated levels of acetone and formaldehyde were observed (Tables 6.23 to 6.27) when the hydrogen peroxide used was greater than 11 mg/L. Thus, although higher level of hydrogen peroxide will enhance the MTBE degradation efficiency, the level of hydrogen peroxide should be optimized so that the balance in MTBE and the degradation by-products level in the effluent are achieved.

Nitrate Conversion

As mentioned previously, in the absence of hydrogen peroxide, nitrate can effectively converted to nitrite during the pulsed UV process. Since nitrite has lower regulatory limit, under some cases, undesirable conversion of nitrate to nitrite may provide conditions where nitrite levels exceed the regulatory limit. The high levels of hydrogen peroxide thus can also contribute in a way by hindering the conversion of nitrate to nitrite by filtering UV light.

Table 6.27
Removal of 2,000 μg/L MTBE in the presence of TBA with pulsed UV of 7 Hz and 32 mg/L H_2O_2

Sample Location	MTBE (μg/L)	TBA (μg/L)	TBF (μg/L)	Acetone (μg/L)	Acetaldehyde (μg/L)	Formaldehyde (μg/L)	Glyoxal (μg/L)	M-Glyoxal (μg/L)	H_2O_2 (mg/L)
Influent	2227	398	NA	139	NA	NA	NA	NA	32
Effluent	63	102	NA	321	3	190	3	130	16
BAC Eff	35	164	NA	173	1	ND	ND	1	2

Table 6.28
Conversion of nitrate to nitrite with pulsed UV of 4 Hz and 1 mg/L H_2O_2

Sample Location	NO_3^- as N (mg/L)	NO_2^- as N (mg/L)	Total N (mg/L)
Influent	4.5	ND	4.5
Effluent	3.0	1.8	4.8
BAC Effluent	1.5	0.6	2.1

Table 6.29
Conversion of nitrate to nitrite with pulsed UV of 7 Hz and 1 mg/L H_2O_2

Sample Location	NO_3^- as N (mg/L)	NO_2^- as N (mg/L)	Total N (mg/L)
Influent	4.4~4.5	ND	4.5
Effluent	3.3~3.4	1.0~1.4	4.4~4.7
BAC Effluent	1.9~2.4	0.4~0.8	2.4~2.9

With 1 mg/L of hydrogen peroxide the amount of nitrite as nitrogen (NO_2^- as N) converted from background nitrate was approximately 1.4 to 1.8 mg/L (as nitrogen; Tables 6.28 and 6.29). In the BAC column, both nitrate and nitrite levels were reduced by as much as 50%. With higher level of hydrogen peroxide, the level of nitrite was lower (Tables 6.30 to 6.35). Although higher amount of nitrate was removed in the BAC when higher hydrogen peroxide was used, controlling nitrite is more important since the regulation is far more stringent for nitrite than nitrate.

Summary of Findings

In general, the applied UV dose (frequency) was less sensitive than the hydrogen peroxide in terms of MTBE removal efficiency. Increased in the pulsed frequency from 4 Hz to 7 Hz did not significantly improve the MTBE removal efficiency under the similar operating conditions. On the contrary, increase in the hydrogen peroxide significantly improved the overall treatment

Table 6.30
Conversion of nitrate to nitrite with pulsed UV of 4 Hz and 4 mg/L H_2O_2

Sample Location	NO_3^- as N (mg/L)	NO_2^- as N (mg/L)	Total N (mg/L)
Influent	4.4	ND	4.4
Effluent	3.5	0.9	4.4
BAC Effluent	1.0	0.2	1.2

Table 6.31
Conversion of nitrate to nitrite with pulsed UV of 7 Hz and 5 mg/L H_2O_2

Sample Location	NO_3^- as N (mg/L)	NO_2^- as N (mg/L)	Total N (mg/L)
Influent	4.4	ND	4.4
Effluent	3.6	0.6	4.2
BAC Effluent	1.5	0.3	1.8

Table 6.32
Conversion of nitrate to nitrite with pulsed UV of 4 Hz and 11 mg/L H_2O_2

Sample Location	NO_3^- as N (mg/L)	NO_2^- as N (mg/L)	Total N (mg/L)
Influent	4.5	ND	4.5
Effluent	3.9	0.6	4.5
BAC Effluent	1.6	0.4	2.0

Table 6.33
Conversion of nitrate to nitrite with pulsed UV of 7 Hz and 10~11 mg/L H_2O_2

Sample Location	NO_3^- as N (mg/L)	NO_2^- as N (mg/L)	Total N (mg/L)
Influent	4.4~4.5	ND	4.4~4.5
Effluent	3.9	0.5	4.4
BAC Effluent	1.6~2.2	0.3~0.4	1.9~2.6

efficiency by generating more hydroxyl radicals, which oxidize MTBE and the by-products. Also, although the biological columns used as the post treatment to complement the pulsed UV system enhanced the overall removal efficiency for both MTBE and the by-products by chemical adsorption and/or biological removal, the removal trend was not consistent under the operating conditions used. Although the pilot scale test demonstrated that BAC alone can effectively remove MTBE in a wide range, depending on the adsorption/desorption phase, the capacity for

Table 6.34
Conversion of nitrate to nitrite with pulsed UV of 4 Hz and 28 mg/L H_2O_2

Sample Location	NO_3^- as N (mg/L)	NO_2^- as N (mg/L)	Total N (mg/L)
Influent	4.4	ND	4.4
Effluent	4.1	0.3	4.4
BAC Effluent	2.7	0.2	2.9

Table 6.35
Conversion of nitrate to nitrite with pulsed UV of 7 Hz and 29~32 mg/L H_2O_2

Sample Location	NO_3^- as N (mg/L)	NO_2^- as N (mg/L)	Total N (mg/L)
Influent	4.4	ND	4.4
Effluent	4.2	0.2~0.3	4.4~4.5
BAC Effluent	2.3~3.3	0.2	2.5~3.5

the BAC may fluctuate and thus may not provide the consistent removal efficiency for the MTBE and the by-products. Also, with fluctuation in the MTBE concentration, it may be difficult to project the performance of the BAC. Thus, in employing the overall treatment process, the primary treatment mechanism, such as pulsed UV should be first optimized to provide sufficient removal of both MTBE and the degradation by-products, and the post treatment using biological GAC should provide secondary treatment for low residuals found the in the AOP effluent or occasional elevated levels of contaminants. Groundwater from Santa Monica, which contained high level of nitrate exhibited overall lower level of removal efficiency than the surface water (SPW) used because nitrate strongly absorbs UV light, thus decreasing the amount of hydroxyl radicals generated during the pulsed UV process. Depending on the operational conditions used, excess amount of nitrite may be formed in the effluent especially when the level of initial hydrogen peroxide is low. Thus for the operational consideration, adequate amount of hydrogen peroxide in the range of 10 mg/L and optimized retention time may provide effluent that meets all the requirements including MTBE, TBA, other by-products, and nitrite.

IMPLICATIONS FOR FULL-SCALE USE

Process Design Criteria

Based on bench and pilot testing it was inferred that a pulsed UV dose of 2,000 kJ with a contact time of 10 minutes is necessary to reduce the influent MTBE levels from 200 μg/L to 10 μg/L or less. The lamp pulse frequency was assumed to be 4 Hz and the lamp life was assumed to be one month. Test data also indicated a peroxide dose of 10 mg/L is necessary to achieve the MTBE oxidation. MTBE and by-products will be further removed by biological GAC treatment. Test data indicated a design loading rate of 2 gpm/sf and a contact time of 20 minutes is necessary for biological GAC to remove MTBE and by-products to very low (e.g., non-detect) levels.

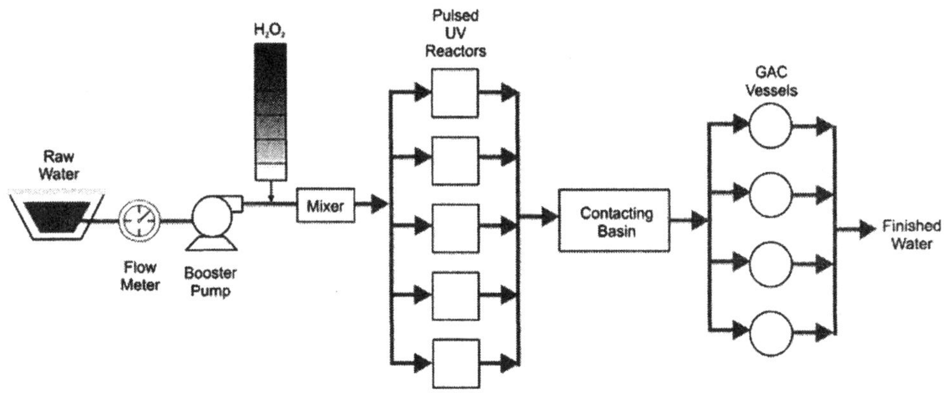

Figure 6.23 Pulsed UV/peroxide schematic

Process Schematics and Layout

Pulsed UV/Hydrogen Peroxide System

A schematic of the pulsed UV/hydrogen peroxide system is shown in Figure 6.23. Raw water supply is metered prior to pumping. Booster pumps provide the hydraulic head required for flow through the entire system. Following the pumps, hydrogen peroxide is added to the water at a concentration of 10 mg/L. The concentration of hydrogen peroxide was determined from pilot testing. Hydrogen peroxide can either be fed through a venturi-type injector device which facilitates mixing or metered into the process stream and mixed downstream of the injection point via a static or dynamic mixing device. The MTBE contaminated water is then directed through the pulsed UV reactors. Pilot testing resulted in a design UV dose of 2000 kJ.

Following the UV reactors is a contacting basin that provides additional residence time such that the total residence time is 10 minutes, determined in pilot testing. The final unit process consists of biologically enhanced granular activated carbon (GAC) polishing to destroy any remaining MTBE and MTBE by-products. Allowing biological growth within the carbon beds enhances the ability of the process to destroy the remaining chemicals and extends the life of the carbon. Pilot testing of GAC resulted in a recommended hydraulic loading of 2 gpm/sf and a contact time of 20 minutes. The design parameters require multiple pre-manufactured GAC vessels, depending on the flow rate through the system.

Based on the process schematic, potential site layouts were prepared for 1 mgd (Figure 6.24) and 10 mgd (Figure 6.25) treatment systems. Major equipment is shown on the site layout. MTBE contaminated water is pumped from the well pump base and boosted through the booster pump stations. From the booster pumps, the water flows through a mixing device where the hydrogen peroxide is added and mixed. For a 1 mgd system, an 8 foot diameter hydrogen peroxide tank will be required and a 10 foot diameter hydrogen peroxide tank will be required for the 10 mgd system. The process water flows through a header system and into the UV reactors. Each UV reactor has a volume of 50 gallons and contains one UV lamp. For the required UV dose, one UV reactor is required for each 200,000 gpd. Each UV reactor has a footprint of approximately 3 feet by 3 feet and an associated power supply with a footprint of approximately 2 feet by 3 feet.

Based on UV dose requirements, 5 UV reactors are required for the 1 mgd system and 50 UV reactors are required for the 10 mgd system. The process water is recombined after the UV

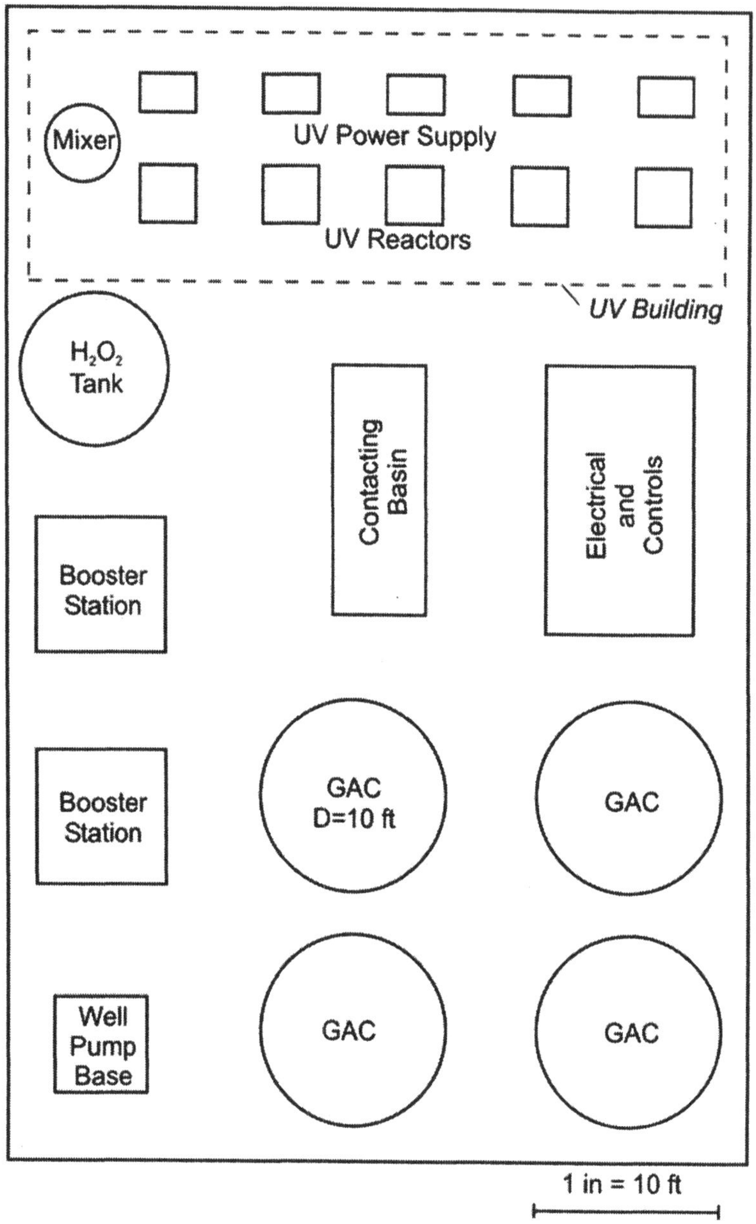

Figure 6.24 Pulsed UV/peroxide layout for 1 mgd system

reactors and flows through the contacting basin. Contacting basins were assumed to be of on-site concrete construction with a length to width ratio of 2.5. Following the contacting basin, process water flows through a distribution header to the GAC vessels. Large pre-manufactured vessels are sold in pairs of 20,000 lb carbon capacity. The size of each vessel is 10 feet in diameter. To meet the hydraulic loading requirement, 4 GAC vessels are required for the 1 mgd system and 40 GAC vessels are required for the 10 mgd system. After determining the number of GAC vessels required to meet the hydraulic loading requirement, the volume of carbon required for each vessel was calculated to meet the required 20 minute contact time. Finally, an electrical control room was included to house electrical components and process control equipment.

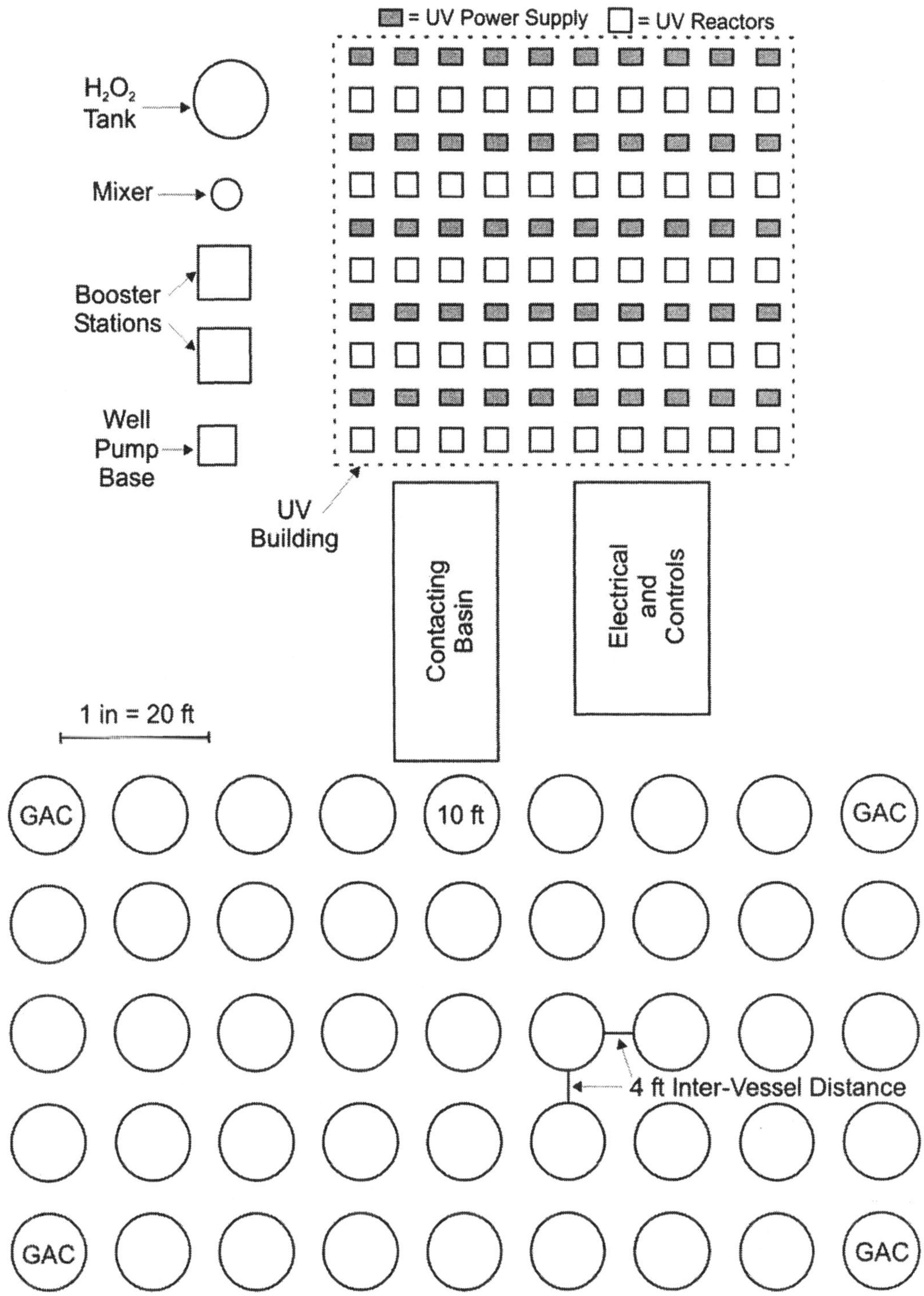

Figure 6.25 Pulsed UV/peroxide layout for 10 mgd system

Technology Costs

Every row of the attached cost table (Table 6.36) is numbered and comments and assumptions for the costs are documented below by row number. For costing purposes, the pulsed UV lamps were assumed to be operated at a pulse frequency of 4 Hz. The lamps were assumed to replaced once a month.

Capital Cost Assumptions

(1) Row 1 shows the design flow in million gallons/day (mgd). Four flow rates were chosen to represent the range of treatment sizes under consideration. The flow rates costed out are 0.5, 1.0, 3.0, and 10.0 mgd, displayed across four columns.
(2) Conversion of mgd to gallons/minute (gpm). mgd is multiplied by 694 to obtain the flow rate in gpm.
(3) Hydrogen peroxide (gallons/day) required for the design concentration of 10 mg/L. These values were calculated assuming 35% volume/volume H_2O_2.
(4) Volume of contacting basin required for 10 minutes residence time. Volume of contacting basin (in gallons) is obtained by multiplying the flow rate in row 2 with 10.
(5) Cost of pulsed UV lamps, sleeves and all other electrical equipment necessary for UV system operation provided by the pulsed UV lamp manufacturer.
(6) Cost of UV reactor chambers provided by the pulsed UV reactor manufacturer.
(7) Peroxide storage tank costs provided by tank vendors.
(8) Peroxide injection system costs provided by vendors.
(9) Monitoring instrumentation costs for hydrogen peroxide equipment plus UV equipment, provided by the above referenced vendors.
(10) Control package costs for hydrogen peroxide equipment plus UV equipment. The vendors provided costs for hydrogen peroxide control equipment. Cost for UV control systems was assumed to be 10% of the UV equipment costs listed in row 5.
(11) Piping and valves costs for hydrogen peroxide and UV equipment are shown in this row. Vendors provided costs for hydrogen peroxide piping and valving. Cost for UV equipment piping and valving was assumed to be 10% of UV equipment plus reactor costs, i.e., 10% of row 5 + row 6.
(12) Pump capital and associated equipment costs are obtained using the Water/Wastewater Cost models (Culp, Wesner and Culp, 1990).
(13) Contacting basin costs are based on required contacting volume (row 4) minus UV reactor volume, multiplied by 1.15 for tank freeboard. These costs are derived from the contactors cost worksheet developed for the EPA Stage-2/LT-2 Technology and Cost document (2002).
(14) Process cost subtotal consists of the sum of all process costs (rows 5 - 13).
(15) Process cost multiplier is 2.5 for flow rates equal to or below 1 mgd and 2.0 for flow rates above 1 mgd. Multiplier accounts for the following items: site work, contractor's overhead and profit, contingencies, engineering and design, mobilization and bonding, legal and administrative, interest during construction, and installation. Direct capital cost is obtained by multiplying the process cost (row 14) with the appropriate multiplier.

Table 6.36
Capital, O&M, and total costs for pulsed UV/peroxide treatment of MTBE

Flow Rates					
Design Flow (mgd)	(1)	0.5	1	3	10
Design Flow (gpm)	(2)	347	694	2082	6940
Sizing Calculations					
Hydrogen Peroxide Required (gal/d)	(3)	13	25	76	252
Contactor Volume (gal)	(4)	3470	6940	20820	69400
Capital Cost Summary					
Process Capital Cost					
UV Lamps, Sleeves	(5)	$120,000	$160,000	$480,000	$1,600,000
Reactor Chamber	(6)	$30,000	$40,000	$120,000	$400,000
Peroxide Storage Tank	(7)	$36,000	$36,000	$36,000	$40,000
Peroxide Injection System	(8)	$9,000	$9,000	$9,000	$10,000
Monitoring Instrumentation	(9)	$8,250	$9,250	$14,250	$32,500
Control Package	(10)	$18,750	$22,750	$54,750	$166,750
Pipes and Valves	(11)	$33,000	$38,000	$78,000	$220,000
Pumping	(12)	$46,703	$74,390	$155,578	$349,232
Contacting Basin	(13)	$19,132	$29,593	$58,641	$124,077
Process Cost Subtotal	(14)	$320,835	$418,983	$1,006,219	$2,942,559
Direct Capital Cost	(15)	$802,088	$1,047,458	$2,012,438	$5,885,118
Indirect Capital Cost					
Piloting	(16)	$10,000	$65,000	$65,000	$65,000
Permitting	(17)	$9,625	$12,569	$30,187	$88,277
Land	(18)	$9,625	$12,569	$30,187	$88,277
Operator Training	(19)	$1,796	$1,796	$1,796	$1,796
Housing	(20)	$34,639	$46,530	$94,094	$260,566
Indirect Cost Subtotal	(21)	$65,686	$138,465	$221,263	$503,916
Total Capital Cost	(22)	$867,774	$1,185,923	$2,233,701	$6,389,034
Annual O&M Summary					
Chemicals	(23)	$25,331	$50,662	$151,986	$506,620
Materials – Lamp Replacement	(23)	$14,400	$24,000	$72,000	$240,000
Electricity	(24)	$13,201	$23,319	$69,989	$233,288
Labor	(26)	$18,221	$36,442	$54,662	$72,883
Total Annual O&M Cost	(27)	$71,153	$134,423	$348,638	$1,052,791
For a project period of 20 years with 7% interest					
Amortized Total Annual Cost	(28)	$153,065	$246,365	$559,483	$1,655,870
Cost/1000 Gallons Treated	(29)	$0.84	$0.67	$0.51	$0.45

(16) Piloting costs were assumed to be \$10,000 for flow rates below 1 mgd and \$65,000 for flow rates equal to and above 1 mgd. For large systems, the piloting costs were calculated at \$65,000 and for small systems it was assumed that limited piloting would be provided for by \$10,000 (EPA Stage-2/LT-2 Technology and Cost Document, 2002).
(17) Permitting costs were assumed to be similar to those calculated for the EPA Arsenic Technology and Cost document (2001). Permit costs equal 3% of the process cost subtotal (row 14), with a minimum of \$2500.
(18) Land costs were assumed to be 3% of the process cost subtotal (row 14).
(19) Operator training costs were based on 40 hours of supervisory training at a rate of \$45/hour.
(20) The building size for housing was estimated from the footprints. Costs were developed from the RS Means Building Construction Cost (2002) data for warehouse/office space and equal to \$62/sq.ft.
(21) Indirect cost subtotal consists of the sum of all indirect process costs (rows 16 - 20).
(22) Total capital cost equals indirect (row 21) and direct (row 15) capital costs.

O&M Cost Assumptions

(23) Chemical costs for hydrogen peroxide was assumed to be \$5.50/gallon.
(24) Materials – Pulsed UV lamp manufacturer specified monthly lamp replacement at a unit cost of \$400 per lamp. There are 3, 5, 15, and 50 lamps required respectively for flow rates of 0.5, 1.0, 3.0, and 10 mgd.
(25) Electricity costs were based on UV system requirements of 4 kW/h per system plus main process pumping costs. Pumping costs were estimated by calculating energy required for pumping the desired flow rate under the assumptions of 50ft of head, 75% pump efficiency, and 85% motor efficiency. The total energy required was then multiplied by 1.1 to account for miscellaneous additional energy requirements and power was estimated at \$0.08/kWh.
(26) Labor costs were estimated at the following amounts:

    0.5 mgd system = 2 hours/day
    1.0 mgd system = 4 hours/day
    3.0 mgd system = 6 hours/day
    10 mgd system = 8 hours/day

    The labor rate was assumed at a typical technician rate of \$25/hour.

(27) Annual O&M subtotal consists of the sum of all O&M costs (rows 23 - 26).

Total Cost Assumptions

(28) The total capital costs (row 22) were amortized over 20 years at a 7% interest rate and then annual O&M (row 27) was added to derive the amortized total annual cost.
(29) Cost per thousand gallons treated was calculated by dividing the amortized total annual cost (row 28) by the flow rate in thousands of gallons and the days in a year.

CHAPTER 7
ELECTRON BEAM TREATMENT

PILOT SCALE TESTING

Introduction

E-Beam pilot tests were performed at the Port Hueneme Naval Base in California. Approximately 1,800 gallons of Santa Monica groundwater was transported to the Naval Base in a clean tanker. Five experiments, each with 300 gallons of groundwater, were performed.

Experimental Plan

In the present study, five separate studies were performed in which the destruction of MTBE was determined. In addition to the destruction of MTBE, the formation of reaction by-products was determined. For each of the five tests performed approximately 300 gallons of Santa Monica groundwater was transferred from a tanker truck to a 500-gallon tank located in the pump room of the mobile electron beam treatment system. Once transferred to the 500-gallon tank, the water matrix was spiked with the appropriate chemicals to yield a final compound target concentration as shown in Table 7.1.

After the appropriate chemicals were added, the resulting solution was mixed through the pump room plumbing (500 tank to PC pump to 500 gallon tank), at a flow rate of approximately 40 gpm, for 30 minutes. The appropriate valves were then opened and closed, and resulting solution was pumped through the electron beam treatment system at a flow rate of 6 gpm. For experiments 1 and 5, a zero dose influent and effluent sample were collected, after 5 minutes of pumping. Additional effluent samples were collected at treatment doses of 1.67, 3.76, 5.85, 8.77, and 10.03 kGy. During dose changes, the treatment system was given 5 minutes of treatment at each dose before the effluent sample was collected. During sampling, the effluent sampling port was left open allowing the treated water to continuously flush the sampling port to reduce cross contamination between differing treatment doses. For experiments 2, 3, and 4 only effluent samples were collected at 0, 5.85 and 10.03 kGy.

Table 7.1
Experimental design for the Electron Beam studies of water obtained from Santa Monica

Experiment	MTBE ($\mu g\ L^{-1}$)	TBA ($\mu g\ L^{-1}$)	NO_3^- ($mg\ L^{-1}$)	Br^- ($mg\ L^{-1}$)
1	200	-	-	-
2	200	200	-	-
3	200	-	36	-
4	200	-	-	0.5
5	200	200	36	0.5

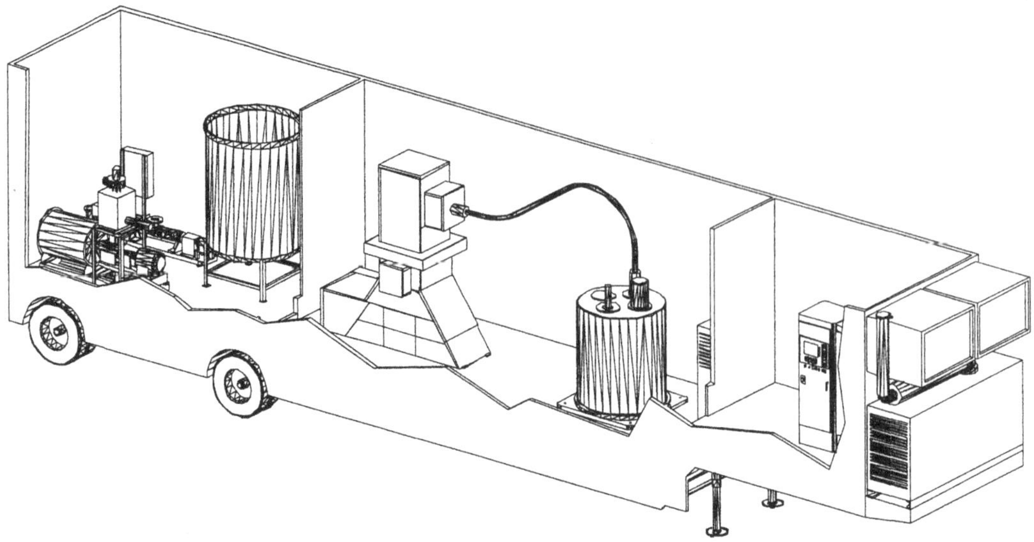

Figure 7.1 Schematic of the mobile electron beam system used for treating water from Santa Monica, CA

Between experiments, the plumbing distribution system was drained as completely as possible. Time constraints did not permit rinsing the entire plumbing system with potable water between tests.

Materials and Methods

All of the analytical data were obtained from others.

All experiments were conducted using a 25 kW mobile electron beam system shown in Figure 7.1 (Nickelsen et al., 1998). The E-beam system is housed in an 8- by 48-foot trailer and is rated for a maximum flow rate of 50 gallons per minute (gpm). The E-beam system includes the following components: a strainer basket, an influent pump, the E-beam unit, a cooling air processor, a turbo blower, and a control console. These components are situated in three separate rooms: the pump room, process room, and control room. The pump room contains all ancillary equipment for the E-beam unit for both water and air handling; the radiation-shielded process room contains the E-beam unit itself; and the control room contains the control console where system-operating conditions are monitored and adjusted.

The E-beam unit is made up of the following components: an electron accelerator, a scanner, a contact chamber, and lead shielding. The electron accelerator is capable of generating an accelerating voltage of 500 kilovolts (kV) and a beam current of between 0 and 40 milliamps (mA). The accelerating voltage determines the E-beam penetration depth, while the beam current determines the concentration or number of electrons that impact the flowing stream.

A pyramid-shaped scanner located beneath the electron accelerator uses magnetic coils to deflect the E-beam, causing it to scan the flowing water (the E-beam scanner operation is similar to the scanner operation in a television set). Contaminated groundwater is pumped through the contact chamber, which is located beneath the scanner. The scanner is operated in such a way that the E-beam contacts the entire surface of the water flowing through the contact chamber. The contact chamber dimensions are proprietary information.

Two 0.0015-inch-thick titanium windows separate the scanner from the contact chamber. The first (primary) window is necessary to maintain a vacuum in the scanner and the second (secondary) titanium window isolates cooling air from the contact chamber; the vacuum is required to minimize E-beam energy losses. As the E-beam passes through the primary titanium window, the window absorbs some of the E-beam's energy. This energy absorption is manifested in the form of heat. Because of the E-beam's high energy, the titanium window experiences significant heating. Therefore, passing-cooling air between the primary and secondary windows cools the titanium window. Cooling air exiting from between the two windows flows through a cooling air processor and is returned to re-cool the primary window by a turbo blower; this is a closed-loop system. The cooling air system includes a turbo blower and a water-jacketed air chiller.

Lead shielding surrounds the E-beam unit to prevent x-ray emissions. X-rays are formed when the E-beam contacts various internal stainless steel surfaces. As an added safety measure, the process room is inaccessible during system operation.

Resistance temperature devices (RTD) were used to measure the temperature of groundwater before and after treatment. As previously discussed the change in water temperature is used to determine the E-beam dose. The E-beam system was configured with two RTDs, one immediately upstream and one RTD immediately downstream from the contact chamber. The RTDs have a sensitivity of 0.01° C.

Results and Discussion

The variables that were chosen for study, in the experimental design, were MTBE, TBA, NO_3^- and Br^- and five experiments were completed. The results of the five experiments are summarized in Tables 7.2 and 7.3. The data in Table 7.2 show that the E-Beam process destroys both MTBE and TBA and that with increasing dose increasing destruction is observed.

At a first glance, the data in Table 7.2 seem to show that both the MTBE and TBA were substantially removed at the zero-dose, indicating removal via physical means (i.e., stripping). However, it has been shown that MTBE and TBA are not removed at all by stripping using this mobile electron beam system by measuring influent and effluent concentrations at zero dose (Cooper et al., 2000a, b).

Upon further review of the data it was determined that the E-beam system was not allowed to come into solution equilibrium, that is, the plumbing system was not flushed with enough of the sample matrix at the zero dose, prior to collecting the effluent sample. Evidence of this is demonstrated by the results of Experiment 3 that showed a high TBA concentration in the zero dose effluent, when in fact TBA was not spiked into the test matrix. This high effluent TBA concentration is as a result of cross contamination from the previous test run. That is, the plumbing system was not sufficiently flushed.

Experiment 1 indicates that MTBE is efficiently removed, from 82 μg/L to <MDL at 1.67 kGy and that TBA is formed but never greater than 7.5 μg/L. Formaldehyde and methylglyoxal were observed at 15 and 21 μg/L, respectively. Bromate ion was observed at the detection limit of 1 μg/L that suggests an analytical error.

Experiment 2 demonstrated that the addition of TBA adversely influenced the removal of MTBE. However at the highest dose, 10 kGy, both MTBE and TBA were reduced to concentrations that are within the suggested regulatory limits of both, in drinking water. The reaction by-product formaldehyde was observed at 250 μg/L while the methylglyoxal was 36 μg/L.

Table 7.2
Measured process parameters for the E-Beam treatment of Santa Monica groundwater

Exp.	Dose kGy	MTBE $\mu g\ L^{-1}$	TBA $\mu g\ L^{-1}$	Nitrate Ion as NO_3^- $mg\ L^{-1}$	Nitrate Ion as NO_3^--N $mg\ L^{-1}$
	MDL →	0.078	0.13	8.8	2
1	0 (Influent)	140	<MDL	17.6	3.99
	0	82.1	<MDL	17.4	3.95
	1.67	<MDL	6.72	17.6	3.99
	3.76	<MDL	7.65		
	5.85	<MDL	7.49		
	8.77	<MDL	4.89		
	10.03	<MDL	<MDL		
2	0	118	247		
	5.85	27.1	100	19.8	4.51
	10.03	<MDL	7.21		
3	0	117	54.1	54.1	12.3
	5.85	4.04	11.7	50.2	11.4
	10.03	<MDL	8.43		
4	0	233	6.28	5.17	22.7
	5.85	4.55	8.61	4.52	19.9
	10.03	<MDL	6.31		
5	0 (Influent)	195	266		
	0	97.2	4.87	24	5.46
	1.67	2.07	18.3	49.7	11.3
	3.76	<MDL	15.9		
	5.85	<MDL	47.8		
	7.94	<MDL	83.3		
	10.03	<MDL	36.5		

Again, bromate ion was detected at 1.6 µg/L in an experiment in which no bromide ion was added, suggesting analytical error.

In experiment 3 the spiked nitrate ion appeared to retard the removal of MTBE and TBA (TBA in experiment 3 at zero does confirmed that the system piping was not properly cleaned between experiments) by acting as a radical scavenger. The nitrate ion reacts rapidly with the aqueous electron to form nitrite ion. The nitrite ion then goes on to react with the hydroxyl radical to ultimately reform nitrate ion. The nitrate and nitrite ions are not destroyed in the process; this continuous reactionary loop serves to scavenge both hydroxyl radicals and aqueous electrons. This scavenging by the NO_3^-/NO_2^- is unique for AOPs, because of the presence of both the reducing, (e^-_{aq}/H·), and oxidizing reactive species (·OH). This is an example of where the presence of both oxidizing and reducing species reduces the effectiveness of the treatment process system by measuring influent and effluent concentrations at zero dose (Cooper et al., 2000a, b).

The data from experiment 4, where bromide ion was added at 0.5 mg L^{-1}, indicated that it had no effect on the removal of either the MTBE or TBA. However, bromate ion was observed at

Table 7.3
Reaction by-product analyses for the E-Beam treatment of Santa Monica groundwater

Exp	Dose kGy	Acetone μg L^{-1}	Acetaldehyde μg L^{-1}	Formaldehyde μg L^{-1}	Glyoxal μg L^{-1}	Methylglyoxal μg L^{-1}	Nitrite NO_2^--N mg L^{-1}	Bromate μg L^{-1}
	MDL →	5.1	1	5	1	1	0.5	1
1	0 (Influent)	27.9	2	7	<MDL	<MDL	<MDL	<MDL
	0	31.3						
	1.67	34.6						
	3.76	51.5						
	5.85	29.4						
	8.77	20.7	5	15	1	21	<MDL	1
	10.03	16.4						
2	0	357						
	5.85	21.7						
	10.03	29.8	1	250	3	36	1.4	1.6
3	0	23.5	<MDL	<MDL	<MDL	<MDL	<MDL	1
	5.85	36.8	2	380	5	20	2.3	1.6
	10.03	57.3						
4	0	<MDL	<MDL	<MDL	<MDL	<MDL	<MDL	<MDL
	5.85	20.4						
	10.03	18.8	1	280	3	22	1.2	3.6
5	0 (Influent)	13.0	<MDL	<MDL	<MDL	<MDL	<MDL	1.1
	0	16.7						
	1.67	85.8						
	3.76	49.9						
	5.85	67.3						
	7.94	90.3	1	410	4	44	2.2	3.4
	10.03	65.1						

a concentration of 3.6 μg/L. Although the concentration of the bromate ion was well below the EPA regulatory level of 10 μg L^{-1}, we feel that these are anomalous results. In concurrent studies for EPA, the destruction of MTBE (1,600 μg/L) has been studied in the presence of BTEX (2,500 μg/L) and 1 mg L^{-1} of bromide, showed no formation of bromate ion. In addition, in previous work the electron beam process was shown to be an effective technology for destroying bromate ion (Siddiqui et. al., 1996a; 1996b).

Experiment 5, in which all of the variables were added, showed efficient removal of the MTBE at the lowest dose (1.67 kGy), that is below the action limit, and the formation of TBA with increasing dose to 83 μg/L at 7.94 kGy. At the highest dose, 10 kGy, the TBA was 36.5 μg/L, and the formaldehyde and methylglyoxal, 410 and 44 μg/L, respectively. The bromate ion was 3.4 μg/L, below the regulatory limit.

The bromate formation results of this study appear to contradict both other work and published data on the reduction of bromate ion with the electron beam process. In a study that has just been completed, (Innovative Technology Evaluation Report for the High Energy Electron

Injection (E-Beam) Technology, prepared for the USEPA, Office of Research and Development, National Risk Management Research Laboratory, Cincinnati, OH) a ground water containing 900 µg/L of bromide was treated for the destruction of MTBE and BTEX with the mobile electron beam system. No bromate ion was observed in any of the studies, an EPA certified contract lab conducted all of the analyses. The lack of the formation of bromate ion in that ground water corroborates studies that have reported the reduction of bromate ion using the electron beam process at the Miami Electronic Beam Research Facility, at 100-120 gallons per minute (Siddiqui et al., 1996a; 1996b).

Summary of Findings

In summary, the electron beam process was show to be effective in destroying MTBE from spiked Santa Monica groundwater. The addition of TBA decreased the efficiency of the removal of MTBE. Because of the limited treatment doses in the experiments where nitrate ion and bromide ion were added, it is difficult to determine the effect on MTBE removal; however, the presence of either nitrate ion or bromide ion, appeared to increase the concentration of the reaction by-product formaldehyde. In the experiment where all of the variables were added, Experiment 5, the removal of the MTBE appeared relatively unaffected, but the concentration of the reaction by-products formaldehyde and methylglyoxal were high than the other experiments. Although bromate ion was observed in the tow experiment in which bromide ion was added, this result is not consistent with other studies that we have conducted at much higher concentrations of either bromide ion or bromate ion. E_{EO} values for MTBE destruction using E beam treatment can be obtained from previous publications of Cooper and Tornatore.

IMPLICATIONS FOR FULL-SCALE USE

The implications of the pilot-scale feasibility study indicate that the electron beam treatment process has direct application in treating waters contaminated with MTBE. Studies have been conducted at large scale, ≥100 gallons per minute (0.40 m^3 per minute). Therefore, there are no impediments engineering or scientific impediments to the implementation of the electron beam process at full scale.

Process Design Criteria

- Contaminant concentration
- Water characteristics, pH, alkalinity, and other potential scavengers
- Treatment objectives
- Process flow rate
- Economics

Safety Issues and Permits

Although the electron beam technology utilizes ionizing radiation as a means of destroying organic contaminants in solution, electron accelerators are not regulated by the Nuclear Regulatory Agency mainly because the technology derives its source for the radiation from electrical means. Thus, the electron accelerator can be easily switched off by simply removing the source of electrical power. The technology is however a machine source of radiation and therefore must be registered as such with State regulatory agencies, much in the same way as medical X-ray machines. The typical costs for machine source registration ranges from \$10 to \$300, depending on the State. For example, the mobile electron beam system that was used in this study has been licensed in Florida, California, New York, Germany and Switzerland.

X-rays are electromagnetic radiation with energies greater than about 0.1 keV. They are produced whenever high-speed electrons are rapidly decelerated, for example, by passing through the electric field of an atomic nucleus, leading to their description as *bremsstrahlung* (i.e., braking radiation). Bremsstrahlung production is an alternative means of energy loss by fast electrons to the production of ionization and excitation; it is favored at high electron energies and by high-Z (i.e., high atomic number) stopping materials. In practice, X-rays are produce by stopping beams of accelerated electrons in high-Z, high-melting, metal targets; in principle, any accelerator producing an electron beam can be used to generate X-rays. Since electron beam treatment systems are primarily composed of structural steel and stainless steel (high-Z metal targets), incident electrons have the capability of generating X-rays. The energy of the resulting X-ray radiation ranges from near zero to the maximum energy of the incident electrons.

Shielding

To ensure the safety of employees, clients, and any visitors, electron beam treatment systems are shielded with lead, concrete or earthen berms. The shielding employed is of sufficient thickness and geometry to attenuate any generated electron or X-ray to background levels.

Radioactivity Residuals

Electron beam treatment systems contain no radioactive isotopes; therefore, the equipment itself is not radioactive. In addition, electron beam treatment systems for environmental applications (10 MeV or less) do not contain enough energy to induce nuclear activation. That is, the material being irradiated does not itself become radioactive. As such, there is no radiation residual left in either the treated waste stream or any structural equipment.

Once the electrical power to the electron beam treatment system has been turned off, the generation of electrons and production of X-ray radiation is terminated.

Process Schematics and Layout

The Electron Beam Research Facility (EBRF) is a typical of a full-scale electron beam treatment facility and it is located at the Miami-Dade Virginia Key (Central District) Wastewater Treatment Plant, Miami, FL (Kurucz et al., 1995). The facility consists of a 1.5 MeV electron accelerator rated at 75 kW and was designed to treat 120 gallons per minute of anaerobically digested sludge prior to land application. A schematic drawing of the facility is shown in Figure 7.2. The electron accelerator is powered by an insulated core transformer (ICT) operating at 1.5 MeV with a continuously variable beam current from 0 to 50 mA. The electron beam is scanned at 200 Hz to give a coverage of 1.2 m wide and 5 cm high. Varying the beam current changes the absorbed dose in a linear fashion, allowing for experimentation at doses from 0 to 8.0 kGy.

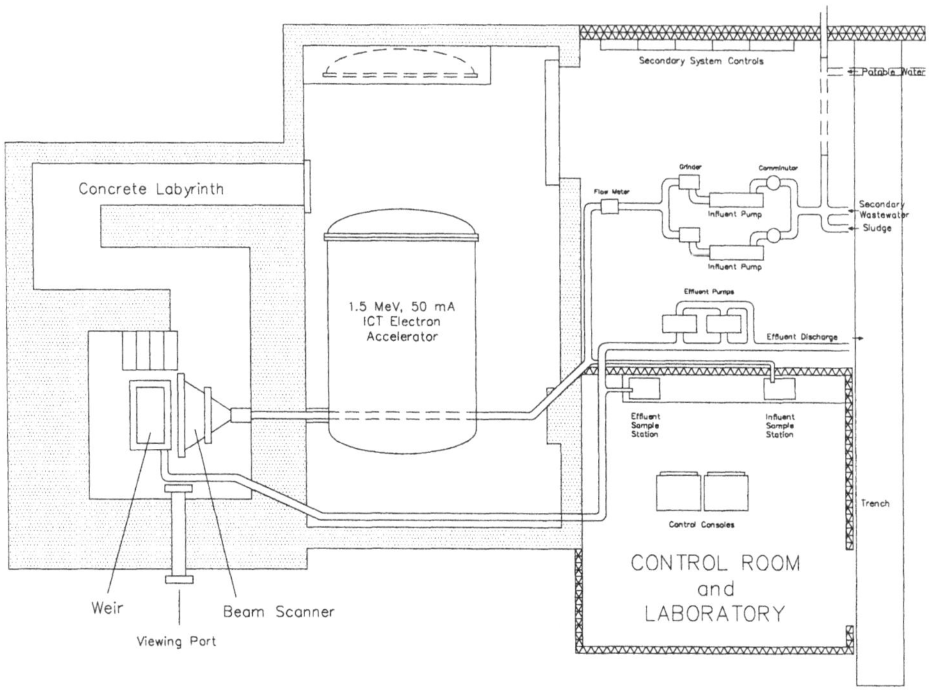

Figure 7.2 Floor plan of the Electron Beam Research Facility, Miami, Florida

Table 7.4
Costs associated with the E-Beam treatment of Santa Monica MTBE contaminated groundwater

MGD	No. of Beams*	Installed Cost	Annual Cost†	Annual Maintenance Cost‡	Annual Electrical Costs§	Annual Labor Cost**
0.5	1	$2,000,000	$140,000	$100,000	$70,000	$50,000
1	2	$3,500,000	$245,000	$175,000	$140,200	$50,000
3	4	$6,000,000	$420,000	$300,000	$280,300	$50,000
10	12	$18,000,000	$1,260,000	$900,000	$841,000	$100,000

MGD	No.of Beams*	Total Annual Costs††	1000 Gallons Per Year Treated	$/1000 Gallons
0.5	1	$360,000	182,500	$1.97
1	2	$610,200	365,000	$1.67
3	4	$1,100,300	1,095,000	$1.00
10	12	$3,101,000	3,650,000	$0.85

* 100 kW E-beam units at a cost of $1,000,000 per unit.
† Based on 20-year amortization at 7% interest rate.
‡ Calculated at 5% of installed cost.
§ Number of beams × 100 kW × $0.08/kWh × 8760 hr/yr.
** One/two full-time operators at $25/hour.
†† Annual Cost + Maintenance + Electrical + Operators.

The facility is designed to treat aqueous streams at 0.46 m^3 min^{-1} liters (120 gpm). Influent streams at the EBRF are presented to the scanned beam in a 4 mm thick falling stream. Since the maximum penetration in water is approximately 7 mm for 1.5 MeV electrons, some electrons pass through the stream and thus not all of the beam energy is transferred to the water. Furthermore, the addition of over-scanning the waste stream to insure that the edges of the stream are irradiated, additional energy is lost resulting in a total efficiency of energy transfer of approximately 60 – 85%. Thus, at 50 mA (75 kW) we have recorded doses of between 6.5 and 8.0 kGy. Total power consumption, including pumps, chillers and other auxiliary equipment is about 125 kW, with a total up time of between 97 and 99%.

The EBRF is instrumented with resistance temperature devices (RTDs) to obtain direct estimates of absorbed dose. Five RTDs are mounted in the influent (2 sensors) and effluent (3 sensors) stream immediately before and after irradiation. The RTDs are connected via an interface to a computer, which continuously reads and records temperatures and the absorbed dose is estimated by converting the observed temperature differences to the energy transferred to the water.

Technology Costs

Costs for E beam treatment are presented in Table 7.4. The footnotes give the assumptions that were used in developing these costs. The E beam treatment costs varied from \$1.97 (for 0.5 mgd) to \$0.85 (for 10 mgd) per 1,000 gallons of water treated.

CHAPTER 8
COMPARISON OF AOP TECHNOLOGIES

OVERVIEW OF BENEFITS AND LIMITATIONS OF AOP TECHNOLOGIES

This study evaluated the following AOP technologies with respect to MTBE removal efficiency, water quality impact and operational issues:

- Ozone/peroxide
- Continuous UV/peroxide
- Pulsed UV/peroxide
- E beam

Some of the key benefits in using AOP technologies for treating potable water are:

- AOPs are destructive processes that generate no MTBE-laden residuals (e.g., spent GAC or resin) or off-gases that may require further treatment
- Several AOPs have disinfectant capabilities and can supplement the existing/proposed disinfection at a water treatment plant
- Some of the individual processes that are part of AOPs (e.g., ozone and UV) are already being used in drinking water treatment
- AOPs are available as modular, package treatment systems that can be installed as wellhead treatment systems or in large, surface water treatment plants
- In addition to MTBE, AOPs can be used to treat other organic contaminants like benzene/toluene/tertiary butyl alcohol/geosmin that may be present in the water

Some of the key limitations of the AOP technologies are:

- Potential for accumulation of by-products. Complete oxidation of MTBE and its by-products may require longer oxidation times or use of polishing treatments like biological activated carbon (BAC). Residual peroxide can serve as an oxygen source for microorganisms and can promote microbial re-growth in the distribution system
- Presence of radical scavengers (e.g., bicarbonates, NOM) and interfering contaminants (e.g., nitrate) can reduce the overall effectiveness of AOPs
- Limited full-scale application of AOP treatment for MTBE removal. There is little information available in terms of full-scale plant operational data for AOPs
- AOPs that rely on UV radiation exhibit a decrease in efficiency as the source water turbidity increases. Turbidity of the water lowers the transmittance and penetration of UV light into the source water
- Ozone/peroxide treatment of waters containing elevated levels of bromide (> 0.1 mg/L) might result in the formation of bromate in excess of its MCL of 10 μg/L
- Ozone/peroxide treatment may require an air permit for ozone emissions in addition to an off-gas treatment system for ozone destruction

All the aforementioned AOP technologies were able to achieve 95% or greater removal of MTBE under varying water quality and operational conditions. Selecting an appropriate AOP technology should give due consideration to costs, qualitative issues (e.g., public acceptance) influent and finished water qualities. In the following sections, the various AOP technologies will be compared with respect to these decision drivers.

AOP TREATMENT COSTS

The AOP treatment costs were developed using the design criteria based on pilot study results and described in Chapters 4 through 7. These costs were developed using similar assumptions and approaches for all the AOP technologies. The design assumptions common to all the technologies are shown in Table 8.1. As shown by the bench and pilot scale results, AOPs can reduce MTBE from 200 μg/L to 10 μg/L or less. Polishing treatment with BAC is assumed to remove the residual MTBE, its by-products and any peroxide. When AOP and BAC sequential treatment is operated under optimized conditions it is possible to achieve non-detect levels (<0.2 μg/L) of MTBE in the treated waters (as shown by the pilot study results).

Some of the key design criteria used in developing the costs for the various AOP technologies are summarized in Table 8.2. These design criteria were developed based on bench/pilot scale test data and best professional judgment. Details on the design criteria are in Chapters 4–7.

Summarized in Table 8.3 are the capital, O&M and total costs for the various AOP technologies and the BAC polishing treatment. These costs were presented for design flows of 0.5 mgd, 1.0 mgd, 3.0 mgd and 10.0 mgd. The capital costs for E beam were the highest, followed by ozone/peroxide or pulsed UV/peroxide, depending on the flow size. The continuous wave LP/MP-UV/peroxide capital costs were the lowest among the AOP technologies considered. The O&M costs for the three AOPs, other than E beam, appear to be more or less similar for system sizes up to 3 mgd. For the 10 mgd size, the O&M costs for LP-UV/peroxide were the lowest followed by pulsed UV/peroxide and MP-UV/peroxide.

The total costs per 1,000 gallons treated were lowest for LP/MP-UV/peroxide and pulsed UV/peroxide processes, followed by ozone/peroxide and E beam (Table 8.3). These costs are in good agreement with costs that were reported in earlier studies for MTBE removal using AOPs (California MTBE Research Partnership, 2000). The total cost per 1,000 gallons treated for MP-UV/peroxide for a 1 mgd system was estimated to be \$0.81 (excluding BAC polishing treatment cost), which is close to \$0.71 (without any polishing treatment cost) reported in the earlier study (California MTBE Research Partnership, 2000). Similarly, the costs for 1,000 gallons treated by ozone/peroxide treatment for a 1 mgd system was found to be \$0.92 (excluding BAC polishing treatment cost), and this is an agreement with \$0.84 (without any polishing treatment cost) in the earlier study (California MTBE Research Partnership, 2000). It should be noted that the approaches used for estimating the costs in the present study are different than those used in the California MTBE Research Partnership study (2000). BAC polishing treatment will add about \$0.20 per 1,000 gallons treated for all AOP + BAC treatment scenarios.

The cost comparisons developed in this report should be used as a guideline. An understanding of the actual costs will require pilot testing to determine site-specific costs. Selecting an AOP treatment for an MTBE contaminated potable water source will more than likely be based on other criteria such as process reliability and flexibility in addition to cost. A comparative discussion of the promising AOP technologies relative to their applicability and effectiveness is presented in the following section.

Table 8.1
Assumptions common to all AOP technology cost development efforts

Design parameter	Design assumptions and goals
Influent MTBE for AOP treatment	200 μg/L
Targeted effluent MTBE from AOP treatment	≤10 μg/L
Influent MTBE for BAC polishing treatment	≤10 μg/L
Targeted effluent MTBE from BAC polishing treatment	≤ 0.2 μg/L (non-detect)
Design flows for which costs were developed	0.5, 1, 3 and 10 mgd

Table 8.2
Design criteria used in developing the costs

Treatment process	Key design criteria
Ozone/peroxide	O_3 dose = 15 mg/L H_2O_2 dose = 15 mg/L Contact time = 25 min
MP-UV/peroxide	E_{EO} = 0.35 kWh/m^3-order H_2O_2 dose = 15 mg/L Lamp replacement = once a year
LP-UV/peroxide	E_{EO} = 0.13 kWh/m^3-order H_2O_2 dose = 15 mg/L Lamp replacement = once a year
Pulsed-UV/peroxide	UV dose = 2,000 kJ H_2O_2 dose = 10 mg/L UV reactor residence time = 10 min Pulse frequency = 4 Hz Lamp replacement = once a month
E beam	E beam dose = 10 kGy or 1,000 krads
BAC polishing	Loading rate = 2 gpm/ft^2 Empty bed contact time = 20 min

Costs Sensitivity

The AOP treatment costs are affected by the:

- Extent of MTBE contamination and
- Presence of (i) co-occurring contaminants such as gasoline additives (e.g., BTEX) (ii) interfering constituents such as nitrates, NOM/TOC or particulate matter and (iii) radical scavengers such as bicarbonates in the source water

If removal of co-occurring contaminants such as BTEX is also a treatment goal then it may be necessary to have larger AOP reactors for more contact time, add more oxidants or have

Table 8.3
Summary of capital, O&M, and total costs for various AOP and BAC polishing treatments

Technology	Design flow (mgd) 0.5	1.0	3.0	10
		Total capital costs ($)		
Ozone/Peroxide	$1,218,000	$1,826,000	$2,384,000	$4,340,000
MP-UV/Peroxide	NA*	$1,448,000	$2,012,000	$2,881,000
LP-UV/Peroxide	NA	$1,090,000	$2,140,000	$4,904,000
Pulsed UV/Peroxide	$868,000	$1,186,000	$2,234,000	$6,389,000
E Beam	$2,000,000	$3,500,000	$6,000,000	$18,000,000
BAC Polishing	$367,000	$779,000	$1,792,000	$5,822,000
		Total O&M cost ($/yr)		
Ozone/Peroxide	$82,000	$164,000	$436,000	$1,337,000
MP-UV/Peroxide	NA	$158,000	$407,000	$1,227,000
LP-UV/Peroxide	NA	$97,000	$246,000	$694,000
Pulsed UV/Peroxide	$71,000	$134,000	$349,000	$1,053,000
E Beam	$360,000	$610,000	$1,100,300	$3,101,000
BAC Polishing	$2,300	$4,600	$13,800	$45,900
		Total cost per 1,000 gallons treated ($)		
Ozone/Peroxide	$1.08	$0.92	$0.60	$0.48
MP-UV/Peroxide	NA	$0.81	$0.54	$0.41
LP-UV/Peroxide	NA	$0.55	$0.41	$0.32
Pulsed UV/Peroxide	$0.84	$0.67	$0.51	$0.45
E Beam	$1.97	$1.67	$1.00	$0.85
BAC Polishing	$0.20	$0.21	$0.17	$0.16

* NA – not available or estimated. Insufficient information was available to develop these costs

more lamps (for UV based processes). All of these modifications will increase the capital and operational costs. Presence of nitrates and particulate matter can interfere with the absorption of UV light by hydrogen peroxide. Therefore, in source waters with high concentrations of nitrates or high turbidities, it may be necessary to use more UV lamps or provide longer UV contact time for oxidation reactions to proceed to completion. If the source water has high (1-2 mg/L) influent MTBE and low bromide concentration then ozone/peroxide treatment would be an economically attractive AOP. Ozone/peroxide treatment is being used in pump-and-treat remediation applications where the concentrations of MTBE are in mg/L range.

For all AOPs considered, other than E beam, costs for peroxide addition form a significant portion of the O&M costs. Pilot and full-scale tests are recommended to optimize the oxidant dosages for MTBE removal using AOPs. For E beam and UV based processes, energy costs also form a significant portion of the O&M costs. The energy costs would be a function of unit rate

($/kWh). Lower energy rates (say $0.04/kWh) would significantly lower the O&M costs for UV/peroxide and E beam processes.

Costs for maintenance will increase the AOP O&M costs. It is difficult to quantify the maintenance costs without understanding all the site-specific issues. Complications from sleeve and UV lamp failures have been reported in prior studies with UV/peroxide treatment. Plugging of ozone diffusers or spargers is also reported in some systems for source waters with high iron/manganese concentrations.

OTHER DECISION DRIVERS

In this section, the various AOP technologies are compared with respect to their:

- Mechanical reliability
- Process reliability
- Flexibility
- Adaptability
- Potential for modifications
- Bromate regulatory compliance
- Energy efficiency
- Public acceptability and
- Ease of implementation

Summarized in Table 8.4 are the scores (high, medium or low) for these qualitative drivers for each AOP technology. The following discussion explains the reasoning behind the scores.

Mechanical Reliability

Reliability of AOP technologies can be addressed under two broad categories, namely, process reliability and mechanical reliability. Technologies with fewer moving or replaceable parts are considered to be more mechanically reliable because they will likely require less frequent maintenance. Consequently, the ozone/peroxide and MP, LP-UV/peroxide treatment that have few moving parts receive the highest rating for mechanically reliability. In the ozone/peroxide process, periodic checking and cleaning of the ozone generator and ozone gas diffusers is required. Fouling of spargers from precipitation of carbonates can lead to inefficient ozone transfer. The pulsed UV/peroxide treatment requires periodic replacement (once a month) of the lamps and therefore received a medium rating for mechanical. E beam has a large number of specialty parts and equipment requiring experts for maintenance and possible replacement and thus, receives a low score for mechanically reliability.

Process Reliability

Process reliability is defined as the ability of a given AOP technology to consistently meet treated water goals. The ozone/peroxide and continuous UV/peroxide have been proven to consistently meet targeted water quality goals and are, thus, considered highly reliable. Presently, the lack of large-scale potable water treatment applications for pulsed UV/peroxide and E beam resulted in low ratings; however, an optimized E beam or pulsed UV/peroxide system will be able

Table 8.4
Comparative analysis of various AOPs based on qualitative decision drivers

AOP Technology	Mechanical Reliability	Process Reliability	Flexibility	Adaptability	Potential for Modifications	Bromate Regulatory Compliance	Energy Efficiency	Public Acceptance	Ease of Implementation	Total Score (/27)
Ozone/ peroxide	High (3)	High (3)	High (3)	High (3)	Medium (2)	Low (1)	High (3)	High (3)	High (3)	24
Continuous UV/peroxide	High (3)	High (3)	High (3)	Medium (2)	Medium (2)	High (3)	Medium (2)	High (3)	High (3)	24
Pulsed UV/ peroxide	Medium (2)	Low (1)	Medium (2)	Medium (2)	Medium (2)	High (3)	Low (1)	Medium (2)	Medium (2)	17
E beam	Low (1)	Medium (2)	Low (1)	Medium (2)	Medium (2)	Medium (2)	Low (1)	Low (1)	Medium (2)	14

to consistently remove MTBE as shown by the pilot study results. For all AOP technologies, including those with high reliability ratings, monitoring and controls are recommended to optimize the treatment process.

Flexibility

Flexibility is defined as the ability of a technology to handle wide fluctuations in the influent water flow rate following design and installation. Occasionally, during the operation of a treatment process, the influent stream flows increase or decrease significantly compared to the design flow. A flexible technology should be able to handle these fluctuations with no major impact on the treatment process outcome. Ozone/peroxide and continuous UV/peroxide when designed with sufficient safety factors can handle a large turndown ratio (i.e., ratio of maximum to minimum allowable flow rates). Also, chemical additions (ozone or peroxide) or UV dose can be changed to respond to changing flow rates. For the aforementioned reasons, high ratings were given to ozone/peroxide and continuous UV/peroxide. Since pulsed UV reactors are commercially-available only in 50 gpm reactor sizes, lot of automation would be necessary to turn-off a few reactors when the plant flow decreases significantly. Therefore, medium rating was given to pulsed UV/peroxide process. E beam may require significant changes to its stream distribution, or spreading system when the water flow rates increase, again suggesting a low rating.

Flexible technologies, when necessary, can also be scaled-up with little or no difficulty. The capacities of the modular AOP technologies, such as ozone/peroxide, continuous UV/ peroxide and pulsed UV/peroxide can be expanded by adding additional reactors either in series (to extend the reaction time) or in parallel (to extend the flow rates).

Adaptability

Adaptability of a technology is defined as its ability to handle fluctuations in water quality conditions, such as influent contaminant concentrations, hardness, alkalinity, and turbidity. All the AOP technologies discussed above can achieve MTBE removal efficiencies that are independent of the influent MTBE concentration, but that vary widely with water quality conditions. If influent MTBE concentrations increase while effluent goals remain unchanged, it will be necessary to increase the contact time or oxidant doses in the reaction chamber to meet effluent goals.

Since oxidation via hydroxyl radicals is the predominant mechanism for MTBE removal for each of the AOPs discussed above, presence of radical scavengers will affect treatment performance, independent of the selected AOP. However, those technologies that have other pathways other than radical route for removing MTBE will be less affected by the presence of radical scavengers. Therefore, E beam process that can generate multiple oxidizing and reducing species receives a medium rating. The effectiveness of ozone/peroxide process can be reduced by the presence of excess particulate matter or scaling parameters that foul the ozone gas diffusers. But, in ozone/peroxide treatment, increasing the dosages of the oxidant is relatively easy and therefore for adaptability criterion it (ozone/peroxide treatment) receives a high rating.

The removal efficiency of UV-based technologies, such as continuous UV/H_2O_2 and pulsed UV/ H_2O_2, are hindered by water quality parameters other than radical scavengers, e.g., excess turbidity (which masks the penetration of the UV light), the presence of nitrate (which absorbs effective UV radiation), and iron and other fouling agents (which scale the quartz sleeves). Therefore, the UV based technologies received a medium rating. The removal efficiency of each AOP technology is strongly dependent on the characteristics of the influent water quality. Hence, in the design of AOP systems, due consideration must be given to the concentrations (and expected fluctuations) of radical scavengers and other interfering compounds.

Potential for Modifications

The potential for modifications is defined as the ability to alter the installed system including addition of any necessary pre- and post-treatments to accommodate changes in the design criteria and conditions (e.g., lowered target concentrations of MTBE, removal of high alkalinity or iron, removing by-products like TBA using other polishing treatments). For example, most AOPs treating source waters with medium to high (150-200 mg/L) alkalinities may require pre-treatment for alkalinity removal, which may include a pH adjustment step followed by CO_2 stripping. Most modular processes (e.g., O_3/H_2O_2, UV/H_2O_2, pulsed UV/H_2O_2) are more easily amenable to changes compared to non-modular processes (e.g., E beam). Also, in modular processes, several modular units in parallel or series can supply additional contact time.

When necessary, all of the AOP technologies evaluated can be supplemented with pre- and post-treatment systems. However, for some AOPs, these pre- or post-treatment systems are mandatory prior to drinking water distribution and thus, these AOPs will receive a lower rating. For waters with high bromide concentrations (>100 μg/L), control of bromate formation will be necessary in ozone based AOP systems (e.g., H_2O_2/O_3), although this effect can be mitigated without pre-treating, as shown by the results presented in Chapter 4. Additionally, ozone/peroxide process may require ozone off-gas treatment, resulting in a medium rating. H_2O_2/UV processes may require a post-treatment temperature adjustment for small systems or pre-treatment to remove turbidity, nitrates, or scaling agents; however, these pre- and post-treatments are not always necessary, suggesting a medium rating. E beam requires no pre- or post- treatment processes, but since it is not a modular process it also gets a medium rating.

Bromate Regulatory Compliance

Bromate is classified by the International Agency for Research on Cancer (IARC) as a possible human carcinogen and strictly regulated under the Stage 1 and Stage 2 D/DBP Rules with a MCL of 10 μg/L. This MCL may become more stringent in future rulemaking. Thus, AOPs that generate bromate (O_3/H_2O_2) receives a lower rating than alternatives; however, as discussed previously, varying chemical doses for the O_3/H_2O_2 process can mitigate bromate formation.

Energy Efficiency

Energy usage is rated medium for AOPs that are based on either O_3 or continuous UV in combination with other oxidants. E beam and pulsed UV processes appear to require significant energy for operation, resulting in a low rating. Among the four AOPs, ozone/peroxide would need the least amount of energy and therefore it is given a high rating.

Public Acceptability

Systems that are widely used in remediation and drinking water treatment applications are rated as highly acceptable to the public; whereas, emerging AOP (pulsed UV/peroxide) with little or no field application are classified as medium. E beam is given a low rating for public acceptability due to its reliance on a radiation source that has some associated safety issues.

Ease of Implementation

The number of field installations was used as a surrogate to determine the ease of implementation. Accordingly, the AOPs that use some combination of O_3, LP/MP UV and H_2O_2 were rated high for ease of implementation; whereas, emerging AOPs such as pulsed UV and E beam with no field applications were given a medium grade.

Total Score

Scores of 1, 2 and 3 were assigned respectively for low, medium and high as indicated in parentheses of Table 8.4. The total score for a technology was obtained by summing the scores for a individual drivers. Ozone/peroxide and continuous UV/peroxide obtained the highest score of 24 based on the above-discussed qualitative factors. The maximum possible score is 27. Pulsed UV/peroxide and E beam received the next higher scores of 18 and 14, respectively.

INFLUENT AND TREATED WATER QUALITY CONSIDERATIONS

In selecting an AOP treatment process, due consideration must be given to the influent and treated water qualities. Summarized in Table 8.5 are the influent water quality parameters that determine the effectiveness of the various AOP treatment processes in MTBE and its by-product removal efficiencies. High alkalinity (> 100 mg/L) will have a negative impact on all the AOP treatment processes, since bicarbonate ions can scavenge hydroxyl radicals. High oxidant dosages or longer oxidation times may be required for AOP treatment of MTBE in highly alkaline waters. Presence of natural organic matter, measured as TOC, also works similar to alkalinity. For waters that have high TOCs (> 3 mg/L), radical scavenging will retard the MTBE and its by-products oxidation rates. E beam process is less sensitive to alkalinity and TOC since there are other mechanisms, other than radicals, that can remove MTBE in the E beam process.

Nitrate and turbidity hinder UV-based AOPs by absorbing light in the wavelength region (200-250 nm) that would be effective in MTBE oxidation. Ozone/peroxide process is not impacted by the presence of nitrate, in the levels that are normally seen in natural waters. Turbidity will also have less impact on ozone/peroxide and E beam treatment processes. Nitrate was found to have minimal or no impact on E beam process.

In the pilot tests with all AOPs, presence of TBA delayed the MTBE oxidation. In waters that contain TBA along with MTBE due consideration must be given to this effect in designing AOP treatment. TBA is one of the by-products of MTBE in natural attenuation processes. Presence of TBA in waters contaminated with MTBE has been reported at several locations in California.

Table 8.5
Comparative analysis of various AOPs based on the impact of influent water quality on the MTBE and its by-products removal efficiency

Influent water quality parameter	Ozone/peroxide	Continuous UV/peroxide	Pulsed UV/peroxide	E beam
Alkalinity	Adverse, radical scavenger	Adverse, radical scavenger	Adverse, radical scavenger	Adverse, radical scavenger
Natural organic matter	Adverse, radical scavenger	Adverse, radical scavenger	Adverse, radical scavenger	Adverse, radical scavenger
Nitrate	No impact	Adverse, absorbs UV light	Adverse, absorbs UV light	Minimal impact
Turbidity	No impact	Adverse, absorbs UV light	Adverse, absorbs UV light	Minimal impact
TBA	Adverse, radical scavenger	Adverse, radical scavenger	Adverse, radical scavenger	Adverse, radical scavenger
pH (6 – 9)	Low pH, preferable to control bromate formation	Minimal impact	Minimal impact	Mechanism not well understood

The influent water ambient pH between 6 and 9 has no impact on MTBE based oxidation by continuous or pulsed UV processes. Low pH is preferred for ozone/peroxide treatment of waters that contain high levels (> 100 μg/L) of bromide. The effect of pH on E beam process is not yet studied and therefore is not well understood.

Summarized in Table 8.6 is a comparison of the various AOP treatments with respect to some of the key treated water parameters. In bench and pilot studies conducted with waters that were spiked with bromide, high levels of bromate were observed in treated waters for ozone/peroxide treatment. In E beam tests conducted with waters spiked with bromide, moderate (3-5 μg/L) levels of bromate formation was observed.

Nitrate gets converted to nitrite by AOP treatment. This conversion was excessive in ozone/peroxide treatment with low levels of peroxide. Conversion of nitrate to nitrite may also occur in waters that have high levels (>20 mg/L as NO_3) of nitrate for UV based and E beam treatment processes. It should be noted that the MCL for NO_2-N is 1 mg/L. Presence of nitrite will also contribute to the decay of chloramines in the distribution system.

All AOP technologies can provide additional benefit of disinfection. Prior studies indicated lower DBP formation potentials for waters that were treated with ozone/peroxide and subsequently chlorinated (Zheng et al., 1999a, 1999b). Prior studies also indicated no significant change in DBP formation potential for waters that were chlorinated after UV/peroxide treatment (Zheng et al., 1999a, 1999b). The impact of pulsed UV/peroxide and E beam treatments on DBP formation potential are not well understood.

Table 8.6
Comparative analysis of various AOPs based on the treated water quality

Treated water quality parameter	Ozone/peroxide	Continuous UV/peroxide	Pulsed UV/peroxide	E beam
Bromate formation	High potential, in excess of regulatory limit for waters containing high levels of bromide	Minimal or no potential	Minimal or no potential	Moderate to low potential
Nitrite formation	Minimal	Minimal	Excessive, if operated with low H_2O_2 dose	Minimal
Microbial	Good	Good	Good	Good, E-beam is used in sterilizing surgical instruments
DBPs formation potential (in chlorinated water after AOP treatment compared to before AOP treatment)	Lower	No significant change	Not known	Not known

SUMMARY

Under optimized operating conditions, the four AOP technologies that were evaluated in this study were successful in achieving 95% or higher MTBE removals. Among the four AOP processes studied, ozone/peroxide and UV/peroxide are better understood than the emerging pulsed UV/peroxide and E beam technologies. Also, there are more number of suppliers for the ozone/peroxide and UV/peroxide systems than for E beam and pulsed UV/peroxide systems. Per the aforementioned discussion on water quality issues, qualitative factors, and costs, the ozone/peroxide and continuous wave UV/peroxide appear to be more feasible for treating potable waters contaminated with MTBE than the other AOP technologies.

APPENDIX A
UV PEROXIDE BENCH RESULTS

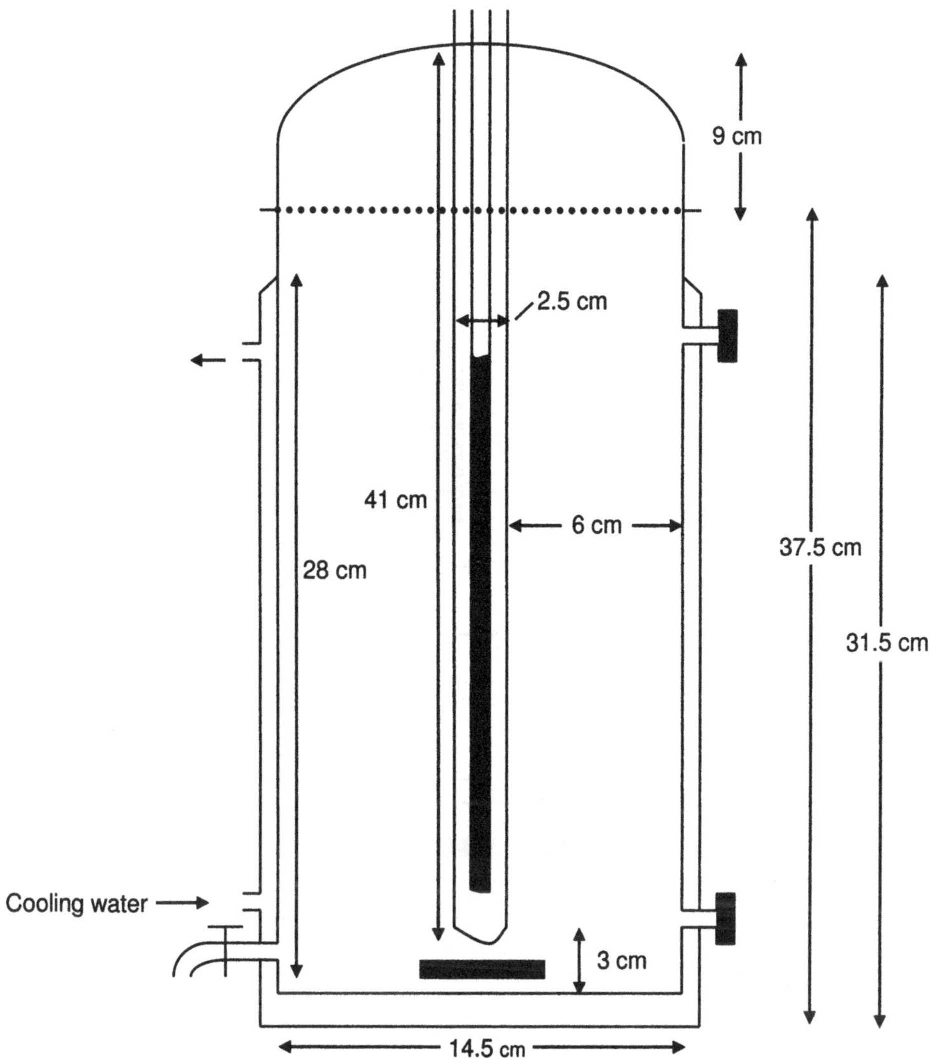

Figure A.1 Schematic of the reactor used in the experiments on the removal of MTBE with LP-UV/H_2O_2 processes

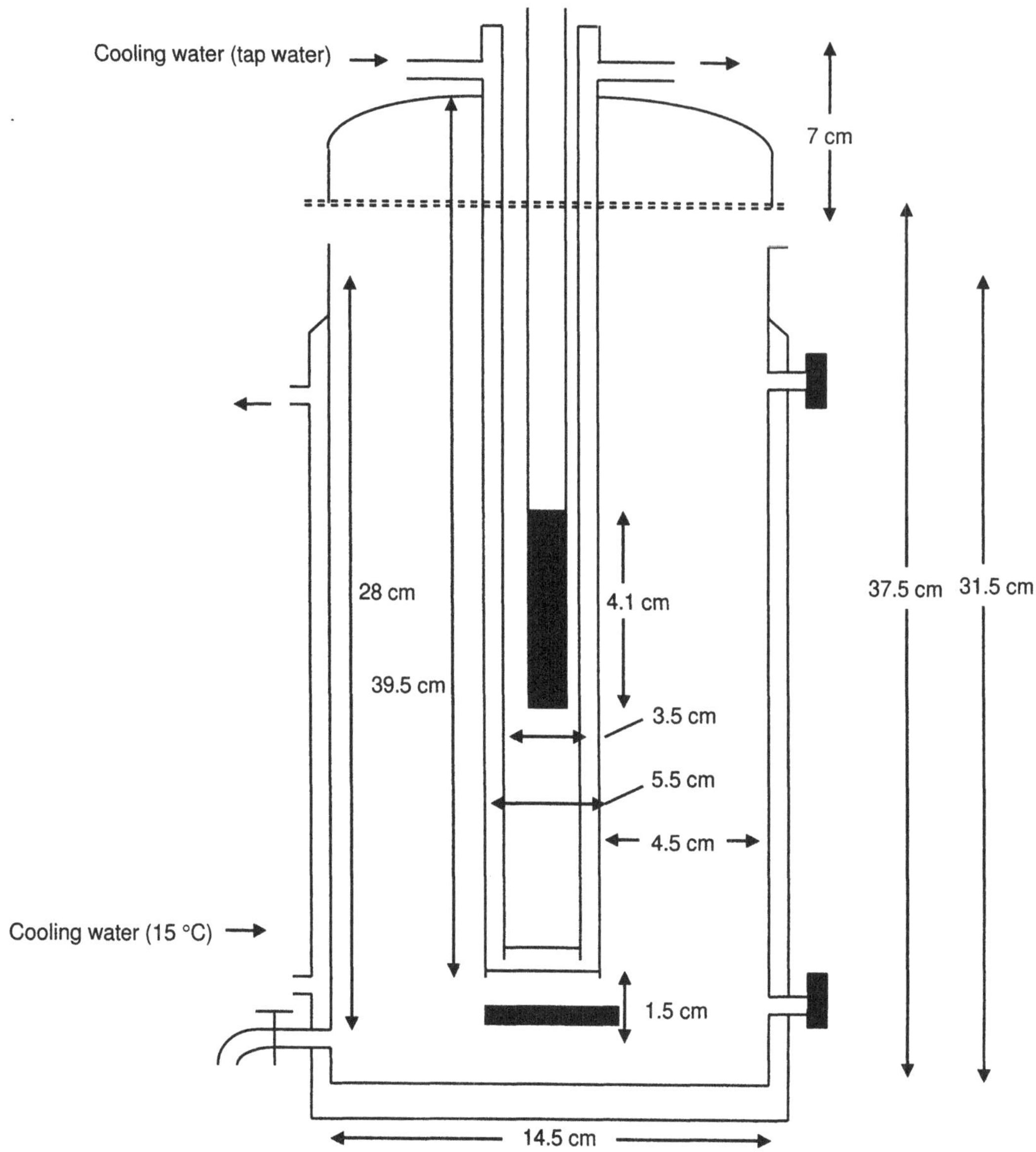

Figure A.2 Schematic of the reactor used in the experiments on the removal of MTBE with MP-UV/H_2O_2 processes

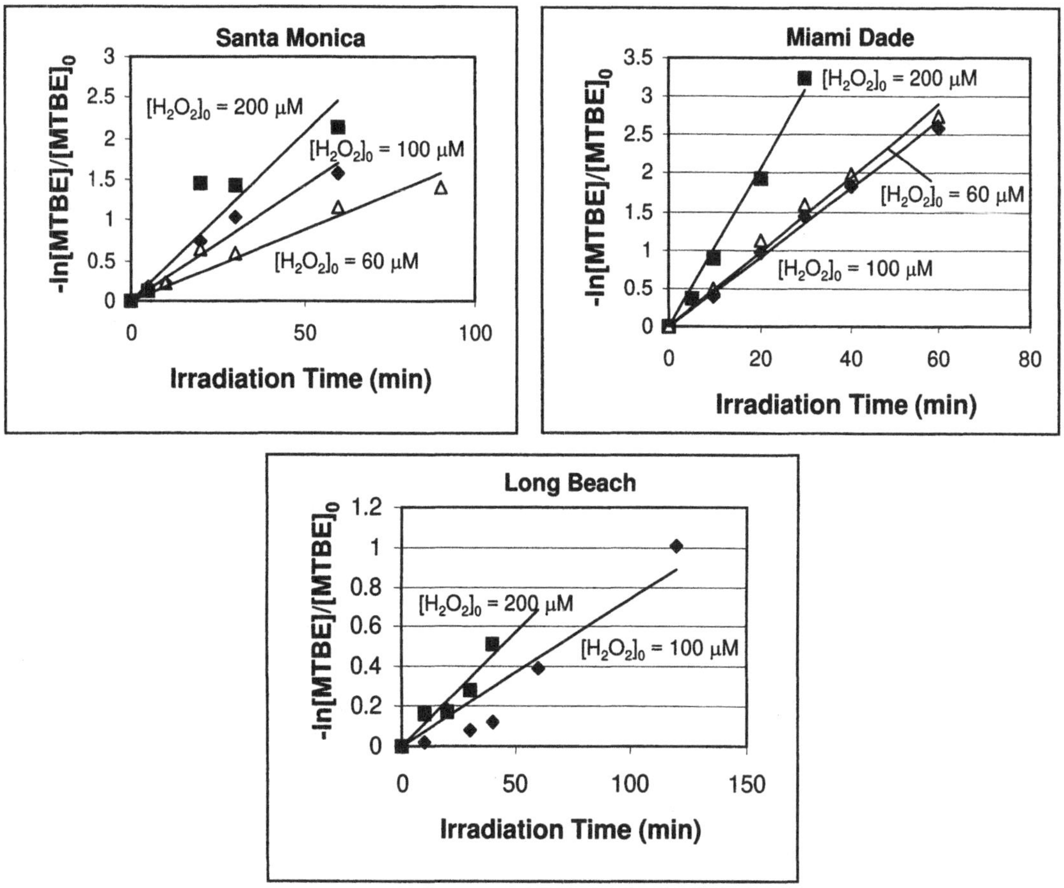

Figure A.3 Pseudo-first order degradation of MTBE in Santa Monica (SM), Suffolk County (SC), Long Beach (LB), and Miami Dade (MD) water samples with LP-UV/H_2O_2 processes at different H_2O_2 concentrations

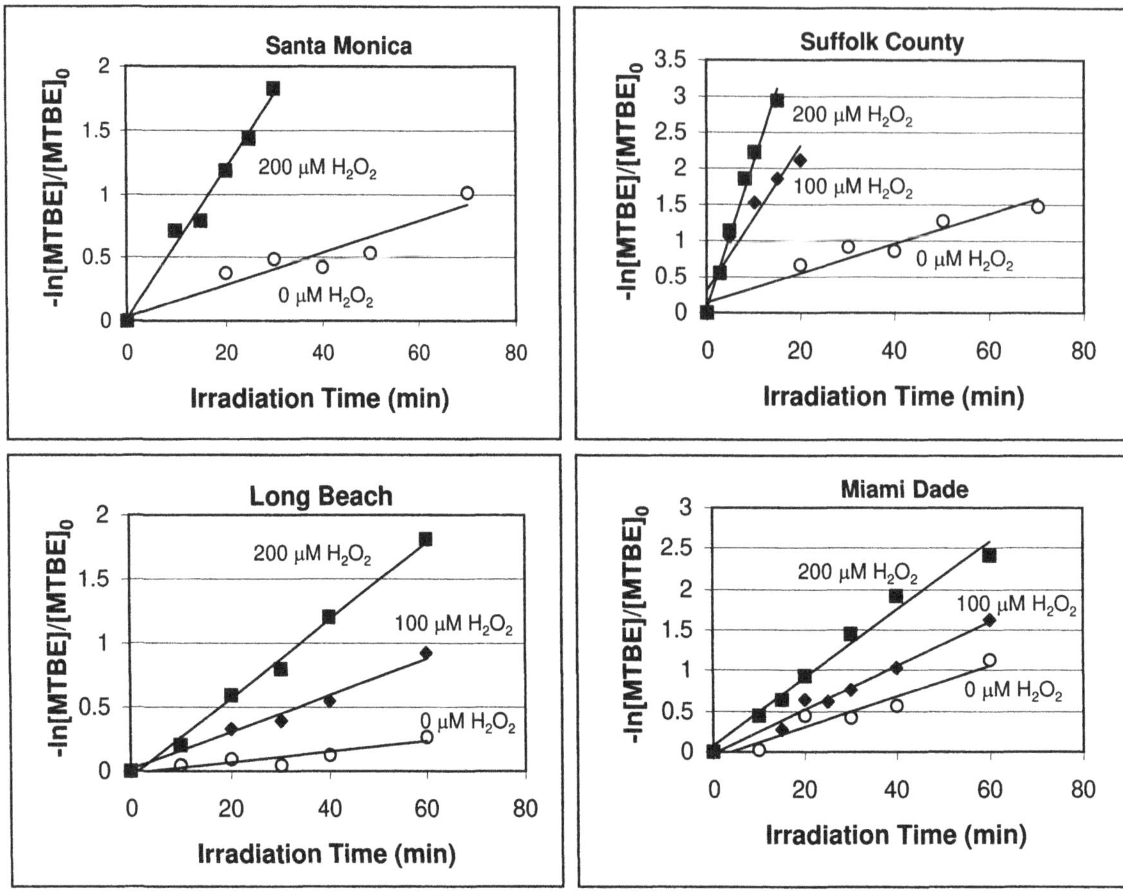

Figure A.4 Pseudo-first order degradation of MTBE in Santa Monica (SM), Suffolk County (SC), Long Beach (LB), and Miami Dade (MD) water samples with MP-UV/H_2O_2 processes at different H_2O_2 concentrations

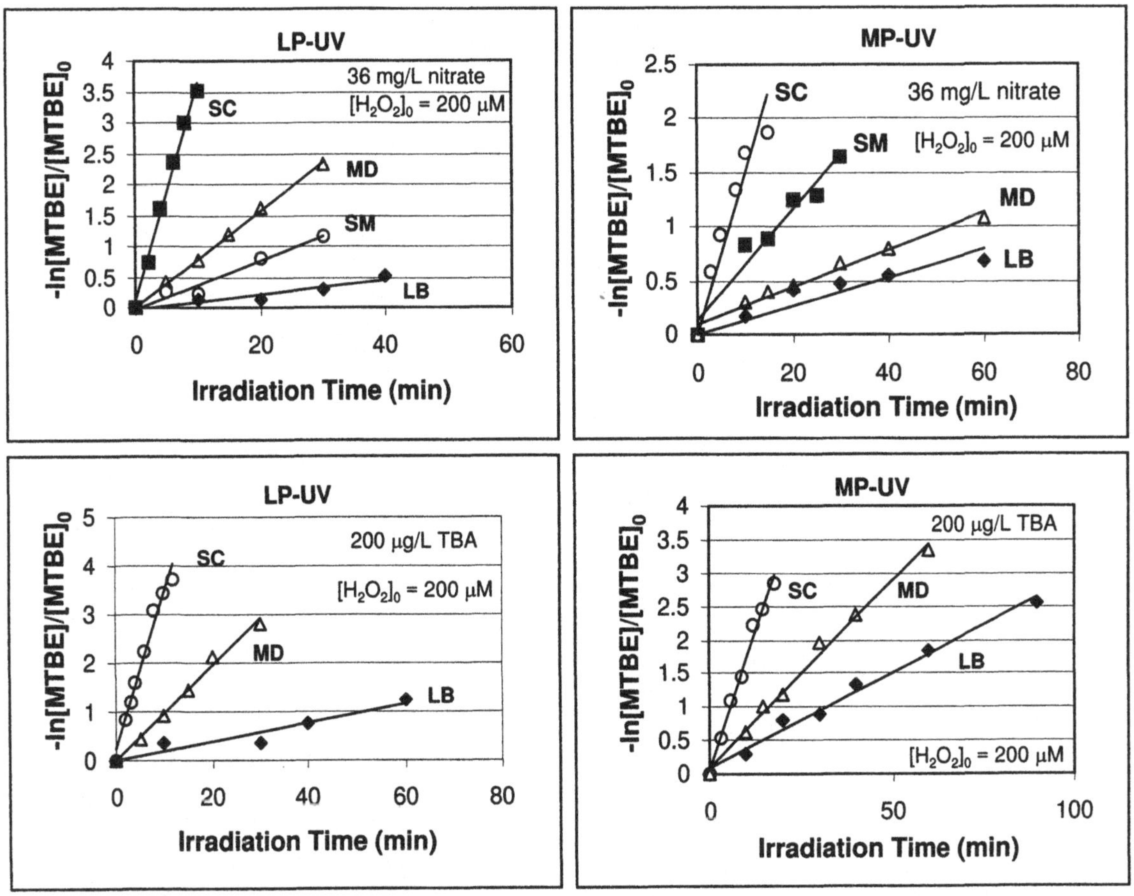

Figure A.5 Pseudo-first order degradation of MTBE in Santa Monica (SM), Suffolk County (SC), Long Beach (LB), and Miami Dade (MD) water samples with LP-and MP-UV/H_2O_2 processes with an addition of 36 mg/L nitrate or 200 µg/L TBA

SUFFOLK COUNTY

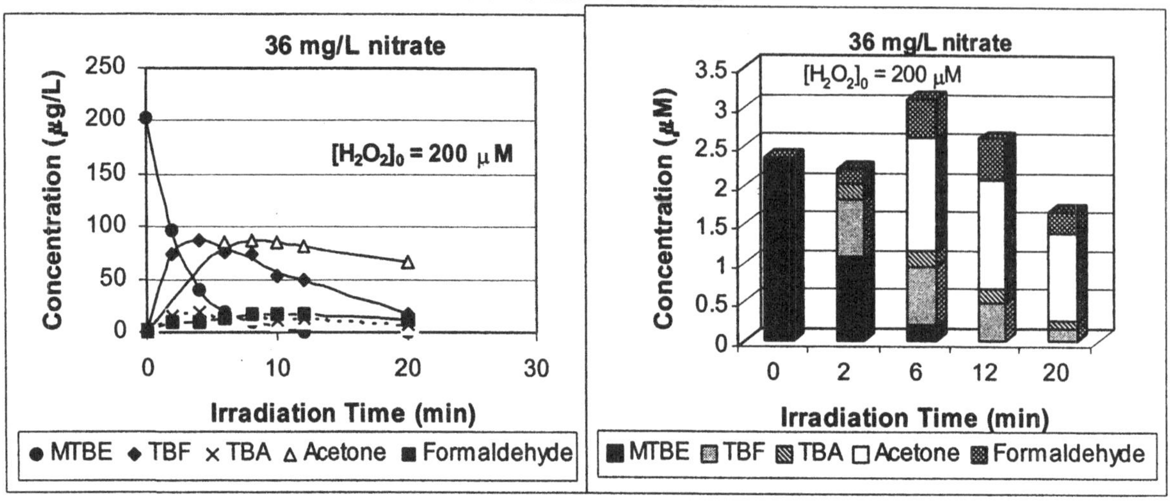

MIAMI DADE

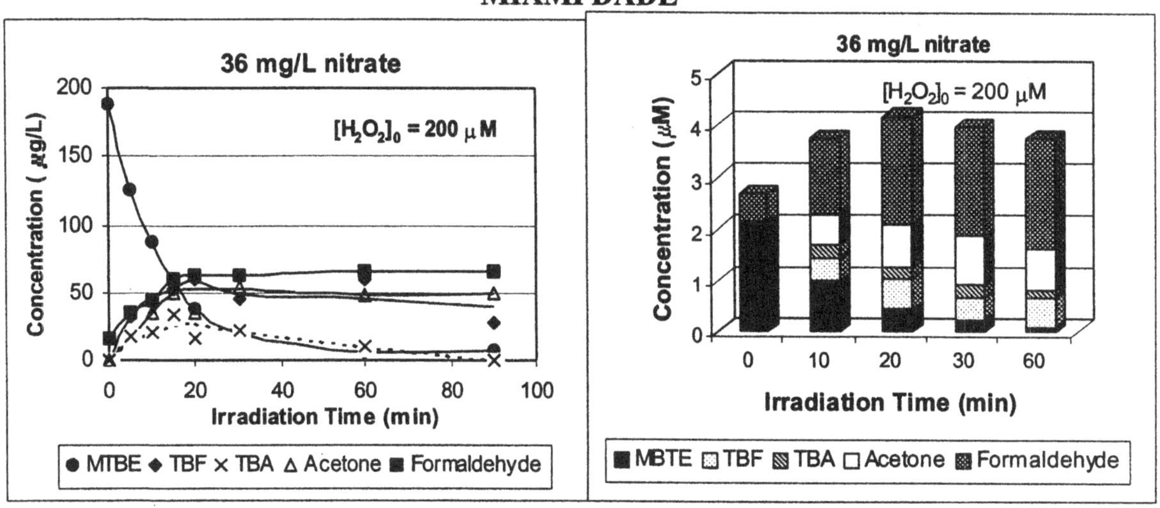

Figure A.6 Decay of MTBE and evolution of by-products in LP-UV/H_2O_2 treated water from Santa Monica and Long Beach with an addition of 36 mg/L nitrate

LONG BEACH

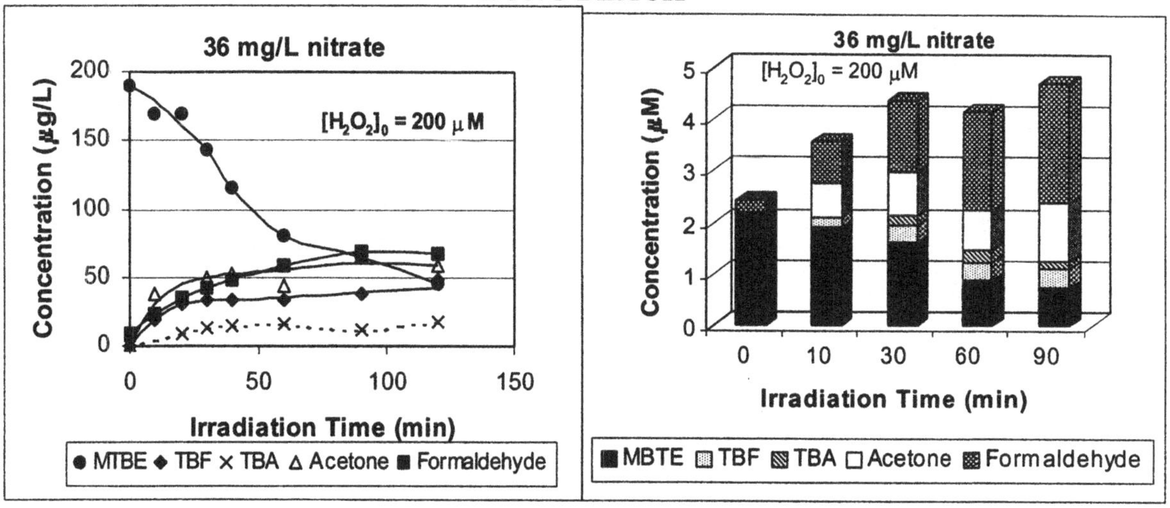

SANTA MONICA

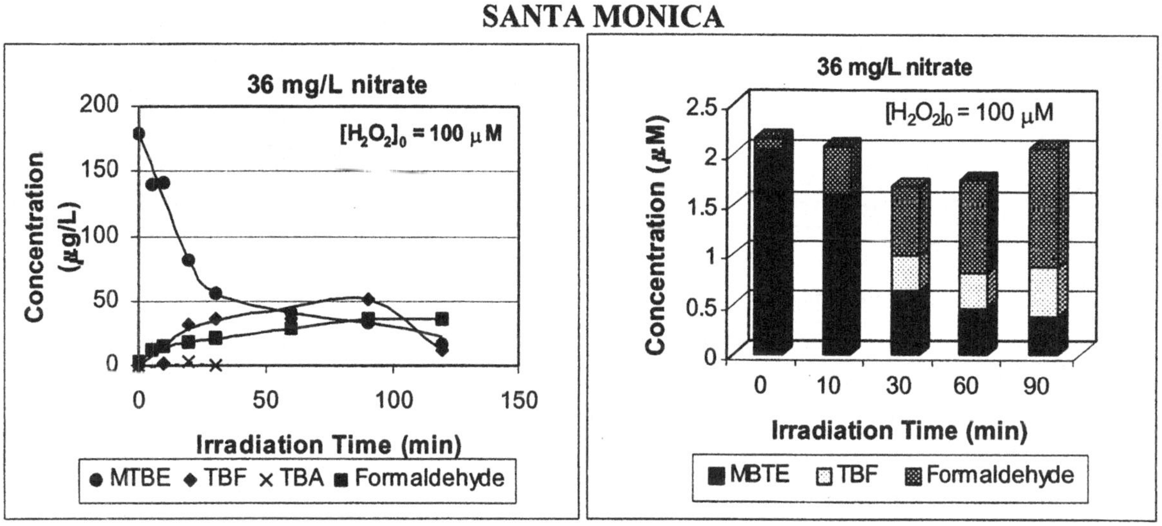

Figure A.7 Decay of MTBE and evolution of by-products in LP-UV/H_2O_2 treated water from Santa Monica and Long Beach with an addition of 36 mg/L nitrate

MIAMI DADE

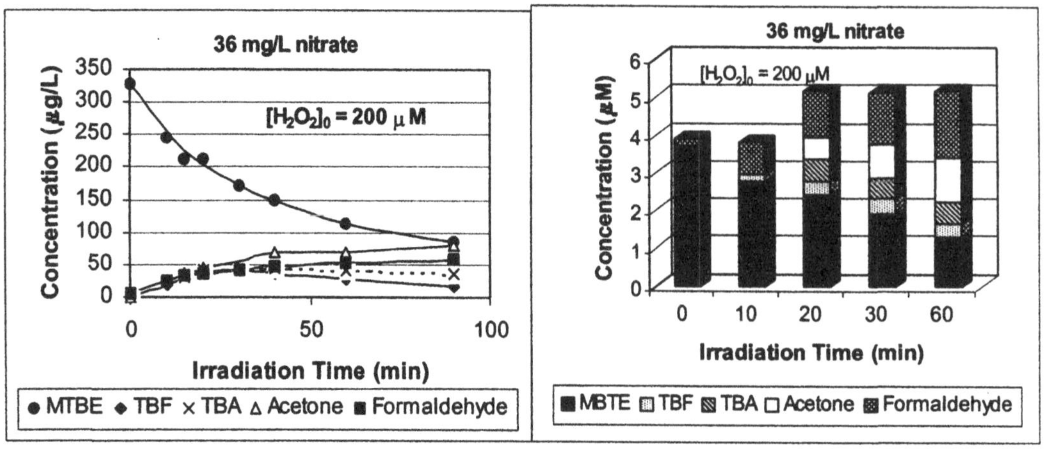

LONG BEACH

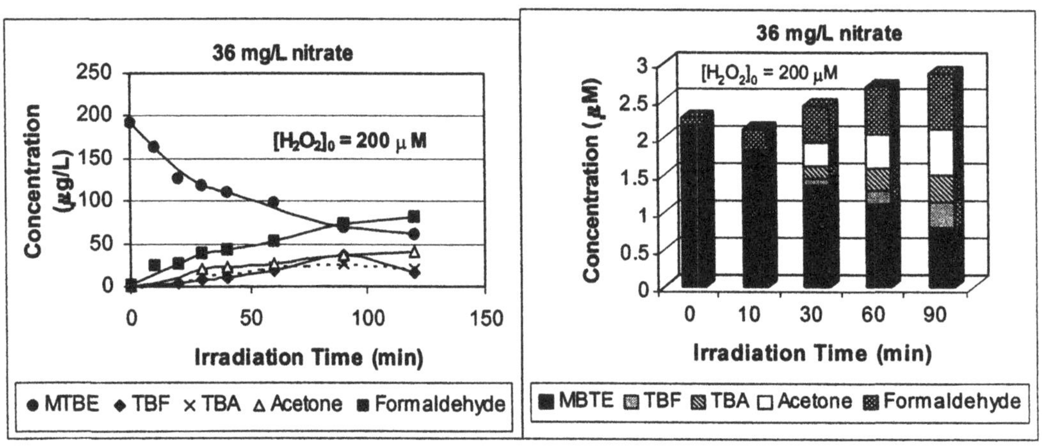

SUFFOLK COUNTY

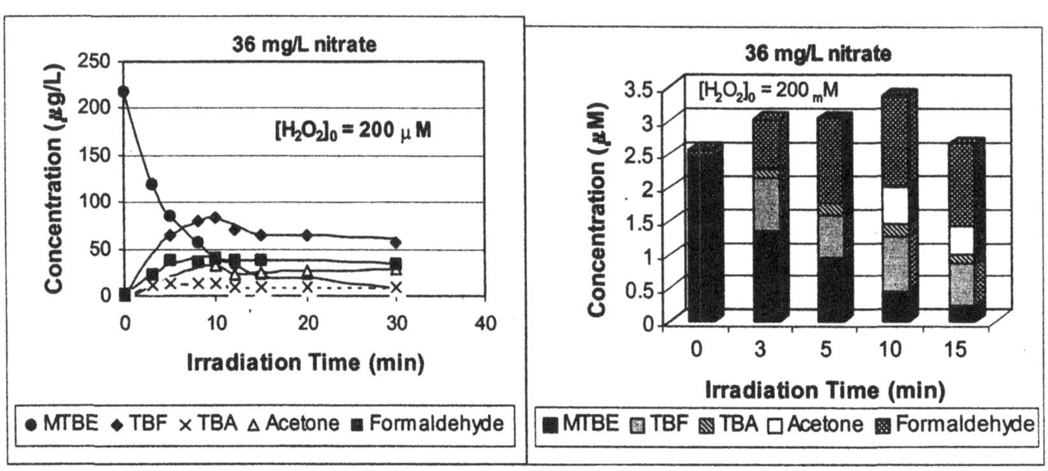

Figure A.8 Decay of MTBE and evolution of by-products in MP-UV/H_2O_2 treated water from Miami Dade, Long Beach, and Suffolk County in the presence of 36 mg/L nitrate

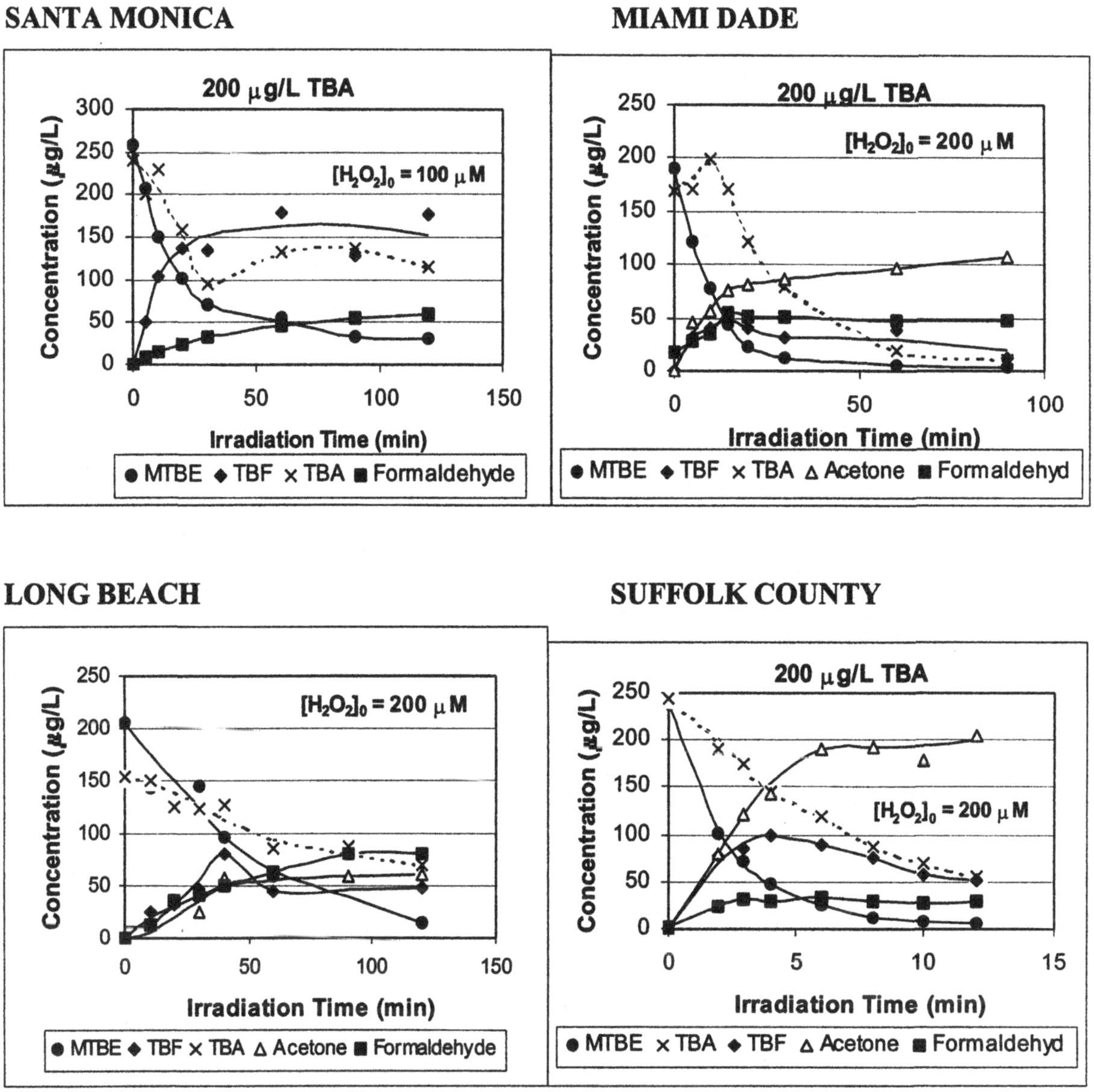

Figure A.9 Decay of MTBE and evolution of by-products with LP-UV/H_2O_2 with the addition of 200 μg/L TBA

LONG BEACH

SUFFOLK COUNTY

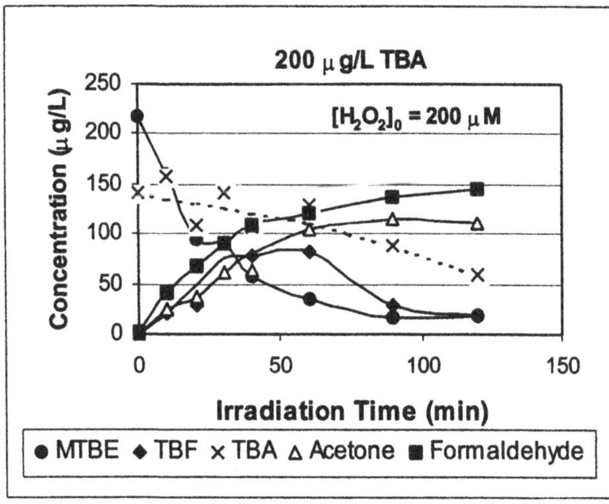

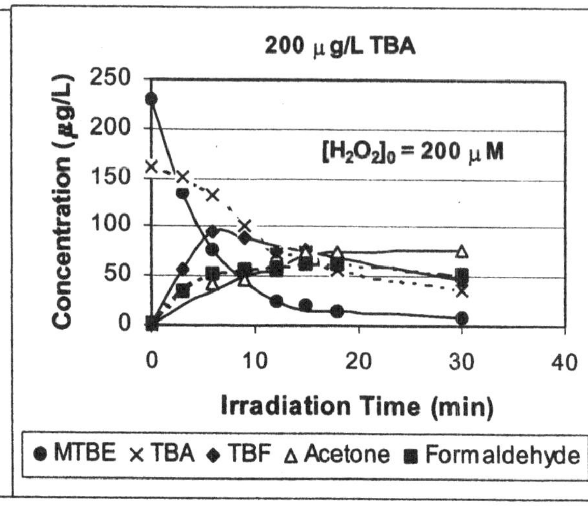

Figure A.10 Degradation of MTBE and evolution of by-products in MP-UV/H_2O_2 treated water from Suffolk County or Long Beach with a 200 μg/L TBA spike. The initial H_2O_2 concentration was 200 μM.

APPENDIX B
MTBE AND BY-PRODUCTS ANALYSIS

METHODS

To analyze MTBE and its degradation by-products (TBF, TBA, and acetone), the solid phase extraction (SPE) method developed by Davis, D. et al., (1999) has been used. Three different cartridges (3 mL) containing 200 mg styrenedivinylbenzene as solid phase were tested; Baker SDB-1, Baker SDB-2, Varian Bond Elut PPL. Of these cartridges, the best results were obtained with the Varian Bond Elut PPL cartridges (Varian Chrompack) and these were consequently used in our further experiments. The extraction procedure was as follows:

One hundred mL water was passed through a cartridge and the retained compounds were eluted with 2 mL of 1-pentanol (99+% ACS reagent, Aldrich). Before the extraction, the cartridges were rinsed with approximately 10 mL of ultrapure water and finally with 3 mL of methanol. The water sample was then passed through the cartridge at a flow rate of 3-4 mL/min. Care was taken not to allow the cartridges to dry both before and during the extraction of the water samples. At the end, a minimum amount of water was left on the solid phase and 2 mL of 1-pentanol was added and eluted in a series of short bursts into pre-weighed collection vials (Notice that a second elution step with 2 mL of fresh 1-pentanol confirmed that the adsorbed compounds were removed from the cartridge during the first elution step). The valve was finally left open for approximately 30 s. The water phase was removed from the vials and the vials were weighed. The extraction yields were then corrected with the appropriate amount of 1-pentanol eluted. Approximately 1.7 mL 1-pentanol was eluted each time. To one mL of the eluted 1-pentanol, five μL of a stock solution of CS_2 (16.7 mM) in 1-pentanol was added as internal standard, and the solution was analyzed by gas chromatography coupled with mass spectrometry (GC/MS).

ANALYTICAL PROCEDURES

The samples were analyzed with a Hewlett Packard mass selective detector combined with a gas chromatograph (GC/MSD) working in the selective ion monitoring (SIM) mode. Table B.1 shows the ions monitored for MTBE, acetone, TBF, and TBA. The compounds were separated on an Alltech polyethylene glycol column of 30 m length and 0.25 mm i.d. with a

Table B.1
Retention times and ions monitored for the compounds analyzed by GC/MS-SIM

Compound	Retention time (min)	Quantitation ion	Qualification ion
MTBE	2.4	73.05	43.00
CS_2	2.8	75.95	-
Acetone	4.0	43.00	58.05
TBF	5.6	59.00	87.05
TBA	6.0	59.00	43.00

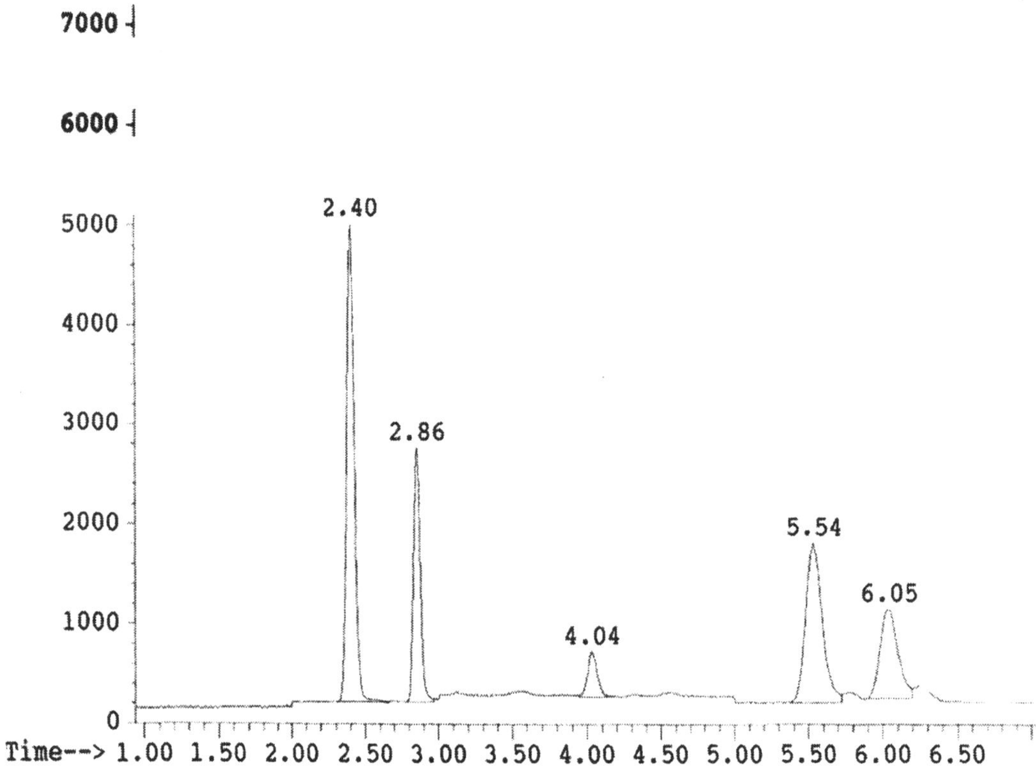

Figure B.1 A GC/MS-SIM chromatogram of a liquid/solid phase extracted water solution containing 200 µg/L MTBE, acetone, TBF, and TBA

stationary phase film thickness of 1 µm. The injector and the interface temperature were set at 150°C and 23°C, respectively. The GC oven temperature was held at 50°C for 10 min and then raised to 200°C at a rate of 50°C/min. The compounds were quantified by determining their respective area relative to that of the standard carbon disulfide. Figure B.1 presents an example of a GC/MS chromatogram of the four compounds (after liquid/solid phase extraction). The detection limits for the compounds with this method was 10 µg/L for MTBE and TBF whereas it was 20 µgL for acetone and TBA.

REPRODUCIBILITY IN ULTRAPURE WATER

To determine the reproducibility of the extraction method, several extractions of aqueous solutions containing known concentrations of MTBE, acetone, TBF, and TBA have been performed. A stock solution with a concentration of 500 ppm of each compound was prepared in ultrapure water and 0.5 mL of this solution was added to 1 L of ultrapure water to get a concentration of ~200 µg/L for each compound. Immediately after the spiking, up to eight fractions (100 mL) of the prepared solution were extracted in parallel as described above and the extraction recovery was determined by comparing the area responses of the compounds with those in a standard solution of the compounds dissolved in 1-pentanol.

Table B.2
Recovery of MTBE, acetone, TBF, and TBA from a 200-μg/L solution by using SPE on Varian Bond Elut PPL cartridges. The recoveries obtained from test 1 and 4 are excluded from the calculated mean value.

	Recovery (%)								
Compound	1	2	3	4	5	6	7	8	Mean ± SD
MTBE	41	79	86	51	86	75	73	87	81 ± 6
Acetone	11	13	13	9	13	11	11	9	12 ± 2
TBF	36	58	66	40	71	61	55	64	63 ± 5
TBA	40	37	43	33	33	49	40	45	40 ± 5

Table B.3
Recoveries of MTBE, acetone, TBF, and TBA from 200 μg/L spiked water solutions from Miami Dade, Long Beach, Santa Monica, and Suffolk County

	Extraction Recovery			
Compound	Miami Dade	Long Beach	Santa Monica	Suffolk County
MTBE	67 ± 9%	84 ± 2%	80 ± 3%	80 ± 3%
Acetone	14 ± 1%	12 ± 1%	10 ± 1%	10 ± 1%
TBF	53 ± 5%	99 ± 2%	73 ± 5%	60 ± 1%
TBA	32 ± 2%	56 ± 4%	40 ± 3%	27 ± 3%

The obtained recoveries together with the standard errors are presented in Table B.2. With the exception of the results of tests 1 and 4 for MTBE and TBF (no explanation can be given to explain such low recoveries) the liquid solid phase extraction protocol provides reasonably good recoveries for MTBE and TBF while the recoveries of TBA and acetone were significantly lower. In addition, the extraction recovery of MTBE or a 10 μg/L MTBE solution was 87 ± 6%.

EXTRACTION RECOVERIES IN THE FIELD WATER SAMPLES

The concentrations of MTBE, TBF, acetone, and TBA in the four field water samples were determined by using the solid phase extraction procedure described above. In order to determine if the different water quality parameters affected the extraction procedure, the extraction recoveries of the compounds were determined in each field water sample three or four fractions (100 mL) of a water sample spiked with 200 μg/L MTBE, TBF, TBA, and acetone were extracted and the extraction efficiencies were determined. The extraction recoveries obtained are presented in Tables B.2 and B.3. The extraction yields of MTBE and acetone were fairly similar in the different field waters whereas the variation in the extraction recoveries of TBF and TBA was much larger. However, because a calibration curve for each compound was prepared in the field water under study the variations in the extraction recovery may be neglected (Table B.4).

Table B.4
Linear regression equations and R^2 values for the calibrations curves of MTBE, TBA, TBF, and acetone in the four water sources

	Santa Monica		Miami Dade	
MTBE	y = 0.0144x - 0.412	r^2 = 0.995	y = 0.014x + 0.0195	r^2 = 0.9826
TBF	y = 0.0099x + 0.0969	r^2 = 0.969	y = 0.0095x + 0.0561	r^2 = 0.9725
TBA	y = 0.0062x + 0.1708	r^2 = 0.967	y = 0.0046x - 0.0187	r^2 = 0.9891
Acetone	y = 0.0016x – 0.0038	r^2 = 0.970	y = 0.0014x – 0.0313	r^2 = 0.9923
	Suffolk County		Long Beach	
MTBE	y = 0.0123x - 0.1317	r^2 = 0.9988	y = 0.0156x - 0.1008	r^2 = 0.9616
TBF	y = 0.0072x - 0.0308	r^2 = 0.9988	y = 0.0107x + 0.1182	r^2 = 0.9954
TBA	y = 0.0035x - 0.0028	r^2 = 0.9966	y = 0.0059x - 0.0123	r^2 = 0.9936
Acetone	y = 0.0007x + 0.0382	r^2 = 0.9957	y = 0.0019x – 0.0126	r^2 = 0.9948

Table B.5
Characteristics of the irradiation sources

	NNI 40/20	TQ 718	TQ 150
Electric power (W)	40	700	150
Total length (mm)	260 ± 2	167.0	165.0
Arc length (mm)	198	97	40
Radiation flux at 254 nm (W)	12	18.6	4.0
Radiation flux at 313 nm (W)	-	20.1	4.3
Radiation flux at 366 nm (W)	-	30.0	6.4
Radiation flux at 546 nm (W)	-	23.9	5.1

The field water samples were spiked with MTBE, TBF, TBA, and acetone and the solutions were extracted and analyzed as described above. Calibration curves were then obtained by plotting the area response versus the corresponding concentration and Table B.4 presents the obtained linear regression equations and R^2 values for each water sample. A new calibration curve was prepared for each set of analyses.

ANALYSIS OF ALDEHYDES

Low-molecular-weight aldehydes, such as formaldehyde, acetaldehyde, and glyoxal were extracted and analyzed according to the procedure described by Glaze, Koga, and Carcilla (1989). In this method, the aldehydes are derivatized by O-(2,3,4,5,6-pentafluorobenzyl) hydroxylamine hydrochloride (PFBOA), extracted with hexane and finally analyzed with GC/ECD. The compounds were analyzed on a Fisons Instrument GC 8180 equipped with a J&W Scientific column of 30 m length, 0.32 mm I.D and with a stationary phase film thickness of 0.25 μm.

REFERENCES

Acero, J.L., Stemmler, K., and von Gunten, U. 2000a. Degradation Kinetics of Atrazine and its Main Metabolites with Ozone and OH Radicals: A predictive Tool for Drinking Water Treatment. *Environmental Science & Technology, 34(4), 591-597.*

Acero, J.L., von Gunten, U., Schmidt, T., and Haderlein, S. 2000b. MTBE Oxidation by Conventional Ozonation and the Combination Ozone/Hydrogen Peroxide: Efficiency and Bromate Formation. Manuscript submitted for publication.

Achten, C. and Puttmann, W. 2000. Determination of Methyl Tert-Butyl Ether in Surface Water by Use of Solid-Phase Microextraction. *Jour. Environmental Science and Technology,* 34(7)1359-1364.

Allen, A.O. 1961. *The Radiation Chemistry of Water and Aqueous Solutions.* 1961. Van Nostrand-Reinhold. Princeton, N.J.

American Petroleum Institute (API). 1994. Odor Threshold Studies Performed With Gasoline and Gasoline Combined With MTBE, ETBE, and TAME. TRC Environmental Corporation. Conducted by TRC Environmental Corp.

American Public Health Association, American Water Works Association, and Water Environment Federation (APHA, AWWA, and WEF). 1995. *Standard Methods for the Examination of Water and Wastewater.* 19th ed. Washington, D.C., American Public Health Association.

American Public Health Association, American Water Works Association, and Water Environment Federation. *Standard Methods for the Examination of Water and Wastewater.* 20th ed. 1998. Washington, D.C., American Public Health Association.

Applebury, T. 1999. Personal Communication. Chief Technical Officer, Applied Process Technologies, Pleasant Hill, California.

ARCO Chemical Company. 1993. The Odor and Taste Threshold Studies Performed with Methyl Tertiary-Butyl Ether (MTBE) and Ethyl Tertiary-Butyl Ether (ETBE). TRC Environmental Corporation. Conducted by TRC Environmental Corp.

Asmus, K.D.; Mockel, H.; Henglein, A. 1973. Pulse Radiolytic Study of the Site of OH Radical Attack on Aliphatic Alcohols in Aqueous Solution. *J. of Physical Chemistry,* 77, 10, 1218-12.

AwwaRF. 1998. Effect of Bicarbonate Alkalinity on Performance of Advanced Oxidation Processes. Prepared by: Peyton, G.R., Bell, O.J., Girin, E., LaFaivre, M.H. and Sanders, J., Denver.

Bader, H. and Hoigné, J. 1982. Determination of Ozone in Water by the Indigo Method: A Submitted Standard Method. *Jour. Ozone: Science and Engineering,* 4(4), 169-176.

Barreto, R.D., Gray, K.A. and Anders, K. 1995. "Photocatalytic Degradation of MTBE in TiO_2 Slurries: A Proposed Reaction Scheme," *Water Research,* 29(5):1243-1248.

Bender, J.B. May 1998. Photolytic Oxidation of Contaminated Water Using the Riptide Water Purification System. Toronto Environmental Trade Show and Conference, Seminar # 2, Toronto, Canada.

Bender, J.B. 1999. Personal Communication. President, Pulsar Environmental Remediation Systems Incorporated.

Bolton, J.R.; Valladares, J.E.; Cooper, W.J.; Waite, T.D.; Kurucz, C.N.; Nickelsen, M.G.; and Kajdi, D.C. 1998. Figures-of-Merit for Advanced Oxidation Processes - A Comparison of Homogeneous UV/H_2O_2, Heterogeneous TiO_2 and Electron Beam Processes. *Jour. Adv. Oxid. Technol.,* 3, 174-181.

Bukhari, Z., Hargy, T.M., Bolton, J.R., Dussert, B. and Clancy, J.L. 1999. Medium-Pressure UV for Oocyst Inactivation. *Jour. AWWA,* 91(3):86-94.

Buscheck, 1998. Occurrence and Behavior of MTBE in Groundwater. Underground Storage Tank Conference, Los Angeles, State of California Water Resources Control, Sacramento.

Buxton, G.V.; Greenstock, C.L.; Helman W.P.; and Ross, A.B. 1988. Critical Review of Rate Constants for Hydrated Electrons, Hydrogen Atoms and Hydroxyl Radicals (OH/O^-) in Aqueous Solution. *Journal Physical and Chemical Reference Data,* 17(2)13.

Calgon AOT Handbook. 1996. Markham, Ontario, Canada: Calgon Carbon Oxidation Technologies.

California MTBE Research Partnership. 2000. Treatment Technologies for Removal of Methyl Tertiary Butyl Ether (MTBE) from Drinking Water: Air Stripping, Advanced Oxidation Processes, Granular Activated Carbon, Synthetic Resin Sorbents (2nd Edition). Editor - Gina Melin, National Water Research Institute, Fountain Valley, CA.

Cater, S. 1999. Personal Communication. Research and Technical Manager, Calgon Carbon Technologies, Markham, Ontario, Canada.

Cater, S.R., Stefan, M.I., Bolton, J.R. and Safarzadeh-Amiri, A. 2000. UV/H_2O_2 Treatment of Methyl Tert-Butyl Ether in Contaminated Waters, *Jour. Environmental Science and Technology,* 34(4)659 – 662.

Chang, P., and Young, T. 1999. Reactivity and By-products of Methyl Tertiary-Butyl Ether Resulting from Water Treatment Processes. University of California Toxic Substances Research & Teaching Program.

Chang, P.B.L., and Young, T.M. 2000. Kinetics of Methyl Tert-Butyl Ether Degradation and Byproduct Formation During UV/hydrogen Peroxide Water Treatment. *Jour. Water Research,* 34(8), 2233-2240.

Chiou, C.-F.; Mariñas, B.J., and Adams, J.Q. 1995. Modified Indigo Method for Gaseous and Aqueous Ozone Analyses. *Jour. Ozone: Science and Engineering,* 17, 329-344.

Church, C.D.; Isabelle, L.M.; Pankow, J.F.; Rose, D.L. and Tratnyek, P.G. 1997. *Jour. Environmental Science Technology,* 31, 3723-3726.

Church, C.D.; Pankow, J.F.; and Tratnyek, P.G. 1999. Hydrolysis of Tert-Butyl Formate: Kinetics, Products and Implications for the Environmental Impact of Methyl Tert-Butyl Ether. *Environmental Toxicology and Chemistry,* 18(12), 2789-2796.

Cooper, W.J. January, 1998. Treatment of MTBE in Drinking Water Supplies Using High Energy Electron Injection: Innovative, Proven, Economical. Orange County ACS Meeting.

Cooper, W.J., and Mezyk, S.P. 2001. MTBE and Priority Contaminant Treatment With High Energy Electron Beam Injection. *Jour. Radiation Physics and Chemistry.*

Cooper, W.J.; Nickelsen, M.G.; Meacham, D.E.; Waite, T.D.; and Kurucz, C.N. 1992a. High Energy Electron Beam Irradiation: An Innovative Process for the Treatment of Aqueous-Based Organic Hazardous Wastes. *J. Environ. Sci. Health,* A27(1), 219-244.

Cooper, W.J.; Nickelsen, M.G.; Meacham, D.E.; Waite, T.D.; and Kurucz, C.N. 1992b. High Energy Electron Beam Irradiation: An Advanced Oxidation Process for the Treatment of Aqueous Based Organic Hazardous Wastes. *Wat. Poll. Res. J. Canada,* 27, 69-95.

Cooper, W.J.; Meacham, D.E.; Nickelsen, M.G.; Lin, K.; Ford, D.B.; Kurucz, C.N.; and Waite, T.D. 1993a. The Removal of Tri- (TCE) and Tetrachloroethylene (PCE) from Aqueous Solution Using High Energy Electrons. *J. Air Waste Manage. Assoc.* 43, 1358-66.

Cooper, W.J.; Cadavid, E.; Nickelsen, M.G.; Lin, K.; Kurucz, C.N.; and Waite, T.D. 1993b. Removing THMs From Drinking Water Using High-Energy Electron-Beam Irradiation. *Jour. AWWA,* 85 (Sept.), 106-112.

Cooper, W.J.; Dougal, R.A.; Nickelsen, M.G.; Waite, T.D.; Kurucz, C.N.; Lin, K.; and Bibler, J.P. 1996a. Benzene Destruction in Aqueous Waste, Part 1. Bench-Scale Gamma Irradiation Experiments. *Rad. Phys. Chem.,* 48: 81-87.

Cooper, W.J.; Sawal, K.L.; Slifker, R.A.; Nickelsen, M.G.; Hoogland, Y.S.; Kurucz, C.N.; and Waite, T.D. 1996b. Precursor Removal from Natural Waters Using an Innovative Treatment Process. In *Disinfection By-Products in Water Treatment: The Chemistry of Their Formation and Control.* R.A. Minear and G.L. Amy, Eds, CRC Press, Inc. Boca Raton, FL, 151-162.

Cooper, W.J.; Leslie, G.; Tornatore, P.M.; Hardison, W.C. and Hajali, P.A. 1999. MTBE and Priority Contaminant Treatment With High Energy Electron Injection." Draft.

Cooper, W.J. and Tornatore, P.M. March, 1999. Destruction of MTBE in Contaminated Groundwater Using Electron Beam Injection. ACS Proceeding, Anaheim, California.

Cooper, W.J.; Tobien, T.; Mezyk, S.P.; Adams, J.W.; Nickelsen, M.G.; O'Shea, K.E.; Inclan, G.; Tornatore, P.M.; Hajali, P.; and Weidman, D.J. 2001a. The Electron Beam Process for the Destruction of MTBE. In *Exploring the Environmental Issues of Mobile, Recalcitrant Compounds in Gasoline,* A.F. Diaz & D.L. Drogos (Eds.). Washington, DC: American Chemical Society Symposium Series.

Cooper, W.J.; Tobien, T.; Mezyk, S.P.; Adams, J.W.; Nickelsen, M.G.; O'Shea, K.E.; Inclan, G.; Tornatore, P.M.; Hajali, P.; and Weidman, D.J. 2001b. The Electron Beam Process For The Destruction of MTBE. In *Oxygenates in Gasoline: Environmental Aspects.* Eds. A. F. Diaz and D. Drogos, ACS Symposium Series 799, American Chemical Society, Washington, DC. P. 153-164

Cooper, W.J.; Nickelsen, M.G.; Mezyk, S.P.; Leslie, G.; Tornatore, P.M.; Hardison, W.; and Hajali, P.A. 2002. MTBE and Priority Contaminant Treatment With High Energy Electron Beam Injection. *Rad. Phys. Chem.* (in press).

Crawford, M. 1999. Personal Communication. Calgon Carbon Incorporated. Suisun City, California.

Crittenden, J., Hand, D., Perram, D.L., and Marchand, E.G. 1997. Environmental Technologies Design Option Tools (ETDOT) for the Clean Process Advisory Systems - Adsorption, Aeration and Physical Properties Software. National Center for Clean Industrial and Treatment Technologies.

Culp, Wesner and Culp. 1990. WaterCost Model: A Computer Program for Estimating Water and Wastewater Treatment Costs. Civic Engineering, San Clemente, CA.

Dale, M.S.; Moylan, M.S.; Koch, B.; Davis, M.K. 1997. MTBE: Taste and Odor Threshold Determinations Using the Flavor Profile Method. Metropolitan Water District of Southern California.

Davidson, J.M., and Creek, D.N. 1998. *Granular Activated Carbon, in Treatment Technologies for Removal of Methyl Tertiary Butyl Ether (MTBE) From Drinking Water: Air Stripping, Advanced Oxidation Processes (AOP), and Granular Activated Carbon (GAC).* Fort Collins, CO: Alpine Environmental, prepared for MTBE Research Partnership.

Davis, S.W., and Powers, S.E. 1999. Alternative Sorbents for Removing MTBE from Gasoline-Contaminated Groundwater. *Journal of Environmental Engineering.*

Davis, D.V.; Sclimenti, M.J.; Pastor, S.J. and Barrett, S.E. 1999. Submitted to *Environmental Science and Technology.*

Deeb, R.A.; Stocking, A.; Alvarez-Cohen, L. and Kavanaugh, M. 2001. Biodegradation of MTBE and TBA: A Current Review, Chapter 16. In *Exploring the Environmental Issues of Mobile, Recalcitrant Compounds in Gasoline.* Diaz, A. and Drogos, D. (Eds.), American Chemical Society Books and Oxford University Press. (In press.)

Delzer, G.C.; Grady S.J.; Zogorski, J.S.; Rowe, B.L. and Koch, B. 2001. National Survey of MTBE, Other Oxygenates, and Other VOCs in Community Drinking-Water Sources, Abstract for AwwaRF Project 2507. Posted on the web site at www.awwa.org/mtbe. January 2001.

Dougal, R.A.; Cooper, W.J.; Nickelsen, M.G.; Lin, K.; Waite, T.D.; Kurucz, C.N. and Bibler, J.P. 1998. Industrial-scale Destruction of Benzene in a Simulated Mixed Waste Using Electron-Beam Irradiation. *Jour. Environmental Applications of Ionizing Radiation.* John Wiley & Sons, Inc. New York, N.Y. 417-427.

Dyksen, J.E.; Ford, R.; Raman, K.; Schaeffer, J.K.; Fung, L.; Schwartz, B; and Barnes, M. 1992. In-line Ozone and Hydrogen Peroxide Treatment for Removal of Organic Chemicals, AwwaRF, Denver, CO.

Eibenberger, J.; Schulte-Frohlinde, D.; and Steenken, S. 1980. One-Electron Oxidation of Mono-alkoxyalkyl Radicals by Tetranitromethane via an Intermediate Adduct. Influence of Radical Structure on Rate of Decomposition of the Adduct. *J. Phys. Chem.,* 84, 704-710.

Elovitz, M.S., and von Gunten, U. 1999. Hydroxyl Radical/Ozone Ratios During Ozonation Processes I. The Rct Concept. *Jour. Ozone Science and Engineering,* 21:239-260.

Elovitz, M.S., von Gunten, U., and Kaiser, H.-P. 2000. Hydroxyl Radical/Ozone Ratios During Ozonation Processes. II. The Effect of Temperature, pH, Alkalinity, and DOM Properties. *Jour. Ozone Science and Engineering,* v. 22, p. 123-150.

EPA Arsenic Technology and Cost Document. 2002. Prepared by Malcolm Pirnie under Cadmus Group Inc. contract to USEPA.

EPA Stage-2/LT-2 Technology and Cost Document. 2002. Prepared by Malcolm Pirnie under Cadmus Group Inc. contract to USEPA.

EPA/600/R-98/048. 1998. Oxygenates in Water: Critical Information and Research Needs. USEPA Office of Research and Development, Washington, D.C.

EPA/600/R-98/160VS. 1999. Environmental Technology Verification Statement, Sentinel™ UV Reactor, USEPA Office of Research and Development, Washington D.C.

Finneran, K. and Lovley, D. 2001. Anaerobic Degradation of Methyl Tert-Butyl Ether (MTBE) and Tert-Butyl Alcohol (TBA). *Environ. Sci. and Technol.,* 35(9): 1785-1790.

Glaze, W.H.; Koga, M.; and Cancilla, D. 1989. Ozonation Byproducts 2: Improvement of Aqueous-Phase Derivatization Method for the Detection of Formaldehyde and Other Carbonyl Compounds Formed by the Ozonation of Drinking Water. *Environmental Science and Technology,* 23: 838.

Gruber, W. 1994. "The Ultrox UV Oxidation System," In *EI Digest – Industrial and Hazardous Waste Management,* p. 7-11.

Gullick, R.W., and LeChevallier, M.W. 2000. Occurrence of MTBE in Drinking Water Sources. *Jour. AWWA,* 92(1):100-113.

Hanson, J.R.; Ackerman, C.E.; and Scow, K.M. 1999. Biodegradation of Methyl Tert-Butyl Ether by a Bacterial Pure Culture. *Jour. Applied and Environmental Microbiology,* 65(11):4788-4792.

Happel, A.M.; Beckenbach, E.H. and Halden, R.U., 1998. An Evaluation of MTBE Impacts to California Groundwater. Lawrence Livermore National Laboratory, Livermore, CA UCRL-AR-130897.

Hardison, D.R.; Cooper, W.J.; Mezyk, S.P. and Bartels, D.M.. 2002. The Free Radical Chemistry of Tert-Butyl Formate: Rate Constants for Hydroxyl Radical, Hydrated Electron and Hydrogen Atom Reaction in Aqueous Solution. *Rad. Phys. Chem.* (In Press.)

Hengesbach, B.; Schoenen, D.; Hoyer, O.; Bernhardt, H.; Mark, G., Schuchmann, H.P., and Sonntag, C. 1993. UV Disinfection of Drinking Water – the Question of Bacterial Regrowth and the Photolytic Degradation of Biogenic High-Molecular Weight Substances. J. WSRT-AQUA. 42(1):13-22.

Himebaugh, W. 1999. Personal Communications. Industrial Wastewater Manager, Signal Hill, California.

Hoigné, J. 1998. "Chemistry of Aqueous Ozone and Transformation of Pollutants by Ozonation and Advanced Oxidation Processes." In *The Handbook of Environmental Chemistry,* Vol. 5, Part C, Quality and Treatment of Drinking Water (edited by J. Hrubec). Springer-Verlag, Berlin.

Hoigné, J. and Bader, H. 1976. The Role of Hydroxyl Radical Reactions in Ozonation Processes in Aqueous Solution. *Jour. Water Research,* 10:377-386.

Hong, S., Duttweiler, C.M. and Lemley, A.T. 1999. *J. Chromat. A.,* v. 857, p. 205-216.

Hurt, K., Wilson, J., Beck, F. and Cho, J. 1999. In *Natural Attenuation of Chlorinated Solvents, Petroleum Hydrocarbons, and Other Organic Compounds,* Alleman, B. and Leeson, A. (Eds.). Battelle Press, Columbus, OH, Vol. 1, pp 103-108.

HVEA (High Voltage Environmental Applications). 1999. Product Brochure, Rochester, New York.

Kang, J-W. and Hoffmann, M.R. 1998. Kinetics and Mechanism of the Sonolytic Destruction of MTBE by Ultrasonic Irradiation in the Presence of Ozone. *Jour. Environmental Science and Technology,* 32(20):3194-3199.

Kang, J-W., Hung, H-M., Lin, A. and Hoffmann, M.R. 1999. Sonolytic Destruction of Methyl Tert-Butyl Ether by Ultrasonic Irradiation: The Role of O_3, H_2O_2, Frequency, and Power Density. *Environmental Science and Technology.* 33(18):3199-3205.

Karpel Vel Leitner, N., Papailhou, A., Croué, J., Peyrot, J., and Doré, M. 1994. Oxidation of Methyl Tert Butyl Ether (MTBE) and Ethyl Tert Butyl Ether (ETBE) by Ozone and Combined Ozone/Hydrogen Peroxide. *Ozone Science and Engineering,* v. 16 (41).

Kim, D.K.; O'Shea, K.E. and Cooper, W.J. 2002. The Degradation of MTBE and Related Gasoline Oxygenates in Aqueous Media by Ultrasound Irradiation. *ASCE – Journal of Environ. Engineer.* (Submitted.)

Kok, G.L.; Thompson, K. and Lazrus, A.L. 1986. Derivatization Technique for the Determination of Peroxides in Precipitation. *Analytical Chemistry,* 58: 1192-1194.

Kormann, C., Bahnemann, D.W., and Hoffmann, M.R. 1988. Photocatalytic Production of H_2O_2 and Organic Peroxides in Aqueous Suspensions of TiO_2, ZnO, and Desert Sand. *Environmental Science and Technology,* 22:798–806.

Krasner, S.T.; Glaze, W.H.; Weinberg, H.S.; Daniel, P.A. and Najm, I.N. 1993. Formation and Control of Bromate During Ozonation of Waters Containing Bromide. *Jour. AWWA,* 85:1:73.

Kuo, C.-Y., Krasner, S.W.; Stalker, G.A. and Weinberg, H.S. 1990. Analysis of Inorganic Disinfection By-Products in *Ozonated Drinking Water By Ion Chromatography.* American Water Works Association Water Quality Technology Conference, San Diego, CA.

Kurucz, C.N.; Waite, T.D., and Cooper, W.J. 1995. The Miami Electron Beam Research Facility: A Large Scale Wastewater Treatment Application. *Rad. Phys. Chem.,* 45, 299-308.

Landmeyer, J., Chapelle, F., Herlong, H. and Bradley, P. 2001. Methyl Tert-Butyl Ether Biodegradation by Indigenous Aquifer Microorganisms under Natural and Artificial Oxic Conditions. *Environ. Sci. Technol.* 35(6): 1118-1126.

Lassiter, J. 1986. ACUCHEM™ (Version 1.4) [Computer software]. Richardson, TX: Acuchem Laboratories.

Lazrus, A.L.; Kok, G.L.; Gitlin, S.N. and Lind, J.A. 1985. Automated Fluorometric Method for Hydrogen Peroxide in Atmospheric Precipitation. *Analytical Chemistry* 57: 917-922.

Liang, S.; Palencia, S.L.; Yates, R.S.; Davis, M.K.; Bruno, J.M., and Wolfe, R.L. 1999a. Oxidation of MTBE by Ozone and Peroxone. *Jour. AWWA,* 91(6), 104-114.

Liang, S.; Yates, R.S.; Davis, D.V.; Pastor S.J.; Palencia, L.S. and Bruno J-M. 1999b. "Treatability of Groundwater Containing Methyl *tertiary*-Butyl Ether (MTBE) by Ozone and PEROXONE and Identification of By-products." (Submitted to JAWWA for Publication.)

Lin, K.; Cooper, W.J.; Nickelsen, M.G.; Durucz C.N. and Waite, T.D. 1995. Decomposition of Aqueous Solutions of Phenol using HEEB Irradiation: a Large Scale Study. *Applied Radiation Isotopes,* 46(12):1307-1316.

Mackay, D.M., Einarson, M.D., Wilson, R.D., Fowler, B., Scow, K., Hyman, M., Naas, C., Schirmer, M., and Durrant, G.D. 1999. Field Studies of In-Situ Remediation of an MTBE Plume at Site 60, Vandenberg Air Force Base, California. In *Proceedings of the 2000 Petroleum Hydrocarbons and Organic Chemicals in Ground Water: Prevention, Detection, and Remediation.*

Mo, K.; Lora, C.O.; Wanken, A.E.; Javanmardian, M.; Yang, X. and Kulpa, C.F. 1997. Biodegradation of Methyl t-Butyl Ether by Pure Bacterial Cultures. *Microbiol. Biotechnol.,* 47: 69-72.

Mak, F.T., Cooper, W.J.; Kurucz, C.N.; Nickelsen, M.G. and Waite, T.D. 1996. Removal of Chloroform from Drinking Water Using High Energy Electron Beam Irradiation, *Disinfection By-Products in Water Treatment: The Chemistry of Their Formation and Control,* R.A. Minear and G.L. Amy, Eds, CRC Press, Inc. Boca Raton, FL, 131-150.

Mak, F.T., Zele, S.; Cooper, W.J.; Kurucz, C.N.; Waite, T.D. and Nickelsen, M.G.. 1997. Kinetic Modeling of Carbon Tetrachloride, Chloroform, and Methylene Chloride Removal from Aqueous Solution using the Electron Beam Process. *Water Research,* 31, 219-228.

Malcolm Pirnie, Inc. June, 1998. Taste and Odor Properties of Methyl Tertiary-Butyl Ether and Implications for Setting a Secondary Maximum Contaminant Level. Prepared for the Oxygenated Fuels Association, Inc.

Metropolitan Water District of Southern California and James M. Montgomery Consulting Engineers. 1991. Pilot-Scale Evaluation of Ozone and PEROXONE, AwwaRF and AWWA. Denver, Colo.

Mezyk, S.P.; Cooper, W.J.; Bartels, D.M.; O'Shea, K.E. and Wu, T. 2001. Radiation Chemistry of Alternative Fuel Oxygenates: Substituted Ethers. *J. Phys. Chem. Part A,* 105: 3521-3526.

Mezyk, S.P.; Cooper, W.J.; Tobien, T.; Nickelsen, M.G.; Asmus, K.-D.; O'Shea, K.E. and Bartels, D.M. 2002. Radiation Chemistry of Methyl Tert-Butyl Ether (MTBE) in Aqueous Solution. *Environ. Sci. Technol.* (In Preparation)

Montgomery, J.M. 1985. *Water Treatment Principles and Design,* James M. Montgomery Consulting Engineers, A Wiley-Interscience Publication, New York.

Moran, M.J.; Zogorski, J S. and Squillace, P.J. 1999. MTBE in Groundwater of the United States-Occurrence, Potential Sources and Long-Range Transport, AWWA Water Resources Conference, Norfolk, Virginia.

Morel, F.M.M. and Hering, J.G. 1993. *Principles and Applications of Aquatic Chemistry.* John Wiley & Sons, New York.

NDWAC. 2001. National Drinking Water Adhoc Committee Report on Costing Methods. EPA.

Nickelsen, M.G.; Cooper, W.J.; Waite, T.D. and Kurucz, C.N. 1992. Removal of Benzene and Selected Alkyl Substituted Benzenes from Aqueous Solution Utilizing Continuous High-Energy Electron Irradiation. *Environ. Sci. Technol,.* 26, 144-152.

Nickelsen, M.G.; Cooper, W.J.; Lin, K.; Kurucz, C.N.; Waite, T.D. 1994. High Energy Electron Beam Generation of Oxidants for the Treatment of Benzene and Toluene in the Presence of Radical Scavengers. *Water Research,* 28, 1227-1237

Nickelsen, M.G.; Kajdi, D.C.; Cooper, W.J.; Kurucz, C.N.; Waite, T.D.; Gensel, F.; Lorenzl, H.; and U. Sparka. 1998. Field Application of a Mobile 20-kW Electron Beam Treatment System on Contaminated Groundwater and Industrial Wastes. In *Environmental Applications of Ionizing Radiation,* W.J. Cooper, R.D. Curry and K.E. O'Shea, Eds., John Wiley and Sons, Inc. N.Y. 451-466.

Nicole, I.; De Laat, J., Doré, M., Duguet, J.P. and Bonnel, C. 1999. *Water Research,* v. 24, 157-168.

O'Shea, K.E.; Wu, T. and Cooper, W.J. 2001. TiO_2 Photocatalysis of Gasoline Oxygenates. Kinetic Parameters and Effects of Catalyst Types and Loading on the Degradation of Methyl Tert-Butyl Ether. In *Oxygenates in Gasoline: Environmental Aspects,* Eds. A.F. Diaz and D. Drogos. ACS Symposium Series 799, American Chemical Society, Washington, DC. p. 165-176.

O'Shea, K.E.; Kim, D.K.; Wu, T.; Cooper, W.J. and Mezyk, S.P. 2002. Degradation of MTBE/ BTEX Mixtures by Gamma Radiolysis: A Kinetic Modeling Study. *Rad. Phys. Chem.* (In Press.)

OSTP Report, 1997. Office of Science and Technology Policy Report. Executive Office of the President. National Science and Technology Council. Committee on Environmental and Natural Resources. Interagency Assessment of Oxygenated Fuels. Washington, D.C.

Pfaff, J.D. and Brockhoff, C.A. 1990. Determining Inorganic Disinfection By-products by Ion Chromatography. *Jour. AWWA,* 82(4): 192.

Pinkernell, U., and von Gunten, U. 2001, in press. Bromate Minimization During Ozonation: Mechanistic Considerations. *Environmental Science & Technology.*

Raese, J.W.; Rose, D.L. and Sandstrom, M.W. 1995. U.S. Geological Survey Laboratory Method for Methyl Tert-Butyl Ether and Other Fuel Oxygenates. In Fact Sheet 219-95; U.S. Geological Survey: pp. 4.

Reckhow, D.A.; Knocke, W.R.; Kearns, M.J. and Parks, C.A. 1991. *Jour. Ozone Science and Engineering,* 13:623.

Rodriguez, Rey. November, 1999. Project Manager for Santa Monica Drinking Water Treatment System. Personal Communication.

Rose, D.L.; Schroeder, M.P. 1995. Methods of Analysis by the U.S. Geological Survey National Water Quality Laboratory—Determination of Volatile Organic Compounds in Water by Purge and Trap Capillary Gas Chromatography/Mass Spectrometry. In Open-File Report 94-708; U.S. Geological Survey.

RS Means. 2001. Building Construction Cost Data 2002. 15th Annual Edition.

Salanitro, J.P.; Diaz, L.A.; Williams, M.P. and Wisniewski, H.L. 1994. Isolation of a Bacterial Culture that Degrades Methyl T-Butyl Ether. *Appl. Environ. Microbiol.*, 60(7): 2593-2596.

Salanitro, J.P.; Spinnler, G.E.; Neaville, C.C.; Maner, P.M.; Stearns, S.M.; Johnson, P.C. and Bruce, C. 1999. Demonstration of the Enhanced MTBE Bioremediation (EMB) In Situ Process. The Fifth International In Situ On-Site Bioremediation Symposium.

Salanitro, J.; Johnson, P.; Spinnler, G.; Maner, P.; Wisniewski, H. and Bruce, C. 2000. Field-Scale Demonstration of Enhanced MTBE Bioremediation Through Aquifer Bioaugmentation and Oxygenation. *Environ. Sci. Technol.*, 34(19): 4152-4162.

Schwarzenbach, R.P., Gschwend, P.M., and Imboden, D.M. 1993. Environmental Organic Chemistry. N.Y.: John Wiley & Sons, Inc.

Sclimenti, M.J., Krasner, S.W.; Glaze, W.H., and Weinberg, H.S. 1990. Ozone Disinfection By-Products: Optimization of the PFBHA Derivatization Method for the Analysis of Aldehydes. American Water Works Association Water Quality Technology Conference, San Diego, Cal., AWWA.

Shen, Y.F.; Yoo, L.J.; Fitzsimmons, S.R.; Yamamoto, M.K. 1998. Threshold Odor Concentrations of MTBE and Other Fuel Oxygenates. Orange County Water District.

Siddiqui, M. S.; Amy, G., Ozekin, K.; Zhai, W., and Westerhoff, P. 1994. Alternative Strategies for Removing Bromate. *Jour. AWWA,* 86(10):81-95.

Siddiqui M.; Amy, G.L.; Cooper, W.J.; Kurucz, C.N.; Waite, T.D. and Nickelsen, M.G. 1996a. Bromate Removal by High Energy Electron Beam Process (HEEB). *Jour. AWWA.* 88(10): 90-101.

Siddiqui, M.; Amy, G.L. and Cooper, W.J. 1996b. Bromate Ion Removal by Electric-Arc Discharge and High Energy Electron Beam Processes, in *Water Disinfection and Natural Organic Matter,* R.A. Minear and G.L. Amy, Eds, American Chemical Society, Symposium Series 649, Washington, D.C. 366 -382.

Simon, E. 2001. Optimization of Hydrogen Peroxide-Ozone Process for MTBE Oxidation and Bromate Minimization for Differing Water Quality Types. Master's Thesis, University of Colorado, Boulder.

Song R.; Westerhoff P.; Minnear, R. and Amy, G. 1997. Bromate Minimization during Ozonation. *Jour. AWWA,* 89(6):69.

Squillace, P.J.; Pankow, J.F.; Korte, N.E., and Zogorski, J.S., 1997. Review of the Environmental Behavior and Fate of Methyl Tert-Butyl Ether. *Environmental Toxicology and Chemistry,* 16(9):1836-1844.

Staehelin, J. and Hoigné, J. 1982. Decomposition of Ozone in Water: Rate of Initiation by Hydroxide Ions and Hydrogen Peroxide. *Jour. Environmental Science and Technology,* 16:676-681.

Stefan, M.I.; Mack, J., and Bolton, J.R. 2000. Degradation Pathways during the Treatment of Methyl Tert-Butyl Ether by the UV/H_2O_2 Process. *Jour. Environmental Science and Technology,* 34(4)650 – 658.

Steffan, R.J.; McClay, K.; Vainberg, S.; Condee, C.W. and Zhang, D. 1997. Biodegradation of the Gasoline Oxygenates Methyl Tert-Butyl Ether, Ethyl Tert-Butyl Ether, and Tert-Amyl Methyl Ether by Propane-Oxidizing Bacteria. *Appl. Environ. Microbiol.,* 63(11): 4216-4222.

Stewart, M.H.; Hwang, C.J.; Hacker, P.A.; Yates, R.S., and Wolfe, R.L. 1993. Microbial and Chemical Implications of Using Ultraviolet Irradiation for Treatment of Biological Filter Effluent. Proceedings AWWA 1993 WQTC. Miami, FL.

Stocking, A.J.; Suffet, I.H.; McGuire, M.J., and Kavanaugh, M.C. 2001. Implications of an MTBE Odor Study for Setting Drinking Water Standards. *Jour. AWWA,* 93(3):95-105.

Stumm, W. and Morgan, J.J. 1996. *Aquatic Chemistry: Chemical Equilibria and Rates in Natural Waters,* 3rd Edition, A Wiley Interscience Publication, John Wiley and Sons, New York.

Tornatore, P. Personal Communication. October 1999. Vice-President, Haley & Aldrich, Rochester, NY.

Tornatore, P.M. and Cooper, W. March. 1999. Remediation of MTBE Contaminated Drinking Water Supplies with High Energy Electron Injection. ACS Proceedings, Anaheim, California.

Tornatore, P.M.; Powers, S.T.; Cooper, W.J., and Isacoff, E.G. 2000. Emerging Treatments for MTBE Synthetic Adsorbents and High Energy Electron Injection. In *Chemical Oxidation and Reactive Barriers,* Eds., G.B. Wickramanayake, A.R. Gavaskar and A..S.C. Chen. 2(7): 57-64 (Battelle Press).

USEPA 1997. Advisory: *Consumer Acceptability Advice and Health Effects Analysis on Methyl Tertiary-Butyl Ether* (MTBE). EPA 822-F-97-008.

USEPA January 1998a. MTBE Fact Sheet #1: Overview. EPA 510-F-97-014. Office of Solid Waste and Emergency Response. Washington, D.C.

USEPA January, 1998b. MTBE Fact Sheet #2: Remediation of MTBE Contaminated Soil and Groundwater. EPA 510-F-97-015. Office of Solid Waste and Emergency Response. Washington, D.C.

USEPA January, 1998c. MTBE Fact Sheet #3, Use and Distribution of MTBE and Ethanol. EPA 510-F97-016. Office of Solid Waste and Emergency Response. Washington, D.C.

USEPA 1998d *Oxygenates in Water: Critical Information and Research Needs.* EPA/600/R-98/048.

USEPA 1999. *Achieving Clean Air and Clean Water: The Report of Blue Ribbon Panel on Oxygenates in Gasoline.* EPA 420-R-99-021.

USEPA 2000. *Control of MTBE in Gasoline,* EPA420-F-00-010.

von Gunten, U. and Hoigné, J. 1994. Bromate Formation during Ozonation of Bromide-Containing Waters: Interaction of Ozone and Hydroxyl Radical Reactions. *Jour. Environmental Science and Technology,* 28:1234-1242.

von Gunten, U. and Oliveras, Y. 1997. Kinetics of the Reaction between Hydrogen Peroxide and Hypobromous Acid: Implication on Water Treatment and Natural Systems. *Jour. Water Research,* 31(4):900-906.

von Gunten, U., and Oliveras, Y. 1998. Advanced Oxidation of Bromide-Containing Waters: Bromate Formation Mechanisms. *Environmental Science & Technology,* 32(1), 63-70.

Wagler, J.L. and Malley, J.P. Jr. 1994. The Removal of MTBE from a Model Groundwater using UV/Peroxide Oxidation. *Journal of New England Water Works Association,* p. 236-243.

Westerhoff, P. 1995. Ozone Oxidation of Bromide and Natural Organic Matter. Ph.D. Dissertation. Civil, Environmental, and Architectural Engineering, University of Colorado at Boulder, 80309.

Westerhoff, P.; Aiken, G.; Amy, G. and DeBroux, J. 1999. Relationships Between the Structure of Natural Organic Matter and its Reactivity Towards Molecular Ozone and Hydroxyl Radical. *Wat. Res.,* 33: 2265-2276.

White, G.C. 1999. *Handbook of Chlorination and Alternative Disinfectants.* A Wiley-Interscience Publication. New York. John Wiley and Sons Inc.

Wu, T.; Cruz, V.; Mezyk, S.P.; Cooper, W.J., and O'Shea, K.E. 2002. Gamma Radiolysis of Methyl Tert-Butyl Ether (MTBE). A Study of Hydroxyl Radical Mediated Reaction Pathways. *Rad. Phys. Chem.* (In Press.)

Young, W.F.; Horth, H.; Crane, R.; Ogden, T., and Arnott, M. 1996. Taste and Odor Threshold Concentrations of Potential Potable Water Contaminants, 30(2):331-340.

Zepp, R. G. 1988. Factors Affecting the Photochemical Treatment of Hazardous Waste. *Environmental Science and Technology,* 22(3):256.

Zheng, M.; Andrews, S.A. and Bolton, J.R. 1999a. Impacts of Medium-Pressure UV on THM and HAA Formation in Pre-UV Chlorinated Drinking Water. AWWA Water Quality Technology Conference, Tampa, Florida.

Zheng, M.; Andrews, S.A. and Bolton, J.R. 1999b. Impacts of Medium-Pressure UV and UV/ H_2O_2 Treatments on Disinfection By-product Formation, AWWA Annual Conference, Chicago.

ABBREVIATIONS

AOPs	advanced oxidation processes
ASTM	American Society for Testing and Materials
AwwaRF	AWWA Research Foundation
CAAA	Clean Air Act Amendments
CDHS	California Department of Health Services
Cl·	chlorine radicals
CRW	Colorado River Water
DCE	dichloroethylene
D/DBP	disinfection/disinfection by-product
DI	deionized (water)
DIPE	di-isopropyl ether
DOC	dissolved organic carbon
DOM	dissolved organic matter
E-beam	Electron beam
EBRF	Electron Beam Research Facility
EE/O	Electrical Energy per Order (of Removal)
EPA	Environmental Protection Agency
ETBE	ethyl tertiary butyl ether
Fe	iron
ft	foot or feet
GAC	granular activated carbon
GC/MSD	Gas Chromatograph/Mass Selective Detector
gpm	gallons per minute
H	hydrogen atom
HAA_9	haloacetic acids
hí	UV photon
HVEA	High Voltage Environmental Applications
Hz	Hertz (cycles per second)
H_2O_2	hydrogen peroxide
·OH	hydroxyl radicals
ICT	insulated-core transformer
In	inch or inches
kGy	kiloGray (= 100 krad)
kV	kiloVolts
kWh	kiloWatt-hours

LOX	Liquid oxygen
LP	low pressure
LP-UV	low-pressure (ultra violet)
M	meter
mA	milliAmperes
MCL	Maximum Contaminant Level
MGD	million gallons/day
MP	medium pressure
MP-UV	medium-pressure (ultra violet)
MTBE	Methyl tertiary-butyl ether
MWDSC	Metropolitan Water District of Southern California
μg/L	micrograms per liter
NAWQA	National Water Quality Assessment
NDMA	Nitroso dimethylamine
NF^-	nitroform anions
NOM	natural organic matter
$O_2.$	superoxide radical
O_3	ozonation
O_3-H_2O_2	peroxone or ozone/peroxide
OCWD	Orange County Water District
OGI	Oregon Graduate Institute
Oxyfuel	Oxygenated Fuel
pCBA	para-chlorobenzoic acid
PCE	tetrachloroethylene or perchloroethylene
POC	particulate organic carbon
ppb	parts per billion
PSA	pressure swing absorption
P-UV	pulsed-UV
RFG	Reformulated Gasoline
RTD	resistance temperature devices
Sec	second
SIM	Selective ion monitoring
SMCL	Secondary Maximum Contaminant Level
SPE	solid phase extraction
SPW	State Project Water
TAME	tertiary amyl methyl ether
TBA	tertiary butyl alcohol
TBF	tertiary butyl formate

TCE	trichloroethylene
THM	trihalomethanes
TOC	total organic carbon
TSCA	Toxic Substances Control Act
USGS	US Geological Society
UV	ultraviolet

CONVERSION TABLE

SI EQUIVALENT UNITS

Parameters	From Customary Units	To SI Units	Multiply By
	To Convert		
Area	mi^2	km^2	2.59
		or m^2	2.59×10^6
Area	acre	ha	4.047×10^{-1}
Chemical dosage	ppm	mg/L	—
Chemical feed rate	lb/d	kg/d	4.536×10^{-1}
Chemical feed rate	gph	L/h	3.785
		or mL/s	1.052
Displacement velocity	ft/s	m/s	3.048×10^{-1}
Filter backwash rate	gpm/ft^2	$L/m^2/s$	6.790×10^{-1}
Filter head loss	ft	kPa	2.989
		or m	3.048×10^{-1}
Filtration rate	gpm/ft^2	m/h	2.444
		or $m^3/m^2/h$	2.444
Gas feeder differential pressure	in. H_2O (4 °C)	kPa	2.491×10^{-1}
Gas feeder supply pressure	psi	kPa	6.895
Head loss	ft	kPa	2.989
Hydrant spacing	ft	m	3.048×10^{-1}
Hydraulic gradient	ft/1,000 ft	mm/m	—
		or m/km	—
Level gauging	ft	m	3.048×10^{-1}
Metering	gph	m^3/h	3.785×10^{-3}
Metering	ft^3/h	m^3/h	2.832×10^{-2}
Pipe cross-section	in.^{2}	mm^2	6.452×10^2
Pipe diameter	in.	mm	2.54×10
Pipe flow velocity	ft/s	m/s	3.048×10^{-1}
Pipe length	ft	m	3.048×10^{-1}
Plant capacity	mgd	M^3/d	3.785×10^3
		or ML/d	3.785
Power	hp	watt	7.46×10^2
Pressure	psi	kPa	6.895
Pump capacity	gpm	m^3/s	6.309×10^{-5}
		or L/s	6.309×10^{-2}

(continues)

Parameters	To Convert		
	From Customary Units	To SI Units	Multiply By
Pump capacity	ft^3/min	m^3/s	4.72×10^{-4}
		or L/s	4.720×10^{-1}
Raw water flow	gpm	m^3/s	6.309×10^{-5}
		or L/s	6.309×10^{-2}
Raw water flow	ft^3/min	m^3/s	4.720×10^{-4}
		or L/s	4.720×10^{-1}
Raw water temperature	°F	°C	(°F – 32) × 5/9
Residual chlorine	ppm	mg/L	—
Retention time	h	h	—
Settling rate	fph	m/h	3.048×10^{-1}
Storage elevation	ft	m	3.048×10^{-1}
Storage volume	gal	m^3	3.785×10^{-3}
		or ML	3.785×10^{-6}
Storage volume	ft^3	m^3	2.832×10^{-2}
		or ML	2.832×10^{-5}
Water consumption	gal	m^3	3.785×10^{-3}
		or L	3.785
		or ML	3.785×10^{-6}
Water consumption	ft^3	m^3	2.832×10^{-2}
		or ML	2.832×10^{-5}
Weir overflow rate	gal/ft/d	L/m/s	1.437×10^{-4}

IWA Publishing's authorised EU representative for General Product Safety Regulations is Diane D'Arras, 15 rue Duret, 75116 Paris, France, e-mail: safety@iwap.co.uk.

Printed and bound by CPI Group (UK) Ltd, Croydon, CR0 4YY
17/09/2026
02218802-0005